Lecture Notes in Computer Science

Lecture Notes in Artificial Intelligence 15248

Founding Editor

Jörg Siekmann

Series Editors

Randy Goebel, *University of Alberta, Edmonton, Canada*
Wolfgang Wahlster, *DFKI, Berlin, Germany*
Zhi-Hua Zhou, *Nanjing University, Nanjing, China*

The series Lecture Notes in Artificial Intelligence (LNAI) was established in 1988 as a topical subseries of LNCS devoted to artificial intelligence.

The series publishes state-of-the-art research results at a high level. As with the LNCS mother series, the mission of the series is to serve the international R & D community by providing an invaluable service, mainly focused on the publication of conference and workshop proceedings and postproceedings.

Rupert Freeman · Nicholas Mattei
Editors

Algorithmic Decision Theory

8th International Conference, ADT 2024
New Brunswick, NJ, USA, October 14–16, 2024
Proceedings

Editors
Rupert Freeman
University of Virginia
Charlottesville, VA, USA

Nicholas Mattei
Tulane University
New Orleans, LA, USA

ISSN 0302-9743　　　　　　　ISSN 1611-3349 (electronic)
Lecture Notes in Artificial Intelligence
ISBN 978-3-031-73902-6　　　ISBN 978-3-031-73903-3 (eBook)
https://doi.org/10.1007/978-3-031-73903-3

LNCS Sublibrary: SL7 – Artificial Intelligence

© The Editor(s) (if applicable) and The Author(s), under exclusive license
to Springer Nature Switzerland AG 2025

This work is subject to copyright. All rights are solely and exclusively licensed by the Publisher, whether the whole or part of the material is concerned, specifically the rights of translation, reprinting, reuse of illustrations, recitation, broadcasting, reproduction on microfilms or in any other physical way, and transmission or information storage and retrieval, electronic adaptation, computer software, or by similar or dissimilar methodology now known or hereafter developed.
The use of general descriptive names, registered names, trademarks, service marks, etc. in this publication does not imply, even in the absence of a specific statement, that such names are exempt from the relevant protective laws and regulations and therefore free for general use.
The publisher, the authors and the editors are safe to assume that the advice and information in this book are believed to be true and accurate at the date of publication. Neither the publisher nor the authors or the editors give a warranty, expressed or implied, with respect to the material contained herein or for any errors or omissions that may have been made. The publisher remains neutral with regard to jurisdictional claims in published maps and institutional affiliations.

This Springer imprint is published by the registered company Springer Nature Switzerland AG
The registered company address is: Gewerbestrasse 11, 6330 Cham, Switzerland

If disposing of this product, please recycle the paper.

Preface

The 8th International Conference on Algorithmic Decision Theory (ADT 2024), held in October 2024 at the Center for Discrete Mathematics and Theoretical Computer Science (DIMACS) at Rutgers University (Piscataway, NJ, USA), continued in the tradition established by previous ADT conferences in providing a unique opportunity for scientific exchange among researchers and practitioners whose goal is to improve the theory and practice of modern decision support. The conference series is multi-disciplinary, with attendees coming from diverse areas of computer science, economics, and operations research. Previous ADT conferences were held in Venice, Italy (2009); Piscataway, NJ, USA (2011); Brussels, Belgium (2013); Lexington, KY, USA (2015); Luxembourg (2017); Durham, NC, USA (2019); and Toulouse, France (2021).

ADT 2024 received 41 submissions, of which 39 were sent for peer review. Each paper was reviewed by three program committee (PC) members in a double-blind fashion on the basis of originality, significance, and exposition, after which 26 papers were accepted for an acceptance rate of approximately 63%. Authors were given the choice of having a full paper or a one-page abstract published; 18 papers appear as full papers in the proceedings while 8 appear as one-page abstracts.

The works accepted for publication in this volume cover most of the major aspects of algorithmic decision theory, such as preference modeling and elicitation, voting, preference aggregation, fair division and resource allocation, coalition formation, game theory, and matching.

The program also featured three invited talks by distinguished researchers in fields adjacent to algorithmic decision theory, namely Tracy Xiao Liu (Tsinghua University), Hervé Moulin (University of Glasgow), and Jennifer Wortman Vaughan (Microsoft Research).

We thank the authors for submitting and presenting their high-quality work at ADT 2024, and the program committee for their exceptional work evaluating the submissions. We are grateful to DIMACS for hosting the conference and in particular to the staff at DIMACS for their logistical support. We also thank Alexis Tsoukias and the conference steering committee for their advice and support, Springer for supporting the publication process, and Microsoft for use of the CMT platform for organizing the submission and reviewing process.

Finally, we thank our sponsors for their generous support: DIMACS, Darden School of Business at the University of Virginia, the EURO Working Group on Preference Handling, and the Artificial Intelligence journal.

October 2024   Rupert Freeman
                                                                                                                            Nicholas Mattei

Organization

General Chairs

David M. Pennock   Rutgers University, USA
Lirong Xia   Rutgers University, USA

Program Committee Chairs

Rupert Freeman   University of Virginia, USA
Nicholas Mattei   Tulane University, USA

Program Committee

Thomas Allen	Centre College, USA
Ben Armstrong	University of Waterloo, Canada
Haris Aziz	University of New South Wales, Australia
Avinash Balakrishnan	IBM Research, USA
Manel Baucells	University of Virginia, USA
Dorothea Baumeister	Federal University of Applied Administrative Sciences, Germany
Robert Bredereck	TU Clausthal, Germany
Markus Brill	University of Warwick, UK
Katarína Cechlárová	P. J. Šafárik University, Slovakia
Michael Curry	Harvard University, USA, and University of Zurich, Switzerland
Ronald de Haan	University of Amsterdam, The Netherlands
Roy Fairstein	Ben-Gurion University of the Negev, Israel
Piotr Faliszewski	AGH University, Poland
Zack Fitzsimmons	College of the Holy Cross, USA
Judy Goldsmith	University of Kentucky, USA
Umberto Grandi	University of Toulouse 1 Capitole, France
Davide Grossi	University of Groningen, The Netherlands
Tatiana Guy	Institute of Information Theory and Automation, Czech Republic
Noam Hazon	Ariel University, Israel
Ayumi Igarashi	University of Tokyo, Japan

David Rios Insua	ICMAT, Spain
Jérôme Lang	CNRS, Université Paris-Dauphine, France
Andrea Loreggia	University of Brescia, Italy
David Manlove	University of Glasgow, UK
Neeldhara Misra	IIT Gandhinagar, India
Arianna Novaro	Université Paris 1 Panthéon-Sorbonne, France
Georgios Papasotiropoulos	University of Warsaw, Poland
Patrice Perny	Sorbonne University, France
Dominik Peters	CNRS, Université Paris-Dauphine, France
Maria Silvia Pini	University of Padua, Italy
Maria Polukarov	King's College London, UK
Jörg Rothe	Heinrich-Heine-Universität Düsseldorf, Germany
Ahti Salo	Aalto University, Finland
Ulrike Schmidt-Kraepelin	TU Eindhoven, The Netherlands
Maria Serna	Universitat Politècnica de Catalunya, Spain
Warut Suksompong	National University of Singapore, Singapore
Zoi Terzopoulou	Saint-Etienne School of Economics, France
Alan Tsang	Carleton University, Canada
Alexis Tsoukias	CNRS, Université Paris-Dauphine, France
Rohit Vaish	IIT Delhi, India
Kristen Brent Venable	IHMC, UWF, USA
Paolo Viappiani	CNRS, Université Paris-Dauphine, France
Toby Walsh	University of New South Wales, Australia
Stefan Woltran	Vienna University of Technology, Austria
Sasa Zorc	University of Virginia, USA

Contents

Voting Theory

Complexity of Candidate Control for Single Nontransferable Vote
and Bloc Voting 3
    Garo Karh Bet, Jörg Rothe, and Roman Zorn

Compiling the Votes of a Subelectorate for Multi-winner Voting Rules 18
    Neel Karia and Jérôme Lang

On Approximately Strategy-Proof Tournament Rules for Collusions
of Size at Least Three 33
    David Mikšaník, Ariel Schvartzman, and Jan Soukup

As Time Goes By: Adding a Temporal Dimension to Resolve Delegations
in Liquid Democracy 48
    Evangelos Markakis and Georgios Papasotiropoulos

Game Theory, Optimization, and Decision Making

Simple Stochastic Stopping Games: A Generator and Benchmark Library 67
    Avi Rudich, Isaac Rudich, and Rachel Rue

Protective and Nonprotective Subset Sum Games: A Parameterized
Complexity Analysis 82
    *Jaroslav Garvardt, Christian Komusiewicz, Berthold Blatt Lorke,
    and Jannik Schestag*

Algorithmic Decision Analysis for Multi-stage Games with Incomplete
Information 98
    J. M. Camacho, Roi Naveiro, and David Ríos Insua

Non-maximizing Policies that Fulfill Multi-criterion Aspirations
in Expectation 113
    Simon Dima, Simon Fischer, Jobst Heitzig, and Joss Oliver

Adversarial Risk Analysis for Automated Lane-Changing in Heterogeneous
Traffic 128
    Roi Naveiro, David Ríos Insua, and William N. Caballero

A Fully Bayesian Approach to Bilevel Problems 144
    Vedat Dogan, Steven Prestwich, and Barry O'Sullivan

Collaborative Information Dissemination with Graph-Based Multi-Agent
Reinforcement Learning 160
    Raffaele Galliera, Kristen Brent Venable, Matteo Bassani,
    and Niranjan Suri

Toward Fair and Strategyproof Tournament Rules for Tournaments
with Partially Transferable Utilities 174
    David Pennock, Ariel Schvartzman, and Eric Xue

Preference Theory

Noise-Tolerant Active Preference Learning for Multicriteria Choice
Problems 191
    Margot Herin, Patrice Perny, and Nataliya Sokolovska

Learning Multiple Multicriteria Additive Models from Heterogeneous
Preferences 207
    Vincent Auriau, Khaled Belahcène, Emmanuel Malherbe,
    and Vincent Mousseau

Independent Relaxed Subproblems for Dominance Testing in CP-Nets 225
    Liu Jiang and Thomas E. Allen

Allocations and Matching

Core Stability and Nash Stability in k-Tiered Coalition Formation Games 243
    Nathan Arnold and Judy Goldsmith

Envy-Free and Efficient Allocations for Graphical Valuations 258
    Neeldhara Misra and Aditi Sethia

Manipulation With(out) Money in Matching Market 273
    Sushmita Gupta and Pallavi Jain

Abstracts

Metric Distortion Under Probabilistic Voting 291
    Sahasrajit Sarmsarkar and Mohak Goyal

A Linear Theory of Multi-Winner Voting 293
    Lirong Xia

Group Fairness in Multi-period Mobile Facility Location Problems 294
    Haris Aziz, Hau Chan, Xingchen Sha, Toby Walsh, and Lirong Xia

Extending Myerson's Optimal Auctions to Correlated Bidders via Neural
Network Interpolation 295
    Mingyu Guo, Jiayuan Liu, and Vincent Conitzer

No-Regret Learning for Stackelberg Equilibrium Computation
in Newsvendor Pricing Games 297
    Larkin Liu and Yuming Rong

Equilibria of Data Marketplaces with Privacy-Aware Sellers
under Endogenous Privacy Costs 298
    Diptangshu Sen, Jingyan Wang, and Juba Ziani

Learning Linear Utility Functions From Pairwise Comparison Queries 299
    Luise Ge, Brendan Juba, and Yevgeniy Vorobeychik

The Fairness-Quality Trade-Off in Clustering 300
    *Rashida Hakim, Ana-Andreea Stoica, Mihalis Yannakakis,
    and Christos H. Papadimitriou*

Author Index 301

Voting Theory

Complexity of Candidate Control for Single Nontransferable Vote and Bloc Voting

Garo Karh Bet, Jörg Rothe, and Roman Zorn

MNF, Institut für Informatik, Heinrich-Heine-Universität Düsseldorf, Düsseldorf, Germany
{garo.karh.bet,rothe,roman.zorn}@hhu.de

Abstract. Electoral control is a scenario where an election chair changes the structure of an election by actions such as adding or deleting either candidates or voters with the goal of either making a favorite candidate win or precluding a despised candidate's victory. Much work has been done on the computational complexity of controlling elections for single-winner voting rules, yet much less work on the control complexity for multiwinner voting rules which aim at electing not only a single winner but a winning committee of candidates. Meir et al. [20] initiated the investigation of electoral control for multiwinner voting rules, including single nontransferable vote (SNTV) and bloc voting. We study these two rules with respect to control by adding, deleting, or replacing candidates.

Keywords: multiwinner voting · control · SNTV · bloc voting

1 Introduction

Computational social choice [4,22] is an interdisciplinary field at the interface of artificial intelligence and theoretical computer science on the one hand and economics and social choice theory on the other, which focuses on the computational aspects of collective decision-making mechanisms such as voting. Application scenarios range over a wide spectrum and include, for instance, electing a leader of a group of people or an organisation, or selecting a meeting time and place for a group of people, or when judges or referees short-list the finalists of a competition based on their performance [13]. Moreover, voting has been used as a decision-making mechanism in various computational settings such as planning [8,9] and collaborative filtering [21], and also in several large-scale computer settings, including web-page rank aggregation and the related spam reduction and similarity-search problems [7,11]. In such scenarios, particular attention has been paid to ways of tampering with the outcome of elections by manipulation, control, and bribery [3,6,12]. We focus on electoral control—a scenario where the structure of an election can be changed by an election chair via control actions such as adding or deleting either candidates or voters. Bartholdi et al. [1] introduced and studied the constructive variant of the problem where the chair's goal is to make a favorite candidate win. Hemaspaandra et al. [15] introduced the destructive counterpart that aims at precluding a despised candidate's victory.

Much work has been done on studying the complexity of electoral control for single-winner voting rules, especially for the standard control actions of adding, deleting, or partitioning either the candidates or the voters (see the book chapters by Faliszewski and Rothe [12] and Baumeister and Rothe [3] for an overview). However, much less is known about the complexity of control for multiwinner voting rules where not only a single winner is to be elected but a winning committee of candidates [2,13]. Our main goal is to study the complexity of control by *replacing* candidates for two fundamental multiwinner voting rules: *single nontransferable vote (SNTV)* and *bloc voting*, the multiwinner analogues of plurality and k-approval voting, respectively. Control by replacing candidates or voters was introduced and first studied by Loreggia et al. [17,18]. This control type—which combines adding and deleting either candidates or voters but with the constraint that the same number of candidates or voters must be added as has been deleted—has only been studied for single-winner voting so far, see, e.g., the recent work of Erdélyi et al. [10] and Maushagen et al. [19]. Meir et al. [20] were the first to explicitly address the complexity of control of *multiwinner* elections, including SNTV and bloc voting. Subsequently, Yang [23] investigated the complexity of manipulation and control as well as the parameterized complexity of control for various approval-based multiwinner voting rules.

We will use a general result by Loreggia [17] (stated here as Theorem 1) that links control by replacing candidates to control by adding or deleting candidates, provided that the voting rule is insensitive to bottom-ranked candidates, a property introduced by Lang et al. [16]. Therefore, we also need to explore the complexity of control by adding or deleting candidates for SNTV and bloc voting. Note that Meir et al. [20] in particular observed such results as a simple consequence of the corresponding results in single-winner elections. However, they model these control problems differently than it was done in the original papers on control [1,15], using additive utility functions that assign integers to the candidates. That is why we formalize these standard control problems in the original model and adapt the original proofs in the single-winner setting to the multiwinner case. Another difference is that we exclude the committee size one (which is the single-winner case), so unlike the work of Meir et al. [20], our results in the multiwinner setting are not immediate consequences of the corresponding single-winner results—we only consider elections with strictly multiple winners.

2 Preliminaries

A *multiwinner election* is a triple $E = (C, V, k)$, where C and V stand for the set of candidates and the list of votes, respectively, and k, $2 \leq k \leq |C| - 1$, refers to the (fixed) target committee size. Mathematically speaking, V is expressed by a (strict) linear order on C, satisfying the properties of connectivity, transitivity, and asymmetry. Since for $k = 1$ one would be back to the case of single-winner elections, and for $k = |C|$ the solution is trivial, we are going to consider only target committee size values of $1 < k < |C|$.

A *multiwinner voting rule* $\mathcal{R}$ is a function that maps a given multiwinner election $E = (C, V, k)$ to a set $\mathcal{R}(E)$ of k-element subsets of C, which are referred to as *winning committees*. This is denoted by $\mathcal{R}(E) = W = \{W_1, \ldots, W_l\}$, where $|W_i| = k$, $1 \leq i \leq l$. A multiwinner voting rule outputs all the winning committees that could end up winning for some way of resolving ties that may occur while executing the rule; we call them *possible winning committees*.

Let $E = (C, V, k)$ be a multiwinner election and $W = \{W_1, \ldots, W_l\}$ the set of all possible winning committees with $|W_i| = k$. We call a candidate $c \in C$ a *certain winner* if c is a member of every possible winning committee, i.e., $c \in W_i$ for each $W_i \in W$. Note that for a target committee size k, there can be at most k certain winners in a given multiwinner election. We call c an *uncertain winner* if c is in some, yet not in all possible winning committees, i.e., $c \in \bigcup_{1 \leq i \leq l} W_i$ and $c \notin W_i$ for some i, $1 \leq i \leq l$. Similarly, we call c a *certain nonwinner* if there is no possible winning committee $W_i \in W$ such that $c \in W_i$, i.e., $c \notin W_i$ for any $W_i \in W$.

We consider the following two multiwinner voting rules: *Single nontransferable vote (SNTV)* is the multiwinner variant of plurality, since for committee size k, it returns the k candidates with the highest plurality scores, where the *plurality score* $\sigma(c)$ of a candidate $c \in C$ is the number of votes in which c is ranked first. *Bloc voting* is the multiwinner variant of k-approval: For committee size k, bloc voting returns the k candidates with the highest k-approval scores, where the *k-approval score* $\alpha_k(c)$ *of a candidate* $c \in C$ is the number of votes in which c is ranked among the first k positions.

Meir et al. [20] model control for multiwinner voting rules via utility functions. In contrast, we follow the original model of Bartholdi et al. [1] and Hemaspaandra et al. [15] for control in single-winner voting when defining the following multiwinner control problems for any voting rule $\mathcal{R}$ and fixed committee size k:

$\mathcal{R}$-CONSTRUCTIVE-CONTROL-BY-DELETING-CANDIDATES ($\mathcal{R}$-CCDC)

Given: A set of candidates C, a list of votes V over C, a distinguished candidate $c \in C$, and a positive integer $r \leq |C|$.

Question: Is it possible to delete at most r candidates from C such that c is a certain winner of the election? That is, is there a $C' \subseteq C$ with $|C'| \leq r$ such that c is a certain winner of the election $(C \setminus C', V, k)$?

$\mathcal{R}$-DESTRUCTIVE-CONTROL-BY-ADDING-CANDIDATES ($\mathcal{R}$-DCAC)

Given: Two sets C and D of candidates, $C \cap D = \emptyset$, a list V of votes over $C \cup D$, a distinguished candidate $c \in C$, and a positive integer $r \leq |D|$.

Question: Is there a subset $D' \subseteq D$ with $|D'| \leq r$ such that c is a certain nonwinner of the election $(C \cup D', V, k)$?

$\mathcal{R}$-CONSTRUCTIVE-CONTROL-BY-REPLACING-CANDIDATES ($\mathcal{R}$-CCRC)

Given: Two sets C and D of candidates, $C \cap D = \emptyset$, a list V of votes over $C \cup D$, a distinguished candidate $c \in C$, and a positive integer r.

Question: Are there subsets $C' \subseteq C$ and $D' \subseteq D$ such that $|C'| = |D'| \leq r$ and c is a certain winner of the election $((C \setminus C') \cup D', V, k)$?

$\mathcal{R}$-DESTRUCTIVE-CONTROL-BY-REPLACING-CANDIDATES ($\mathcal{R}$-DCRC)

Given: Two sets C and D of candidates, $C \cap D = \emptyset$, a list V of votes over $C \cup D$, a distinguished candidate $c \in C$, and a positive integer r.

Question: Are there subsets $C' \subseteq C$ and $D' \subseteq D$ such that $|C'| = |D'| \leq r$ and c is a certain nonwinner of the election $((C \setminus C') \cup D', V, k)$?

In all these problems, changes in the candidate set are mirrored in the votes of V; for example, when deleting candidates from C, the votes in V over C are restricted accordingly by canceling these candidates out.

A (multiwinner) voting rule $\mathcal{R}$ is said to be *susceptible* to some control type if the chair can successively exert control for at least some instance of the corresponding control problem; otherwise, $\mathcal{R}$ is said to be *immune* to it. If $\mathcal{R}$ is susceptible to some control type, it is said to be *vulnerable* to it if the corresponding control problem is in P (i.e., solvable in *deterministic polynomial time*), and $\mathcal{R}$ is said to be *resistant* to this control type if the corresponding control problem is NP-hard (where NP is the complexity class *nondeterministic polynomial time*). Slightly abusing notation, we use abbreviations such as DCRC both for the control problem defined above and for describing the corresponding control type. For example, we write some rule $\mathcal{R}$ is "resistant to DCRC" as a shorthand for it being "resistant to destructive control by replacing candidates," which means the control problem $\mathcal{R}$-DCRC is NP-hard.

In our proofs, we will use the following standard NP-complete problems [14]:

HITTING-SET

Given: A set $B = \{b_1, \ldots, b_m\}$, a family $\mathcal{S} = \{S_1, \ldots, S_n\}$ of subsets $S_i \subseteq B$, and a positive integer r.

Question: Is there a subset $B' \subseteq B$, $|B'| \leq r$, such that each $S_i \in \mathcal{S}$ is *hit* by B', i.e., $S_i \cap B' \neq \emptyset$ for all $S_i \in \mathcal{S}$?

EXACT-COVER-BY-THREE-SETS (X3C)

Given: A set $B = \{b_1, \ldots, b_m\}$ with $m = 3r$ and $r \geq 1$ and a family $\mathcal{S} = \{S_1, \ldots, S_n\}$ of subsets $S_i \subseteq B$ with $|S_i| = 3$, for each i, $1 \leq i \leq n$.

Question: Is there a subfamily $\mathcal{S}' \subseteq \mathcal{S}$ such that every element of B appears in exactly one subset of $\mathcal{S}'$?

In our resistance proofs for constructive and destructive control by replacing candidates, we will make use of voting rules being *insensitive to bottom-ranked candidates (IBC)*, a notion due to Lang et al. [16].

Definition 1. *A voting rule $\mathcal{E}$ is IBC if its set of winners does not change after adding or deleting a subset of candidates at the bottom of the preference profile.*

Loreggia [17] shows how this property can be used to relate control by adding or deleting candidates to control by replacing candidates.

Theorem 1 (Theorems 3.3.3 and 3.3.4 in the PhD thesis of Loreggia [17]).

1. *Every voting rule that is IBC and resistant to CCDC is also resistant to CCRC.*
2. *Every voting rule that is IBC and resistant to DCAC or DCDC is also resistant to DCRC.*

3 Results for SNTV

We start with SNTV. To prove its resistance of constructive and destructive control by replacing candidates, Theorem 1 will be used. To this end, SNTV's insensitivity to bottom-ranked candidates will be shown first.

Lemma 1. *SNTV is insensitive to bottom-ranked candidates.*

Proof. This is clear from the definition, as SNTV only considers the first rank of any votes. Adding or deleting candidates at the bottom of a vote has no influence on the winning committees. ☐

Additionally, we now need to show that SNTV is resistant to CCDC and DCAC. Meir et al. [20] already observed that SNTV is resistant to CCDC and DCAC by arguing that this immediately follows from the corresponding resistance results in the single-winner case [1, 15]. However, we exclude the case of $k = 1$ from our definition, so this argument does not apply here; further, based on the original definitions, we have modeled our control problems somewhat differently. We now show that SNTV is resistant to CCDC.

Theorem 2. *SNTV is resistant to constructive control by deleting candidates.*

Proof. Our proof modifies the construction by which Bartholdi et al. [1] show that plurality is resistant to CCDC via a (polynomial-time many-one) reduction from X3C. Let (B, S) be a given instance of X3C, with $B = \{b_1, \ldots, b_m\}$, $m = 3r$ for $r \geq 5$ (which can be assumed without loss of generality because excluding instances with $r < 5$ does not change the complexity of the problem), and a family $S = \{S_1, \ldots, S_n\}$ of three-element subsets of B. Let b_i^1, b_i^2, b_i^3 denote the three elements of S_i. Construct a multiwinner election $E = (C, V, k)$ as follows. Let $C = \{c\} \cup A \cup B \cup D$ be the candidate set, where c is the distinguished candidate and $A = \{a_1, \ldots, a_k\}$ and $D = \{d_1, \ldots, d_r\}$ are auxiliary candidates. For each i, $1 \leq i \leq n$, there is one candidate s_i corresponding to the set $S_i \in S$. The list V of votes is divided into five voter groups as shown in Table 1.

Table 1. Voter groups for SNTV-CCDC in the proof of Theorem 2

Voter group	Number of votes	Preference
V_1	1 for each i, $1 \leq i \leq n$	$s_i\ c\ \cdots$
V_2	1 for each i, $1 \leq i \leq n$	$s_i\ b_i^1\ D\ \cdots$
	1 for each i, $1 \leq i \leq n$	$s_i\ b_i^2\ D\ \cdots$
	1 for each i, $1 \leq i \leq n$	$s_i\ b_i^3\ D\ \cdots$
V_3	r	$a_1\ D\ \cdots$
V_4	$r - 1$ for each q, $2 \leq q \leq k$	$a_q\ D\ \cdots$
V_5	$r - 2$ for each j, $1 \leq j \leq m$	$b_j\ D\ \cdots$

Here, a set D in a preference indicates that the candidates from D occur in an arbitrary order and "$\cdots$" indicates that the remaining candidates from C follow in some arbitrary order. Now, based on those votes, the plurality score of each of the candidates can be calculated:

$$\sigma(s_i) = 4 \text{ for } 1 \leq i \leq n,$$
$$\sigma(a_1) = r,$$
$$\sigma(a_q) = r - 1 \text{ for } 2 \leq q \leq k,$$
$$\sigma(b_j) = r - 2 \text{ for } b_j \in B, \text{ and}$$
$$\sigma(c) = 0.$$

Since there are at least k candidates who have a higher plurality score than c, c can never be a member of any winning committee and thus is a certain nonwinner of the election.

We now show that $(B, \mathcal{S})$ is a yes-instance of X3C if and only if c can be made a certain winner of the election resulting from E by deleting at most r candidates from it.

Suppose that $(B, \mathcal{S})$ is a yes-instance of X3C, so there is an exact 3-cover $\mathcal{S}' \subseteq \mathcal{S}$ of B. Delete the candidates s_i corresponding to the sets $S_i \in \mathcal{S}'$. Since the exact 3-cover has a cardinality of r, the plurality score of the candidates will be altered as follows (we denote these scores by σ' in the modified election after $|\mathcal{S}'| = r$ candidates s_i were removed from it):

$$\sigma'(s_i) = \sigma(s_i) = 4 \text{ for } i, \ 1 \leq i \leq n, \text{ with } S_i \notin \mathcal{S}',$$
$$\sigma'(a_1) = \sigma(a_1) = r,$$
$$\sigma'(a_q) = \sigma(a_q) = r - 1 \text{ for } 2 \leq q \leq k,$$
$$\sigma'(b_j) = r - 1 \text{ for } b_j \in B, \text{ after receiving one vote each in voter group } V_2, \text{ and}$$
$$\sigma'(c) = r.$$

Note that candidates c and a_1 have the highest plurality score among the remaining candidates, so they are now members of every winning committee of the election, i.e., certain winners, since $k \geq 2$.

Conversely, assume that there exists a subset C' of no more than r candidates, whose deletion would make c a certain winner of election $E = (C, V, k)$. Based on our construction, c can get votes only from voter group V_1, because c's position in the other groups is lower than r. That being said, only candidates s_i can be deleted to reach our goal and c can get *no more* than r votes. However, c should also receive *no less* than r votes, since that would tie him with the a_q's, which will result in c being an *uncertain* winner at best (and not a certain winner who is in all winning committees of size k). Thus c must receive *exactly* r votes, which can only be achieved by deleting r candidates s_i in group V_1.

Moreover, the sets $S_i \in \mathcal{S}'$ which the candidates s_i correspond to must comprise an exact 3-cover for the X3C instance. To show this, for a contradiction assume otherwise. After deleting the candidates and since $|S_i| = 3$ for all i, $1 \leq i \leq n$, there would be some b_j who receives two (instead of one) additional votes, giving b_j a plurality score of $r - 2 + 2 = r$. This would tie b_j with c and a_1, which leads to c being an uncertain winner only, as c would not be included in all winning committees of size $k = 2$. This means that the initial assumption was wrong and the sets S_i that the deleted candidates s_i correspond to must form an exact 3-cover for the instance $(B, \mathcal{S})$. ☐

Corollary 1. *SNTV is resistant to constructive control by replacing candidates.*

Proof. This follows directly from Theorem 1, as SNTV is IBC according to Lemma 1 and resistant to CCDC by Theorem 2. ☐

Now, to show that SNTV is resistant to DCRC, we start by showing that it is resistant to DCAC.

Theorem 3. *SNTV is resistant to destructive control by adding candidates.*

Proof. For this proof, we modify the construction of Hemaspaandra et al. [15] from their proof that plurality is resistant to DCAC by a reduction from HITTING-SET. Given an instance $(B, \mathcal{S}, r)$ of HITTING-SET, where $B = \{b_1, b_2, \ldots, b_m\}$ is a set, $\mathcal{S} = \{S_1, S_2, \ldots, S_n\}$ is a family of subsets S_i of B, and $r \leq m$ is a positive integer, we construct an instance of SNTV-DCAC as follows. Let $E = (C, V, k)$ be a multi-winner election with registered candidates $C = \{c\} \cup A$, where $A = \{a_1, a_2, \ldots, a_k\}$, unregistered candidates B, and the list V of votes that is divided into four voter groups as shown in Table 2.

Table 2. Voter groups for SNTV-DCAC in the proof of Theorem 3

Voter group	Number of votes	Preference
V_1	$2(m-r) + 2n(r+1) + 3k$	$c\ a_1\ \cdots$
V_2	$2(r+1)$ for each i, $1 \leq i \leq n$	$S_i\ c\ \cdots$
V_3	$2n(r+1) + 3(k+1)$ for each $q, 1 \leq q \leq k$	$a_q\ c\ \cdots$
V_4	2 for each j, $1 \leq j \leq m$, and each q, $1 \leq q \leq k$	$b_j\ a_q\ \cdots$

It is important to note here that in voter groups V_2 and V_4, the candidates $b_j \in B$ are initially not registered. Hence, the sets $S_1, \ldots, S_n$ are initially empty and candidates c and a_q are ranked first in V_2 and V_4, respectively. Now, based on the votes in Table 2, the plurality score of each registered candidate can be calculated:

$$\sigma(c) = 2(m-r) + 2n(r+1) + 3k + 2n(r+1) = 2(m-r) + 4n(r+1) + 3k,$$
$$\sigma(a_q) = 2m + 2n(r+1) + 3(k+1) \text{ for } q, 1 \leq q \leq k.$$

Based on the above plurality scores, it can be seen that initially candidate c is a certain winner of the election (C, V, k).

We now show that $(B, \mathcal{S}, r)$ is a yes-instance of HITTING-SET if and only if c can be made a certain nonwinner of the election resulting from E by adding at most r unregistered candidates.

Indeed, if B' is a hitting set of size r for S, then registering the candidates of B' would cause the plurality scores to change as follows (again, we denote these scores by σ' in the modified election):

$$\sigma'(c) = 2(m-r) + 2n(r+1) + 3k, \text{ after losing } 2n(r+1) \text{ votes from } V_2,$$
$$\sigma'(a_q) = 2(m-|B'|) + 2n(r+1) + 3(k+1) \text{ for } q, \ 1 \le q \le k, \text{ and} \quad (1)$$
$$\sigma(b_j) \le 2k + 2n(r+1) \text{ for } b_j \in B', \quad (2)$$

(1) holds since each a_q loses votes whenever some $b_j \in B'$ was added in voter group V_4, and (2) holds since $b_j \in B'$ may not necessarily be the first element in the S_i in voter group V_2, yet every such b_j has exactly $2k$ votes from group V_4.

Note now that $\sigma'(c) < \sigma'(a_q)$ for q, $1 \le q \le k$, i.e., there are k candidates who have a plurality score higher than c, so any winning committee of size k would consist of candidates in A only. This means that c can never be a member of any winning committee. Thus c has been turned into a certain nonwinner of the election $(C \cup B', V, k)$.

Conversely, suppose that c is a certain nonwinner (not a member of any winning committee) of the election $(C \cup B', V, k)$ for any subset $B' \subseteq B$ of unregistered candidates. This means that there are at least k candidates in $(C \cup B') \setminus \{c\}$ who have a higher plurality score than c. In $(C \cup B', V, k)$, we have:

$$\sigma'(c) = 2(m-r) + 2n(r+1) + 3k + 2(r+1)\ell,$$
$$\sigma'(a_q) = 2(m-|B'|) + 2n(r+1) + 3(k+1) \text{ for } q, \ 1 \le q \le k, \text{ and}$$
$$\sigma'(b_j) \le 2k + 2n(r+1) \text{ for } j, \ 1 \le j \le m,$$

where ℓ is the number of sets in S that have not been hit by B'. That is, if there exists a set $S_i \in S$ not hit by B' (i.e., $S_i \cap B' = \emptyset$), c gains $2(r+1)$ additional votes from voter group V_2 because $S_i = \emptyset$ in the vote "$S_i \ c \ \cdots$," so it takes the form "$c \ \cdots$." In order to guarantee that c is a certain nonwinner of the election $(C \cup B', V, k)$, $\ell = 0$ needs to hold. Recall that there are at least k candidates who have a higher plurality score than c. Based on the plurality scores above, notice that these candidates must be all $a_q \in A$, and that $\sigma'(c) < \sigma'(a_q)$ holds only when $\ell = 0$. However, we have

$$2(m-r) + 2n(r+1) + 3k + 2(r+1)\ell < 2(m-|B'|) + 3(k+1) + 2n(r+1)$$
$$2(m-r) + 3k + 2(r+1)\ell < 2(m-|B'|) + 3k + 3$$
$$2m - 2r + 2(r+1)\ell < 2m - 2|B'| + 3$$
$$2r - 2(r+1)\ell > 2|B'| - 3$$

and for the smallest ℓ such that $\ell \ne 0$, we have $2r - 2r - 2 > 2|B'| - 3$, which is equivalent to $1/2 > |B'|$, a contradiction. This means that only for the value $\ell = 0$ the first equation is valid, which implies that B' is a hitting set of size at most r.

Summing up, we have shown that S has a hitting set of size less than or equal to r if and only if destructive control by adding candidates can be executed for the constructed election (C, V, k) with unregistered candidates B. ◻

It can now easily be concluded that SNTV is also resistant to DCRC.

Corollary 2. *SNTV is resistant to destructive control by replacing candidates.*

Proof. This follows directly from Theorem 1, as SNTV is IBC according to Lemma 1 and resistant to DCAC by Theorem 3. ❑

4 Results for Bloc Voting

We will now turn our attention towards bloc voting and will again start by showing that bloc voting is also IBC.

Lemma 2. *Bloc voting is insensitive to bottom-ranked candidates.*

Proof. Just like SNTV, this is clear from the definition, as for a target committee size of k, bloc voting considers the k candidates ranked in the first k positions of each vote. Adding or deleting candidates at the bottom of the profile has no influence on the winning committees, because target committee size values of only $2 \leq k \leq |C| - 1$ are considered. Hence, the addition or deletion would take place at the $(|C|+1)$-st position. ❑

Theorem 4. *Bloc voting is resistant to constructive control by deleting candidates.*

Proof. This proof uses a similar approach as that of Theorem 2 to show resistance of bloc voting to CCDC, again by modifying the approach of Bartholdi et al. [1] showing that plurality is resistant to CCDC by a reduction from X3C. Let (B, S) be a given instance of X3C, where $B = \{b_1, \ldots, b_m\}$ is a set with $m = 3r$ elements (again assuming $r \geq 5$, without loss of generality) and $S = \{S_1, \ldots, S_n\}$ is a family of three-element subsets of B. Let b_i^1, b_i^2, b_i^3 denote the elements of S_i. Construct a multiwinner election $E = (C, V, k)$ as follows. Define the set of candidates by

$$C = \{c\} \cup \left(\bigcup_{i=1}^{n} A_i\right) \cup B \cup D \cup \left(\bigcup_{j=1}^{m} E_j\right) \cup G \cup H \cup \left(\bigcup_{i=1}^{n} \{s_i\}\right),$$

where c is the distinguished candidate, the candidates s_i correspond to the sets $S_i \in S$, and

$$A_i = \{a_i^1, \ldots, a_i^{k-1}\}, \quad 1 \leq i \leq n,$$
$$D = \{d_1, \ldots, d_k\},$$
$$E_j = \{e_j^1, \ldots, e_j^{k-1}\}, \quad 1 \leq j \leq m,$$
$$G = \{g_1, \ldots, g_{k-1}\}, \text{ and}$$
$$H = \{h_1, \ldots, h_r\}.$$

The list V of votes is divided into five voter groups as shown in Table 3, where an arrow sitting on top of a set of candidates indicates that the candidates in this set are ordered by increasing indices.

Table 3. Voter groups for CCDC in bloc voting

Voter group	Number of votes	Preference
V_1	1 for each i, $1 \leq i \leq n$	$s_i \overrightarrow{A_i}\ c\ \cdots$
V_2	$r-1$	$\overrightarrow{D}\ H\ \cdots$
V_3	1	$d_1\ \overrightarrow{G}\ H\ \cdots$
V_4	1 for each i, $1 \leq i \leq n$	$s_i\ \overrightarrow{A_i}\ b_i^1\ H\ \cdots$
	1 for each i, $1 \leq i \leq n$	$s_i\ \overrightarrow{A_i}\ b_i^2\ H\ \cdots$
	1 for each i, $1 \leq i \leq n$	$s_i\ \overrightarrow{A_i}\ b_i^3\ H\ \cdots$
V_5	$r-2$ for each j, $1 \leq j \leq m$	$b_j\ \overrightarrow{E_j}\ H\ \cdots$

Based on these votes, the k-approval scores of the candidates can now be calculated:

$\alpha_k(s_i) = 4$ for $1 \leq i \leq n$,
$\alpha_k(a_i^t) = 4$ for $1 \leq i \leq n$ and $1 \leq t \leq k-1$,
$\alpha_k(d_1) = r$ and $\alpha_k(d_t) = r-1$ for $2 \leq t \leq k$,
$\alpha_k(b_j) = r-2$ for $1 \leq j \leq m$,
$\alpha_k(e_j^t) = r-2$ for $1 \leq j \leq m$ and $1 \leq t \leq k-1$,
$\alpha_k(g_t) = 1$ for $1 \leq t \leq k-1$ and $\alpha_k(h_t) = 0$ for $1 \leq t \leq r$, and
$\alpha_k(c) = 0$.

Note again that there are at least k candidates who have a higher k-approval score than the distinguished candidate c, so c can never be a member of any winning committee and thus is a certain nonwinner of the election.

Now, we show that c can be made a certain winner by deleting at most r candidates if and only if the X3C instance $(B, \mathcal{S})$ has a solution.

Assume that $(B, \mathcal{S})$ is a yes-instance of X3C, and let $\mathcal{S}' \subseteq \mathcal{S}$ be an exact 3-cover of B. Delete the candidates s_i corresponding to the sets $S_i \in \mathcal{S}'$. Since $|\mathcal{S}'| = r$, the k-approval scores of the candidates change as follows:

- The distinguished candidate c gets r votes in total from voter group V_1.
- Candidates b_j will each get one additional vote from voter group V_4 due to the deletion of candidates s_i, changing their k-approval score to $r-1$ for $1 \leq j \leq m$.
- There are $n-r$ candidates s_i who keep their positions as well as their k-approval score of four votes each, since they are not deleted.
- All other candidates also keep their k-approval scores.

Note that candidates c and d_1 now both have the highest k-approval scores. Hence, c has been turned into a certain winner of the election, since $k \geq 2$.

Conversely, assume again that there exists a subset $C' \subseteq C$ of no more than r candidates whose deletion would make candidate c a certain winner of the election $E' = (C \setminus C', V, k)$. Based on the above construction, c would need to get votes from voter group V_1, because c's position in the other groups is lower than $r+k$. That being said, only candidates s_i or elements of the corresponding set A_i can be deleted to reach our goal and in that way, c would get *no more* than r votes. This would have no effect on the k-approval scores of the $k-1$ candidates in A_i that are not deleted. However, c should also receive *no less* than r votes, since that would tie c with the candidates in $D \setminus \{d_1\}$ and in B, which will result in c not being a certain winner anymore but only an uncertain winner or even a certain nonwinner. This is because in the first case, candidate d_1 would have r votes, and there would be more than $k-1$ candidates with $r-1$ votes, and in the second case, there would be more than k candidates who each have a k-approval score higher than c's. Thus c must receive *exactly* r votes, which can only be achieved by deleting exactly r candidates s_i.

Moreover, the sets $S_i \in \mathcal{S}'$ which the candidates $s_i \in C'$ correspond to, must comprise an exact 3-cover for the X3C instance. To see this, for a contradiction assume otherwise. After deleting the candidates and since $|S_i| = 3$ for all i, $1 \leq i \leq n$, there would be some b_j who receives two (instead of one) additional votes, giving b_j a k-approval score of $r - 2 + 2 = r$. This would tie b_j with c and d_1, which leads to c being an uncertain winner only, as c would not be included in all winning committees of size $k = 2$. This means that the initial assumption was wrong and the sets S_i that the deleted candidates s_i correspond to must form an exact 3-cover for the instance $(B, \mathcal{S})$. □

It can now be easily concluded that bloc voting is also resistant to CCRC.

Corollary 3. *Bloc voting is resistant to constructive control by replacing candidates.*

Proof. This follows immediately from Theorem 1, as bloc voting is IBC according to Lemma 2 and resistant to CCDC by Theorem 4. □

Now, a similar approach will show that bloc voting is resistant to DCRC.

Theorem 5. *Bloc voting is resistant to destructive control by adding candidates.*

Proof. As in the proof of Theorem 3, we modify the construction of Hemaspaandra et al. [15] from their proof that plurality is resistant to DCAC by a reduction from HITTING-SET. Given an instance $(B, \mathcal{S}, r)$ of HITTING-SET, with a set $B = \{b_1, b_2, \ldots, b_m\}$ and a family $\mathcal{S} = \{S_1, S_2, \ldots, S_n\}$ of subsets S_i of B, and $r \leq m$ is a positive integer, construct an instance of our control problem as follows.

Let $E = (C, V, k)$ be a multiwinner election with registered candidates

$$C = \{c\} \cup A \cup G \cup \bigcup_{i=1}^{n} E_i,$$

where $A = \{a_1, \ldots, a_{k-1}\}$, $G = \{g_1, \ldots, g_k\}$, and $E_i = \{e_i^1, \ldots, e_i^{k-1}\}$, with unregistered candidates B, and with the list V of votes that is divided into four voter groups as shown in Table 4.

Note again that in voter groups V_2 and V_4, the candidates $b_j \in B$ are initially not registered. Hence, the sets $S_1, \ldots, S_n$ are initially empty and the distinguished candidate c is ranked in the k-th position in V_2 and g_1 is ranked first in V_4. Based on the votes in Table 4, the k-approval scores of each registered candidate can now be calculated:

Table 4. Voter groups for DCAC in bloc voting for the proof of Theorem 5

Voter group	Number of votes	Preference
V_1	$2(m-r) + 2n(r+1) + 3$	$c \overrightarrow{A} \cdots$
V_2	$2(r+1)$ for each i, $1 \leq i \leq n$	$\overrightarrow{E_i} S_i\, c \cdots$
V_3	$2n(r+1) + 4$	$\overrightarrow{G} \cdots$
V_4	2 for each j, $1 \leq j \leq m$	$b_j \overrightarrow{G} \cdots$

$$\alpha_k(c) = 2(m-r) + 2n(r+1) + 3 + 2n(r+1) = 2(m-r) + 4n(r+1) + 3,$$
$$\alpha_k(a_t) = 2(m-r) + 2n(r+1) + 3 \text{ for all } a_t \in A,$$
$$\alpha_k(e_i^t) = 2(r+1) \text{ for all } e_i^t \in E_i, 1 \leq i \leq n, \text{ and}$$
$$\alpha_k(g_t) = 2m + 2n(r+1) + 4 \text{ for all } g_t \in G.$$

Note that c has the highest k-approval score. This makes c a certain winner of the election $E = (C, V, k)$, that is, without registering any candidates from B.

Now, we show that $(B, \mathcal{S}, r)$ is a yes-instance of HITTING-SET if and only if c can be made a certain nonwinner by adding at most r unregistered candidates.

Indeed, if B' is a hitting set of size r for $\mathcal{S}$, then registering the candidates from B' will alter the k-approval scores of the candidates as follows (denoting their k-approval scores by α'_k in the modified election):

- Since $B' \cap S_i \neq \emptyset$ for all i, $1 \leq i \leq n$, c is pushed beyond the first k positions of the votes in V_2. In total, c thus loses $2n(r+1)$ votes in V_2, so $\alpha'_k(c) = 2(m-r) + 2n(r+1) + 3$.
- For each registered $b_j \in B'$, some votes in V_4 change, and we have $\alpha'_k(g_k) = 2(m - |B'|) + 2n(r+1) + 4$.
- It may not always be the case that a registered $b_j \in B'$ is ranked in the k-th position of the votes in V_2. However, every $b_j \in B'$ has at least two votes from voter group V_4, so we have $\alpha'_k(b_j) \leq 2 + 2n(r+1)$ for each $b_j \in B'$.
- The candidates in A and E_i, $1 \leq i \leq n$, keep their previous k-approval scores.
- The candidates in $G \setminus \{g_k\}$ keep their k-approval score of $2m + 2n(r+1) + 4$ as well.

Since $|B'| \leq r$, we have $\alpha'_k(c) < \alpha'_k(g_k) < \alpha'_k(g_1) \leq \cdots \leq \alpha'_k(g_{k-1})$. Hence, there are k candidates—namely, those in G—who have k-approval scores higher than c's after the candidates from B' have been added. Thus c cannot be a member of the (unique) winning committee G. This means that adding the candidates from B' to the election has turned c into a certain nonwinner.

Conversely, let $B' \subseteq B$ and assume that c is not a member of any winning committee, i.e., a certain nonwinner of the election $(C \cup B', V, k)$. This means there are at least k candidates who definitely have higher k-approval scores than c. The candidates with higher k-approval scores than c have to be the candidates in G, since no candidate from a group other than G is close enough in points to c.

Now recall from the construction above that after registering the candidates from B', we have the following k-approval scores in the modified election (again denoting them by α'_k and letting ℓ be the number of sets in $\mathcal{S}$ that have not been hit by B'):

$$\alpha'_k(c) = 2(m-r) + 2n(r+1) + 3 + 2(r+1)\ell,$$
$$\alpha'_k(g_k) = 2(m-|B'|) + 2n(r+1) + 4,$$
$$\alpha'_k(g_t) = 2m + 2n(r+1) + 4 \text{ for } t, \ 1 \leq t \leq k-1,$$
$$\alpha'_k(b_j) \leq 2 + 2n(r+1) \text{ for } b_j \in B',$$
$$\alpha'_k(a_t) = 2(m-r) + 2n(r+1) + 3 \text{ for } t, \ 1 \leq t \leq k-1, \text{ and}$$
$$\alpha'_k(e^t_i) = 2(r+1) \text{ for } e^t_i \in E_i, 1 \leq i \leq n.$$

Based on the above scores, if the k candidates in the winning committee are the candidates in G, then $\alpha'_k(c) < \alpha'_k(g_t)$ for all $g_t \in G$. However, in order for $\alpha'_k(c) < \alpha'_k(g_k)$ to hold, $\ell = 0$ must hold, too. We have the following:

$$2(m-r) + 2n(r+1) + 3 + 2(r+1)\ell < 2(m-|B'|) + 2n(r+1) + 4$$
$$2m - 2r + 2(r+1)\ell < 2m - 2|B'| + 1$$
$$2r - 2(r+1)\ell > 2|B'| - 1$$

and for the smallest ℓ such that $\ell \neq 0$, we have $2r - 2r - 2 > 2|B'| - 1$, which is equivalent to $-1/2 > |B'|$, a contradiction. This means that only for the value $\ell = 0$ can it hold that $\alpha'_k(c) < \alpha'_k(g_k)$. Hence, all sets in $\mathcal{S}$ are hit by B', so $B' \subseteq B$ is a hitting set of size less than or equal to r for $\mathcal{S}$. □

Finally, we can easily conclude that bloc voting is also resistant to DCRC.

Corollary 4. *Bloc voting is resistant to destructive control by replacing candidates.*

Proof. This follows directly from Theorem 1, because bloc voting is IBC according to Lemma 2 and resistant to DCAC by Theorem 5. □

5 Conclusions and Future Work

We have studied the complexity of control of two of the most popular multiwinner voting systems: single nontransferable vote and bloc voting. We have shown that these two rules are resistant to constructive control by deleting and by replacing candidates and to destructive control by adding and by replacing candidates. This complements previous results by Meir et al. [20] who use a somewhat different framework to model control via utility functions. Note further that we excluded the committee size $k = 1$ (i.e., the single-winner case), so our results do not immediately follow from the corresponding single-winner results. We have shown them by using the notion of IBC and applying a general result of Loreggia [17] by appropriately modifying previous reductions of Bartholdi et al. [1] and Hemaspaandra et al. [15] adapted to our setting.

For future work, we propose to study control by replacing candidates or voters (and further control types, e.g., control by partitioning them) also for other prominent multiwinner voting rules such as the Chamberlin–Courant rule [5].

Acknowledgments. We thank the anonymous reviewers for helpful comments. This work was supported in part by Deutsche Forschungsgemeinschaft (DFG) under grants RO-1202/21-1 and RO-1202/21-2 (project number 438204498).

References

1. Bartholdi, J., III., Tovey, C., Trick, M.: How hard is it to control an election? Math. Comput. Model. **16**(8/9), 27–40 (1992)
2. Baumeister, D., Faliszewski, P., Rothe, J., Skowron, P.: Multiwinner voting. In: Rothe, J. (ed.) Economics and Computation. An Introduction to Algorithmic Game Theory, Computational Social Choice, and Fair Division. Classroom Companion: Economics, 2nd edn., chap. 6, pp. 403–465. Springer, Cham (2024). https://doi.org/10.1007/978-3-031-60099-9
3. Baumeister, D., Rothe, J.: Preference aggregation by voting. In: Rothe, J. (ed.) Economics and Computation. An Introduction to Algorithmic Game Theory, Computational Social Choice, and Fair Division. Classroom Companion: Economics, 2nd edn., chap. 4, pp. 233–367. Springer, Cham (2024). https://doi.org/10.1007/978-3-031-60099-9_4
4. Brandt, F., Conitzer, V., Endriss, U., Lang, J., Procaccia, A. (eds.): Handbook of Computational Social Choice. Cambridge University Press, Cambridge (2016)
5. Chamberlin, J., Courant, P.: Representative deliberations and representative decisions: proportional representation and the Borda rule. Am. Polit. Sci. Rev. **77**(3), 718–733 (1983)
6. Conitzer, V., Walsh, T.: Barriers to manipulation in voting. In: Brandt, F., Conitzer, V., Endriss, U., Lang, J., Procaccia, A. (eds.) Handbook of Computational Social Choice, chap. 6, pp. 127–145. Cambridge University Press (2016)
7. Dwork, C., Kumar, R., Naor, M., Sivakumar, D.: Rank aggregation methods for the web. In: Proceedings of the 10th International World Wide Web Conference, pp. 613–622. ACM Press (2001)
8. Ephrati, E., Rosenschein, J.: The Clarke Tax as a consensus mechanism among automated agents. In: Proceedings of the 9th National Conference on Artificial Intelligence, pp. 173–178. AAAI Press (1991)
9. Ephrati, E., Rosenschein, J.: A heuristic technique for multi-agent planning. Ann. Math. Artif. Intell. **20**(1–4), 13–67 (1997)
10. Erdélyi, G., Neveling, M., Reger, C., Rothe, J., Yang, Y., Zorn, R.: Towards completing the puzzle: complexity of control by replacing, adding, and deleting candidates or voters. J. Auton. Agents Multi-Agent Syst. **35**(2), 41:1–41:48 (2021)
11. Fagin, R., Kumar, R., Sivakumar, D.: Efficient similarity search and classification via rank aggregation. In: Proceedings of the 2003 ACM SIGMOD International Conference on Management of Data, pp. 301–312. ACM Press (2003). Corrigendum: https://doi.org/10.1145/1376616.1376778
12. Faliszewski, P., Rothe, J.: Control and bribery in voting. In: Brandt, F., Conitzer, V., Endriss, U., Lang, J., Procaccia, A. (eds.) Handbook of Computational Social Choice, chap. 7, pp. 146–168. Cambridge University Press (2016)
13. Faliszewski, P., Skowron, P., Slinko, A., Talmon, N.: Multiwinner voting: a new challenge for social choice theory. In: Endriss, U. (ed.) Trends in Computational Social Choice, chap. 2, pp. 27–47. AI Access Foundation (2017)
14. Garey, M., Johnson, D.: Computers and Intractability: A Guide to the Theory of NP-Completeness. W. H. Freeman and Company (1979)
15. Hemaspaandra, E., Hemaspaandra, L., Rothe, J.: Anyone but him: the complexity of precluding an alternative. Artif. Intell. **171**(5–6), 255–285 (2007)

16. Lang, J., Maudet, N., Polukarov, M.: New results on equilibria in strategic candidacy. In: Vöcking, B. (ed.) SAGT 2013. LNCS, vol. 8146, pp. 13–25. Springer, Heidelberg (2013). https://doi.org/10.1007/978-3-642-41392-6_2
17. Loreggia, A.: Iterative voting, control and sentiment analysis. Ph.D. thesis, University of Padova (2016)
18. Loreggia, A., Narodytska, N., Rossi, F., Venable, B., Walsh, T.: Controlling elections by replacing candidates or votes (extended abstract). In: Proceedings of the 14th International Conference on Autonomous Agents and Multiagent Systems, pp. 1737–1738. IFAAMAS (2015)
19. Maushagen, C., Niclaus, D., Nüsken, P., Rothe, J., Seeger, T.: Toward completing the picture of control in Schulze and ranked pairs elections. In: Proceedings of the 33rd International Joint Conference on Artificial Intelligence. ijcai.org (2024)
20. Meir, R., Procaccia, A., Rosenschein, J., Zohar, A.: Complexity of strategic behavior in multi-winner elections. J. Artif. Intell. Res. **33**, 149–178 (2008)
21. Pennock, D., Horvitz, E., Giles, C.: Social choice theory and recommender systems: analysis of the axiomatic foundations of collaborative filtering. In: Proceedings of the 17th National Conference on Artificial Intelligence, pp. 729–734. AAAI Press (2000)
22. Rothe, J. (ed.): Economics and Computation. An Introduction to Algorithmic Game Theory, Computational Social Choice, and Fair Division. Classroom Companion: Economics, 2nd edn. Springer, Cham (2024)
23. Yang, Y.: Complexity of manipulating and controlling approval-based multiwinner voting. In: Proceedings of the 28th International Joint Conference on Artificial Intelligence, pp. 637–643. ijcai.org (2019)

Compiling the Votes of a Subelectorate for Multi-winner Voting Rules

Neel Karia[1] and Jérôme Lang[2]

[1] Columbia University, New York, USA
nmk2154@columbia.edu
[2] CNRS, LAMSADE, PSL, Université Paris Dauphine, Paris, France
lang@lamsade.dauphine.fr

Abstract. Compiling the votes of a subelectorate is a well-known problem in computational social choice. The goal is to store the information contained in the votes cast by a subelectorate in a space-efficient way, such that when the rest of the votes become available, the winners can be accurately ascertained. This problem has been studied for single-winner voting rules. We provide a comprehensive compilation complexity landscape for several ordinal and approval-based multi-winner voting rules.

Keywords: Voting · Compilation Complexity · Social Choice

1 Introduction

In a usual voting setting, it is assumed that the votes cast by the agents arrive simultaneously. A voting rule is then applied to elect a winner (or a set of co-winners). However, in most real-life scenarios, the votes are not obtained at the same time or at the same place. It is then beneficial to compile the votes that are already available, that is, to store the information contained in these votes, using as little space as possible, in such a way that when the rest of the votes are known, the winner(s) can be determined. The compilation complexity of a voting rule is the worst-case size of the most succinct compilation. Compilation has several advantages: first, the votes of the subelectorate can be stored succinctly. second, the compilation provides a synthetic, understandable view of the votes of the electorate, which makes verification easier. Lastly, it may help us develop dynamic programming algorithms for elections under uncertainty, by storing partial results (refer to [13]).

Compilation of single-winner rules has been studied in [4] and [19]. The compilation complexity of voting rules is related to their communication complexity (initiated in [6]), with a major difference: communication complexity allows several communication rounds, while compilation allows only one round (and is thus related to one-round communication complexity). It is also related to the possible and necessary winners problems (initiated in [15]). More recently, there has been related work on sample complexity for approximate winner predictions

[7] and algorithms for the approximate compilation of information from vote streams for some multi-winner voting rules [8]; a major difference of our work from [8] being that ours considers an offline and exact version of compilation.

We study the compilation complexity of multi-winner voting rules. The interpretation of compilation can be either temporal or spatial. We illustrate both with an example. Assume we are considering multi-winner approval voting (MAV) – voter ballots are subsets of candidates, and the winners are the candidates with the highest $k = 2$ scores (using tie-breaking if necessary). We have four candidates a, b, c, d, out of which we want to select two winners.

For the temporal interpretation, assume that votes come in two stages: say today's votes are two votes (abc)[1], one vote (abd), one (bc), and one (cd); we will have more votes tomorrow (but we do not know how many). Which information should we keep from these 5 votes received today? The answer is intuitive enough (and proven formally in the paper): it is sufficient and necessary to store the information that both b and c have one more approval than a, which in turn has one more approval than d. Once we know the late votes, it will then be easy to determine the winners: for instance, if we get two votes (abd) and one vote (a), then we know that the winners for MAV are a and b.

For the spatial interpretation, assume that votes are collected at two different polling stations (say, two towns). At station 1 we have two votes (abc), one vote (abd), one (bc), and one (cd); at station 2 we have two votes (abd) and one vote (a). We need to publish local results, for several reasons: the people would like to know how many votes the candidates got in their town, and the local observers want to check that the results of ballot counting in their polling station (which they assisted in) correspond to the official figures published the day after the election. In most cases, it is undesirable to publish anonymized approval ballots: this could reveal too much information and would take up too much space if there are many candidates and voters. Instead, for each polling station we can simply publish the list all candidates ranked by non-increasing scores, together with their relative differences of the approval scores. This information is all we need to compute the winners (and it is also necessary, as we will later show).

For both interpretations, it is useful to have an order of magnitude estimate of the space needed to store the information coming for a subelectorate.

In Sect. 2 we give the necessary background on compilation functions, compilation complexity, and multi-winner voting rules. In Sect. 3 we extend the compilation framework from single-winner to multi-winner rules. In Sect. 4 we show results on optimal compilation equivalence conditions and compilation complexity for various multi-winner rules with ordinal input. In Sect. 5 we do the same for multi-winner rules with approval-based inputs.

---

[1] For approval-based rules, by (abc) we indicate that a, b, c are approved in the vote. For ordinal rules we sometimes use (abc) to indicate the vote $a \succ b \succ c$.

2 Background

Let A be a set of candidates, with $|A| = m$. Let $[m] = \{1, \ldots, m\}$. Let $\mathcal{P}_A$ be the space of *votes*, which, depending on the rule used, is either the set of linear orders over A (ranked/ordinal ballots) or the set of subsets of A (approval ballots). $\mathcal{P}_A^n$ represents the space of n votes (either ordinal or approval). A *preference profile* $P \in \mathcal{P}_A^n$ is a collection $(V_1, \ldots, V_n)$ of n votes. For any given n it belongs to $\mathcal{P}_A^* = \bigcup_{n \geq 1} \mathcal{P}_A^n$ (the set of profiles with an arbitrary number of votes).

An *irresolute* multi-winner voting rule is a function g that maps any profile P and any $k \in [m]$ to a nonempty subset of k-committees $g(P, k) \in \mathcal{S}_k(A)$. The restriction of an irresolute multi-winner rule to $k = 1$ is an irresolute single-winner rule. Note that all our results use irresolute rules when unspecified.

2.1 Compilation Complexity and Compilation Functions for Single-Winner Voting Rules

For single-winner voting rules, compilation complexity was introduced in [4], and further studied in [3,19]. We recall the setting for single-winner rules.

Let f be a single-winner voting rule defined for a profile of rankings or approvals over candidates. Consider any two profiles $P, Q \in \mathcal{P}_A^n$, that contain the votes of subelectorates composed of n voters. We say that P and Q are f-*equivalent* [4], which we denote by $P \sim_f Q$, if for each $t \geq 0$ and each profile $T \in \mathcal{P}_A^t$, we have $f(P \cup T) = f(Q \cup T)$. Clearly, $\sim_f$ is an equivalence relation. This notion of f-equivalence is further generalised in [19], where the number t of new votes is known beforehand; we will not use this notion in this paper.

A function $\sigma : \mathcal{P}_A^n \to \{0, 1\}^*$ is a *compilation function for f* if there exists a function $\rho : \{0, 1\}^* \times \mathcal{P}_A^* \to A$ such that for all $P \in \mathcal{P}_A^n$, $t \geq 0$ and $T \in \mathcal{P}_A^t$, we have $\rho(\sigma(P), T) = f(P \cup T)$. By $size(\sigma)$, we denote the number of bits needed to represent $\sigma(P)$. The *compilation complexity* of f, denoted $C(f)$, is the minimum value of $size(\sigma)$ over all compilation functions for f. It is known that $C(f)$ is the upper integer part of the logarithm (in base 2) of the number of equivalence classes for $\sim_f$ (*Proposition 1* of [4]).

2.2 Multi-winner Voting Rules

Ordinal Multi-winner Rules. We start with rules whose input is a profile consisting of ranked ballots (*ordinal* rules), for which a good overview is in [11]. All of them but one (sequential plurality) are defined via *scores*: given a profile $P = (V_1, \ldots, V_n)$, a score $sc(S, P)$ is associated with each committee $S \in \mathcal{S}_k(A)$, and the winning committee W is the one that maximizes $sc(S, P)$. If P contains a single vote V then we use $sc(S, V)$ instead of $sc(S, P)$. The multi-winner irresolute rule that selects all committees maximizing sc is denoted by g^{sc}.

For the first four rules below, the score of a k-committee $S \in \mathcal{S}_k(A)$ is the sum of the scores $sc(S, V_i)$ that it gets from all votes $V_1, \ldots, V_n$.

- **Single Non Transferable Vote (SNTV):** $sc(S, V_i) = 1$ if S contains the top candidate of V_i; else $sc(S, V_i) = 0$.

- **Bloc:** $sc(S, V_i)$ is the number of candidates in S ranked in the first k positions of V_i.
- **k-Borda:** $sc(S, V_i)$ is the sum of the Borda scores of the candidates in S; the Borda score of a candidate ranked in position j in a vote is $m - j$.
- **Chamberlin-Courant (β-CC):** $sc(S, V_i)$ is the Borda score w.r.t. V_i of the best candidate in S according to V_i.
- **Gehrlein Stable Rules** [12]: We say that S is a weak Condorcet set if for every candidates c in S and d in $A \setminus S$, at least half of the voters prefer c to d. A multi-winner rule is Gehrlein stable if it outputs a weak Condorcet set whenever there exists one. Two Gehrlein-Stable rules are NED (*number of external defeats*) and SEO (*size of external opposition*) [5], which can be seen as the respective multi-winner counterparts of the Copeland and maximin single-winner rules. The NED rule outputs committees S that maximize the number of pairs $(x, y) \in S \times A \setminus S$ such that x majority-beats y. The SEO rule outputs committees S that maximise $\min_{x \in S, y \in A \setminus S} |\{i : x \succ_i y\}|$.
- **Theta-Winning Sets (θ)** [10] – also called LSE-maximin [1]. A k-committee S is a θ-winning set if, for every candidate d not in S, more than a θ-fraction of the voters prefer some member of S to d. A winning committee W is a k-committee being is a θ-winning set for the largest value of θ.
- **Sequential Plurality (SeqPl)** [2]: We proceed in rounds. Initially, $W = \emptyset$. The candidate ranked first by the largest number of votes is added to W, removed from the profile, and the procedure is repeated k times (breaking ties if resolute, or considering all possibilities if irresolute). W is the output.

SNTV, Bloc, k-Borda and β-CC belong to the larger family of *committee scoring rules* [9,18], for which (1) $sc(S, P) = \sum_{V_i \in P} sc(S, V_i)$, and (2) $sc(S, V_i)$ is a function of the vector containing the ranks of the elements of S in V_i. Moreover, SNTV, Bloc and k-Borda are *decomposable* committee scoring rules: each candidate x has a score function $sc(x, V)$, such that $sc(S, V) = \sum_{x \in S} sc(x, V)$.

An important ordinal rule that we do not consider is Single Transferable Vote (STV), as it seems quite challenging to characterise its compilation complexity.

Approval-Based Multi-winner Voting Rules. For these three rules, the input is a profile consisting of approval ballots. Refer to [16] for a detailed survey.

- **Multiwinner Approval Voting (MAV):** The winning committee consists of the k candidates that are approved most frequently.
- **Approval-based Chamberlin-Courant (α-CC):** $sc(S, V_i) = 1$ if V_i votes for at least one candidate of S; else $sc(S, V_i) = 0$.
- **Proportional Approval Voting (PAV):** $sc(S, V_i) = \sum_{x=1}^{j} \frac{1}{x}$ if V_i approves j candidates of S.

For the three approval-based rules above, $sc(S, P) = \sum_{V_i \in P} sc(S, V_i)$. These three rules belong to a family of voting rules called the *Thiele's Optimization Method* [14], where $sc(S, V_i) = h(|S \cap V_i|)$ where h is a function from $\mathbb{N}_0$ to $\mathbb{R}$ and $h(1)$ is normalised to 1. We focus on a subset of these rules for which

$h(|S \cap V_i|) \leq |S \cap V_i|$ (which include all the three rules above for $h(r) = r$, $h(r) = 1$, and $h(r) = \sum_{i=1}^{r} \frac{1}{r}$ respectively).

Two important approval-based rules that we are not considering are Phragmén [14] and the Method of Equal Shares [17] because we do not have characterisations of their compilation complexity. This is left for future work.

3 Compilation for Multi-winner Voting Rules

The concepts of equivalence classes, compilation complexity and compilation functions used for single-winner rules can easily be extended to multi-winner rules (ordinal or approval-based) as defined below.

Definition 1. *Let g be a multi-winner voting rule. Let $k \in [m]$ be fixed. Consider any two profiles $P, Q \in \mathcal{P}_A^n$, that contain the votes of subelectorates composed of n voters. We say that P and Q are g-equivalent, which we denote by $P \sim_g Q$, if for each $t \geq 0$ and each profile $T \in \mathcal{P}_A^t$, we have $g(P \cup T, k) = g(Q \cup T, k)$. Clearly, $\sim_g$ is an equivalence relation.*

Definition 2. *A function $\sigma : \mathcal{P}_A^n \times [m] \to \{0,1\}^*$ is a compilation function for g if there exists a function $\rho : \{0,1\}^* \times \mathcal{P}_A^* \times [m] \to A$ such that for all $P \in \mathcal{P}_A^n$, $t \geq 0$ and $T \in \mathcal{P}_A^t$, we have $\rho(\sigma(P), T, k) = g(P \cup T, k)$. By $size(\sigma)$ we denote the number of bits needed to represent $\sigma(P)$.*

Definition 3. *The compilation complexity of g, denoted $C(g)$, is the minimum value of $size(\sigma)$ over all compilation functions for g.*

Proposition 1 of [4] for single-winner voting rules continues to hold for multi-winner rules as well, implying that $C(g)$ is the upper integer part of the logarithm (in base two) of the number of equivalence classes for $\sim_g$.

4 Results for Ordinal Rules

Let $k \in [m]$ be fixed. It is easy to derive a sufficient condition for all the committee scoring rules:

Lemma 1. *Let g_C^{sc} be a committee scoring rule with score function sc. If for all committees $S \in \mathcal{S}_k(A)$ we have $sc(S, P) = sc(S, Q)$, then $P \sim_{g_C^{sc}} Q$.*

Proof. Let $sc(S, P) = sc(S, Q)$ for all $S \in \mathcal{S}_k(A)$. For any profile T, $sc(S, P \cup T) = sc(P) + sc(T)$ and $sc(S, Q \cup T) = sc(Q) + sc(T)$. Hence $sc(S, P \cup T) = sc(S, Q \cup T)$, which shows that $P \sim_{g_C^{sc}} Q$. ∎

A weaker sufficient condition holds for the decomposable committee scoring rules. For $P = (V_1, \ldots, V_n)$, let $sc(x, P) = \sum_i sc(x, V_i)$.

Lemma 2. *Let g_{DC}^{sc} be a decomposable committee scoring rule with score function sc. If for each candidate $x \in A$ we have $sc(x, P) = sc(x, Q)$, then $P \sim_{g_{DC}^{sc}} Q$.*

Proof. Decomposability implies $sc(S, V) = \sum_{x \in S} sc(x, V)$. Hence, $sc(S, P) = \sum_{x \in S} sc(x, P)$. Now, for all $a \in A$, let $sc(a, P) = sc(a, Q)$. This implies that $sc(S, P) = sc(S, Q)$, and $P \sim_{g_{DC}^{sc}} Q$ using Lemma 1, as a decomposable committee scoring rule is a committee scoring rule too. ∎

The condition in Lemma 1 is also necessary for anonymous, neutral, and non-constant committee scoring rules (including SNTV, k-Borda, Bloc and Chamberlin-Courant). By non-constant, we mean that for a given vote V there exist $S, S' \in \mathcal{S}_k(A)$ such $sc(S', V) \neq sc(S, V)$.

Theorem 1. *Let g_{AC}^{sc} be an anonymous, neutral, and nonconstant committee scoring rule. $P \sim_{AC} Q$ holds if and only if for all $S, S' \in \mathcal{S}_k(A)$, $sc(S, P) = sc(S, Q)$.*

Proof. ($\Longleftarrow$) follows from Lemma 1. For ($\Longrightarrow$), the proof is as follows. First we show that if $P \sim_{AC} Q$, then for any S and S' in $\mathcal{S}_k(A)$, we have

$$sc(S, P) - sc(S, Q) = sc(S', P) - sc(S', Q) \qquad (1)$$

Assume Eq. 1 fails for some S and S'. Let $sc(S, P) = \alpha$, $sc(S, Q) = \beta$, $sc(S', P) = \gamma$, and $sc(S', Q) = \delta$. Without loss of generality, $\alpha - \beta > \gamma - \delta$.

Let $S_1 = S \cap S'$, $S_2 = S \backslash S'$, and $S_3 = S' \backslash S$. Let $\mathcal{N}$ be an arbitrary bijective mapping from S_2 to S_3. We build a set a collection of votes T_1 obtained from P by swapping the positions of each candidate a in S_2 and $\mathcal{N}(a)$ in S_3. By neutrality, $sc(S, T_1) = \gamma$ and $sc(S', T_1) = \alpha$. Hence, $sc(S, P \cup T_1) = sc(S', P \cup T_1) = \alpha + \gamma$. However, $sc(S, Q \cup T_1) = \beta + \gamma < \alpha + \delta = sc(S', Q \cup T_1)$. In words, S' beats S in profile $Q \cup T_1$, but S' ties with S in profile $P \cup T_1$.

We define a voter set U_1 where the votes consist of a set of all permutations of S_2 followed by all permutations of S_1 followed by all permutations of S_3 followed by all permutations of the rest of the candidates (in cross multiplication, so that makes $|S_1|!|S_2|!|S_3|!(|A \backslash (S_1 \cup S_2)|!$ votes). We take u copies of each vote in U_1, for a large enough u, and call it T_2. Likewise, we define a voter set U_2 where the votes consist of a set of all permutations of S_3 followed by all permutations of S_1 followed by all permutations of S_2 followed by all permutations of the rest of the candidates. We take u copies of U_2 and call it T_3.

We consider a voter set W_1 where the votes consist of a set of all permutations of S_1 followed by all permutations of S_2 followed by all permutations of S_3 followed by all permutations of the rest of the candidates (in cross multiplication). We consider w copies of the above (for a large enough w) and call it as T_4. Likewise, we consider a voter set W_2 where the votes consist of a set of all permutations of S_1 followed by all permutations of S_3 followed by all permutations of the rest of the candidates (in cross multiplication). We consider w copies of W_2 and call it as T_5. Note that if S_1 is empty, then we do not need W_1 and W_2 because they are identical to U_1 and U_2 respectively.

For large enough u and w, adding votes $T_2 \cup T_3 \cup T_4 \cup T_5$ ensures that S and S' beat the rest of the committees in $\mathcal{S}_k(A)$, but increase their own scores by equal amounts (this uses non-constantness). Let $T = T_1 \cup T_2 \cup T_3 \cup T_4 \cup T_5$. For $P \cup T$, the winners are S and S', but for $Q \cup T$, the winner is solely S'. This implies (1).

Then we show that $\sum_{S \in \mathcal{S}_k(A)} sc(S, P)$ is a constant irrespective of the choice of $P \in \mathcal{P}_A^n$ (2). This must hold since for any vote V, whatever preference order they have, the committees cover all possible sets of size k. Hence, by neutrality, we see that $\sum_{S \in \mathcal{S}_k(A)} sc(S, V)$ is a constant, say ρ. $\sum_{S \in \mathcal{S}_k(A)} sc(S, P)$ is $n\rho$, using anonymity, and therefore a constant.

From Eq. 1, if we had $sc(S, P) - sc(S, Q) = sc(S', P) - sc(S', Q) = c > 0$ (without loss of generality), then summing any side over $S \in \mathcal{S}_k(A)$ would contradict (2). Hence, we must have $sc(S, P) = sc(S, Q)$. ∎

As an immediate corollary, we see that the condition in Lemma 2 is also necessary for any anonymous decomposable committee (ADC) scoring rule (which includes SNTV, k-Borda, and Bloc).

Corollary 1. $P \sim_{g_{ADC}^{sc}} Q$ holds if and only if for every candidate $x \in A$ we have $sc(x, P) = sc(x, Q)$.

4.1 Single Non Transferable Vote (SNTV)

As SNTV is an ADC scoring rule, Corollary 1 tells us that two profiles P and Q are SNTV-equivalent if and only if for each candidate x, the number of votes with x on top is the same in P and Q. For instance, consider a profile P with ranked ballots $\{(abc), (abc), (bca)\}$ and a profile Q with ranked ballots $\{(abc), (bca), (bca)\}$. Let $k = 2$. $P \not\sim_{SNTV} Q$, as $sc(a, P) = 2$ but $sc(a, Q) = 1$. If we add $T = \{(cab), (cab)\}$ to P and Q, the winning committee will be (a, c) for $P \cup T$, but it will be (b, c) for $Q \cup T$.

Corollary 2. $C(SNTV) = \Theta\left(m \log\left(1 + \frac{n}{m}\right) + n \log\left(1 + \frac{m}{n}\right)\right)$.

The above result follows from the bounds of the plurality rule in Corollary 1 of [4] because we have the same equivalence classes for plurality and SNTV.

4.2 k-Borda

As k-Borda is an ADC scoring rule, Corollary 1 gives us its equivalence classes. For instance, consider a profile P with ranked ballots $\{(abc), (abc), (bac)\}$ and a profile Q with ranked ballots $\{(abc), (bca), (bac)\}$. Let $k = 2$. $P \not\sim_{k-Borda} Q$, as $sc(a, P) = 5$ but $sc(a, Q) = 3$. If we add $T = \{(cba), (bca)\}$ to P and Q, the winning committee will be (a, b) for $P \cup T$, but it will be (b, c) for $Q \cup T$.

Corollary 3. $C(k - Borda) = \Theta(m \log nm)$.

The above result follows from the bounds of Borda rule in Corollary 2 of [4] as we have the same equivalence classes for Borda and k-Borda rules.

4.3 Bloc

As Bloc is an ADC scoring rule, Corollary 1 gives us a characterisation of its equivalence classes. As an example, consider a profile P with ranked ballots $\{(abc),(abc),(abc)\}$ and a profile Q with ranked ballots $\{(abc),(abc),(acb)\}$. Let $k=2$. $P \not\sim_{Bloc} Q$, as $sc(b,P)=3$ but $sc(b,Q)=2$. If we add $T=\{(cba),(bca)\}$ to P and Q, the winning committee will be (a,b) for $P \cup T$, but it will be (b,c) for $Q \cup T$.

Corollary 4. $C(Bloc) = \Theta\left(m \log\left(1 + \frac{n\hat{k}}{m}\right) + n\hat{k} \log\left(1 + \frac{m}{n\hat{k}}\right)\right)$ where $\hat{k} = \min(k, m-k)$.

The equivalence classes of Bloc rule are identical to those of l-Approval and hence the bounds are obtained from Theorem 1 of [19].

4.4 Chamberlin-Courant (β-CC)

As Chamberlin-Courant is an AC scoring rule, Theorem 1 characterises its equivalence classes. For instance, consider a profile P with ranked ballots $\{(abc),(abc),(bca)\}$ and a profile Q with ranked ballots $\{(abc),(bca),(bca)\}$. Let $k=2$. $P \not\sim_{\beta-CC} Q$, as $sc(bc,P)=4$ but $sc(bc,Q)=5$. If we add $T=\{(cab),(cab)\}$ to P and Q, the winning committee will be (a,c) for $P \cup T$, but it will be (b,c) for $Q \cup T$.

We obtain the following bounds by using some counting arguments (the proofs of these bounds and those that follow for the rest of the voting rules are skipped for brevity; refer to the full version at https://shorturl.at/jJpe4).

Corollary 5. $C(\beta - CC) = O\left(\binom{m}{k} \log(n(m-k))\right)$.

Corollary 6. $C(\beta - CC) = \Omega\left(\log(n(m-k))\right)$.

Unlike the results we obtained for the decomposable committee scoring rules, our lower bound and upper bound results for β-CC are loose. This is because it is difficult to count the number of functions from $\mathcal{S}_k(A)$ to $\mathbb{N}$ corresponding to $sc(.,P)$ for some profile P for each of these rules, because of the dependencies between the scores of intersecting committees. As an example, take $k=2$. If $sc_B(.,i)$ denotes the Borda score of a candidate for vote V_i, then $sc(bc,\{V_i\}) = \max(s_B(b,i), s_B(c,i)) \leq \max\left(sc(ab,\{V_i\}), sc(ac,\{V_i\})\right)$. Summing this inequality over all votes, we obtain that $sc(bc,P) \leq sc(ab,P)+sc(ac,P)$. More generally, if $X \subseteq X_1 \cup \ldots \cup X_q$ then $sc(X,P) \leq \sum_{j=1}^{q} sc(X_j,P)$.

4.5 Gehrlein-Stable Rules (GehrSta)

For any two candidates c and d, let $N_P(c,d)$ be the number of voters in P that prefer c to d. Let $\mathcal{M}_P$ represent the tournament over candidates in A with voters P. We get the following equivalence result which can be proved using a similar idea to the proof of *Lemma 4* of [4].

Theorem 2. $P \sim_{GehrSta} Q$ holds if and only if $\mathcal{M}_P = \mathcal{M}_Q$, where GehrSta is a Gehrlein Stable rule.

For instance, consider the rule NED, a profile P with ranked ballots $\{(abc), (abc), (abc)\}$ and a profile Q with ranked ballots $\{(bac), (bac), (bca)\}$. Let $k = 2$. $P \not\sim_{GehrSta} Q$, as $\mathcal{M}_P \neq \mathcal{M}_Q$ (a's out-degree is different in $\mathcal{M}_P$ and $\mathcal{M}_Q$). If we add $T = \{(cba), (cba)\}$ to P and Q, the winning committee will be (a, b) for $P \cup T$, but it will be (b, c) for $Q \cup T$.

Corollary 7. $C(GehrSta) = O(m^2 \log n)$, and there exists a constant $q > 0$ such that $C(GehrSta) = \Omega\left(m^2 \log\left(\left\lfloor \frac{n}{qm} \right\rfloor - 2\right)\right)$.

The result follows from *Proposition 8* of [4] and *Proposition 5* of [19], as the equivalence classes are identical to those of a Condorcet Consistent WMG Rule.

4.6 θ-Winning Sets (θ)

We know that the winning committee W is the θ-winning set for the largest value of θ (after tie-breaking). Let $N_P(d, S)$ be the number of voters in P that prefer d to any candidate in S. Let the score of S for P be $sc(S, P) = n - \max_{d \in A \setminus S}(N_P(d, S))$. Then, the maximum θ for S is $\frac{sc(S, P)}{n}$.

Theorem 3. $P \sim_\theta Q$ holds if and only if for all $d \in A \setminus S$ and for all $S \in \mathcal{S}_k(A)$, we have $N_P(d, S) = N_Q(d, S)$.

Proof. ($\Longleftarrow$) is straightforward from how the rule is defined. For ($\Longrightarrow$), the proof can be obtained as follows.

Let S_x^P be the xth highest scoring k-committee for P according to sc. Choose the largest i such that $N_P(d, S_j^P) = N_Q(d, S_j^P)$ for all $d \in A \setminus S_j^P$ and for all $j \in [i-1]$. Initially, we consider $i \neq 1$ and denote the winner as W.

First consider the case where $sc(S_i^P, P) \neq sc(S_i^P, Q)$. Assume without loss of generality that $sc(S_i^P, P) > sc(S_i^P, Q)$. Here, we add voters T_1, each of whom contains every permutation of S_i one by one in the first k places until S_i^P becomes the winner for $P \cup T_1$ but not for $Q \cup T_1$. Hence it is not equivalent.

Next consider the case where $sc(S_i^P, P) = sc(S_i^P, Q)$ but there exists a $c \in A \setminus S_i^P$ such that $N_P(c, S_i^P) \neq N_Q(c, S_i)$. Assume without loss of generality that $N_P(c, S_i^P) < N_Q(c, S_i^P)$. Consider profile T_2 consisting of $n - sc(S_i^P, Q) + 1 - N_Q(c, S_i^P)$ voters where each voter votes in the order c followed by S_i in some order followed by the rest of the candidates in some order. This will increase S_i^P's score by $n - sc(S_i^P, Q) + 1 - N_Q(c, S_i^P)$ and $n - sc(S_i^P, Q) - N_Q(c, S_i^P)$ for $P \cup T_2$ and $Q \cup T_2$ respectively. Now since $sc(S_i^P, P) > s_Q(S_i^P, Q)$, T_1 can be obtained similarly as above to prove lack of equivalence.

Now consider the case where $i = 1$. Here, if $sc(W, P) \neq sc(W, Q)$, assume without loss of generality that $sc(W, P) < sc(W, Q)$. Take the highest scoring committee S_y^Q for Q for which $sc(S_y^Q, P) \geq sc(S_y^Q, Q)$ and add a set of voters T_3 each of whom keeps a permutation of S_y one by one at the top of their list. Continue this until S_y^Q wins for $P \cup T_3$ but not for $Q \cup T_3$.

If $sc(W, P) = sc(W, Q)$, that implies that there exists a c in $A\setminus W$ such that $N_P(c, W) \neq N_Q(c, W)$. Adding a T_4 to both P and Q which is similar to T_2, we reach a similar position as that required for adding T_3, which shows a lack of equivalence. Hence, ($\Longrightarrow$) must be true. ∎

For instance, consider a profile P with ranked ballots $\{(abc), (abc), (bca)\}$ and a profile Q with ranked ballots $\{(abc), (abc), (cab)\}$. Let $k = 2$. $P \not\sim_\theta Q$, as $N_P(c, ab) = 0$ but $N_Q(c, ab) = 1$. If we add $T = \{(cab)\}$ to P and Q, the winning committee will be (a, b) for $P \cup T$, but it will be (a, b) or (a, c) for $Q \cup T$ depending on the tie-breaking rule.

Corollary 8. $C(\theta) = O\left(\binom{m}{k}(m-k)\log n\right)$.

Corollary 9. $C(\theta) = \Omega((m-k)\log n)$.

Again we get a gap between the lower bound and upper bound results, for the same reason as in *Subsect.* 4.4.

4.7 Sequential Plurality (SeqPl)

For each vote V, let V^j be the top-j truncation of V, where $j \in [k]$. For instance, if $V = (abcd)$ then $V^2 = (ab)$. Given a profile P, for each ordered sequence of j candidates λ^j, let $N(P, \lambda^j)$ be the number of votes V in P such that $V^j = \lambda^j$. For instance, if $P = \{(abcd), (abdc), (acbd), (dabc), (dabc)\}$ and $k = 2$, $N(P, ab) = N(P, da) = 2$, $N(P, ac) = 1$, and $N(P, \lambda^k) = 0$ for $\lambda^k \neq ab, ac, da$.

Theorem 4. $P \sim_{SeqPl} Q$ *if and only if* $N(P, \lambda^k) = N(Q, \lambda^k)$ *for each ordered sequence of k candidates λ.*

Proof. ($\Longleftarrow$) is straightforward as the voters are anonymous and to select the jth winner for each $j \in [k]$, we need no more information than that in the first j places of each voter's ordering.

For ($\Longrightarrow$), the proof is as follows. It is a necessary condition for $k = 1$, as in that case, it reduces to the Plurality rule for which it is a necessary condition. From now on, assume that $k > 1$.

Suppose it was not a necessary condition for equivalence. There would be a $j \in [k]$ such that $N(P, \lambda^j) \neq N(Q, \lambda^j)$ for some j-sized ordered sequence λ^j. We will show that by taking the smallest such j satisfying the above condition, we can add voters T to both P and Q such that their winning committees will have different candidates at the jth place and the same candidates in all other places.

The first $j-1$ winners (call them C_1) are the same for P and Q as $N(P, \lambda^i) = N(Q, \lambda^i)$, for each ordered sequence of i candidates for all $i \in [j-1]$. First, consider a group of voters V_a^P and V_a^Q for P and Q whose first $j-1$ positions contain the first $j-1$ winners in some fixed order (which could be different from the final ranking) and their j^{th} member is some c which is not the jth member of the winning committee, which is d. Consider the first case where we assume without loss of generality that $|V_a^P| > |V_a^Q|$. Here we add voters T_1 who have their preference order until $j-1$ same as in V_a^P and with c in the jth position and in the rest of the positions the $k-j$ candidates, call them C_2 (apart from c and d and the first $j-1$ candidates that we want as the last $k-j$ winners, in the order needed. Continue this until c is the jth winner for $P \cup T_1$ but it is still d for $Q \cup T_1$.

After this, if the last $k-j$ winners are not C_2 for $P \cup T_1$ and $Q \cup T_1$ both, add the profile T_2 which is constructed as follows. The first $j-1$ candidates for each added voter are C_1. Put the next $k-j+1$ preferences as follows: 2 voters having c followed by C_2, 2 voters having d followed by C_2 and 1 voter each having C_2 followed by c and C_2 followed by d respectively. Add the above collection of 6 voters one by one repeatedly until the last $k-j$ winners are C_2 for both $P \cup T_1 \cup T_2$ and $Q \cup T_1 \cup T_2$. Hence the winning committee is $C_1 \cup \{c\} \cup C_2$ for $P \cup T_1 \cup T_2$ and $C_1 \cup \{d\} \cup C_2$ for $Q \cup T_1 \cup T_2$.

Note that if c was the jth winner for P, consider voters V_b^P and V_b^Q with the same $j-1$ candidates at the top as above and a d in the jth position such that $|V_b^P| < |V_b^Q|$ and such a d must exist given the assumption on V_a above. We can make $C_1 \cup \{c\} \cup C_2$ as the winning committee for $P \cup T_1 \cup T_2$ and $C_1 \cup \{d\} \cup C_2$ as the winning committee for $Q \cup T_1 \cup T_2$, by choosing profiles similar to T_1 and T_2 using the above method.

Now, for the second case where there is a difference in $|V_c^P|$ and $|V_c^Q|$, where V_c^P and V_c^Q are the sets of voters whose first $j-1$ candidates C_1' are not the first $j-1$ members of the winning committee (in any order). We can make C_1' the first $j-1$ winners by adding a profile T_3 consisting of these C_1' in the first $j-1$ positions and all possible permutations for the following $k-j+1$ places. Add the above multiple times if needed until C_1' become the first $j-1$ winners for both $P \cup T_3$ and $Q \cup T_3$. Now it again reduces to one of the cases above profiles similar to T_1 and T_2 can be added to show the lack of equivalence.

Hence ($\implies$) must be true. ∎

For instance, consider a profile P with ranked ballots $\{(abc), (abc), (abc)\}$ and a profile Q with ranked ballots $\{(abc), (abc), (cba)\}$. Let $k=2$. $P \not\sim_{SeqPl} Q$, as $N(P, ab) = 3$ but $N(Q, ab) = 2$. If we add $T = \{(cba), (cba)\}$ to P and Q, the winning committee will be (a, b) for $P \cup T$, but it will be (b, c) for $Q \cup T$.

Corollary 10. $C(SeqPl) = \Theta\left(\frac{m!}{(m-k)!}\log\left(1 + \frac{nm!}{(m-k)!}\right) + n\log\left(1 + \frac{m!}{n(m-k)!}\right)\right)$.

Proof. For each of the $\frac{m!}{(m-k)!}$ possible k-orderings we can have a certain fixed number $[0, n]$ of voters. There are n voters in total, which yields the result. ∎

5 Results for Approval-Based Rules

We begin with a general result for the equivalence classes of Thiele's Optimization Method for voting rules for which score from a vote V, $sc(S, V) = h(|S \cap V|) \leq |S \cap V|$. We refer to this class of rules as *ThiOpt*. Let $sc(S, P)$ below be the *ThiOpt* score received by S from P.

Theorem 5. $P \sim_{ThiOpt} Q$ holds if and only if for all $S, S' \in \mathcal{S}_k(A)$, $sc(S', P) - sc(S, P) = sc(S', Q) - sc(S, Q)$.

Proof. ($\Longleftarrow$) is straightforward because if the relative differences between scores of all pairs of committees are equal within P and Q, they will remain equal after adding new voters, hence the winning committee W will remain the same for both. For ($\Longrightarrow$), the idea of the proof is as follows.

First, we show that ($\Longrightarrow$) holds for Thiele rules with $h(r) < r$ for all $r > 1$ Assume that the right-hand side of the equation in the theorem is not satisfied for P and Q. Then, there exist $S_x, S_y \in \mathcal{S}_k(A)$ such that $sc(S_x, P) - sc(S_y, P) > sc(S_x, Q) - sc(S_y, Q)$ without losing generality. First, we add a set of voters T_1 with approval ballots such that $sc(S_x, Q \cup T_1) = sc(S_y, Q \cup T_1)$ (Note that this can be done by interchanging the candidates in S_x and S_y from the votes in Q, just like in the proof of *Theorem 1*). So, $sc(S_x, Q \cup T_1) - sc(S_y, Q \cup T_1) = 0$, whereas $sc(S_x, P \cup T_1) - sc(S_y, P \cup T_1) > 0$.

Then, add a profile T_2 consisting of a large number (t_2) of voters, each of them voting for every 'single' candidate in $S_x \cap S_y$ in a round-robin manner. Note that if the intersection is empty, then $T_2 = \phi$, and we skip this step. After this, S_x and S_y beat all committees for $P \cup T_1 \cup T_2$ and $Q \cup T_1 \cup T_2$ except possibly those which contain all candidates of $S_x \cap S_y$.

Consider a third set of voters T_3 with t_3 voters who vote for all pairs (c_1, c_2) for each $c_1 \in S_x \setminus S_y$ and for each $c_2 \in S_y \setminus S_x$ in a round-robin manner. Also, note that we must have $t_2 \gg t_3 \gg n$. After adding these voters, it can be shown that S_x and S_y also beat those committees that contain $S_x \cap S_y$ for both $P \cup T_1 \cup T_2 \cup T_3$ and $Q \cup T_1 \cup T_2 \cup T_3$, using the fact that $h(2) < 2$. As an example, if the committees S_x and S_y are (ab) and (cd), then in T_3 we add copies of votes for $(ac), (ad), (bc), (bd)$. This produces an increase in score of $2h(1) + h(2)$ to each one of $(ac), (ad), (bc), (bd)$. But (ab) and (cd) each get $4h(1)$. Since $h(2) < 2 = 2h(1)$, S_x and S_y will eventually beat the other committees.

After adding the profile $T = T_1 \cup T_2 \cup T_3$, we see that S_x continues to be the sole winner for $P \cup T$, whereas there is a tie between S_x and S_y for $Q \cup T$, demonstrating a lack of equivalence. Hence we must have $sc(S', P) - sc(S, P) = sc(S', Q) - sc(S, Q)$ for all $S, S' \in \mathcal{S}_k(A)$, implying that ($\Longrightarrow$) is true if $h(r) < r$ for all $r > 1$.

Next, for ($\Longrightarrow$), we consider the case where $h(r) = r$ for every $r \in [k]$. This corresponds to multiwinner approval voting. This needs an alternate analysis, where we consider an arbitrary tie-breaking rule over the committees (this does not conflict with the rule being irresoluteness, because tie-breaking is used to simplify the proof). There are $\kappa = \binom{m}{k}$ committees of size k. For a given preference profile P, arrange them in the decreasing order of the scores they receive from P from S_1 to S_κ. Given that $P \sim_{ThiOpt} Q$ for some Q, the winning committee W must be identical for P and Q.

Now, assume without loss of generality that for all $i \in [x-1]$, $sc(W, P) - sc(S_i, P) = sc(W, Q) - sc(S_i, Q)$, and $sc(W, P) - sc(S_x, P) < sc(W, Q) - sc(S_x, Q)$. From now on, we refer to S_x as S'. Let S'' be the next higher-ranked committee for P (here, S_{x-1}). Choose an arbitrary candidate $a \in S' \setminus S''$ and add just enough copies of votes (call the profile T_1) approving only a to both P and Q so that S' scores 1 higher than S'' when considering voter $P \cup T_1$, but S' scores equal to or less than S'' when considering votes $(Q \cup T_1)$. Iterate this process till we get $S'' = W$, and let the final added vote profile be T_τ for some $\tau \geq 1$. Let $T = \bigcup_{j=1}^{\tau} T_j$ Now, S' is the sole winner for the profile $P \cup T$, but at best the shared winner for $Q \cup T$, hence demonstrating lack of equivalence. ∎

5.1 Multiwinner Approval Voting (MAV)

Let $sc(a, P)$ be the approval score of $a \in A$ received from P, and let $sc(S, P)$ be the approval score of $S \in \mathcal{S}_k(A)$ from P. Then, the following result follows directly from Theorem 5.

Corollary 11. $P \sim_{MAV} Q$ holds if and only if for all $a, b \in A$, $sc(a, P) - sc(b, P) = sc(a, Q) - sc(b, Q)$.

Take the following example. Let there be a profile P with approval ballots $\{(a), (a), (b)\}$ and a profile Q with approval ballots $\{(a), (b), (b)\}$. Let $k = 2$. $P \not\sim_{AV} Q$, as $sc(a, P) = 2$ but $sc(a, Q) = 1$. If we add $T = \{(c), (c)\}$ to P and Q, the winning committee will be (a, c) for $P \cup T$ but it will be (b, c) for $Q \cup T$.

Corollary 12. $C(MAV) = \Theta(m \log n)$.

5.2 Approval-Based Chamberlin-Courant Rule (α-CC)

Theorem 5 immediately characterises the equivalence classes for Approval-based Chamberlin-Courant. For example, consider a profile P with approval ballots $\{(a), (ab)\}$ and a profile Q with approval ballots $\{(b), (ab)\}$. Let $k = 2$. $P \not\sim_{AV} Q$, as $sc(bc, P) = 1$ but $sc(bc, Q) = 2$. If we add $T = \{(c)\}$ to P and Q, the winning committee will be (a, c) for $P \cup T$, but it will be (b, c) for $Q \cup T$.

Corollary 13. $C(\alpha - CC) = O\left(\binom{m}{k} \log n\right)$.

Corollary 14. $C(\alpha - CC) = \Omega\left(\lfloor \frac{m}{k} \rfloor \log n\right)$.

The lower bound and upper bound do not match, for the same reason as for β-CC discussed in *Subsect.* 4.4.

5.3 Proportional Approval Voting (PAV)

We get the equivalence classes for PAV directly from Theorem 5. Consider the following example. There is a profile P with approval ballots $\{(a), (ab)\}$ and a profile Q with approval ballots $\{(b), (ab)\}$. Let $k = 2$. $P \not\sim_{PAV} Q$, as $sc(ac, P) = 2$ but $N(ac, Q) = 1$. If we add $T = \{(c)\}$ to P and Q, the winning committee will be (a, c) for $P \cup T$ but it will be (b, c) for $Q \cup T$.

Corollary 15. $C(PAV) = O\left(\binom{m}{k}(k \log k + \log(n \log n))\right)$.

Corollary 16. $C(PAV) = \Omega\left(\lfloor \frac{m}{k} \rfloor (\log(n \log n))\right)$.

Note that our upper bound and lower bound results do not match for the same reason as for β-CC discussed in *Subsect.* 4.4 (Table 1).

Table 1. A summary of results of compilation complexity for multi-winner voting rules.

Rule	Equiv. Condition	Upper Bound	Lower Bound
SNTV	$sc(x,P) = sc(x,Q)$	$\Theta\left(m \lg\left(1 + \frac{n}{m}\right) + n \lg\left(1 + \frac{m}{n}\right)\right)$	
k-Borda		$\Theta(m \lg(nm))$	
Bloc		$\Theta\left(m \lg\left(1 + \frac{n\hat{k}^\dagger}{m}\right) + n\hat{k} \lg\left(1 + \frac{m}{n k}\right)\right)$	
β-CC	$sc(S,P) = sc(S,Q)$	$O\left(\binom{m}{k} \lg(n(m-k))\right)$	$\Omega(\lg(n(m-k)))$
GehrSta	$\mathcal{M}_P = \mathcal{M}_Q$	$O(m^2 \lg n)$	$\Omega\left(m^2 \lg\left(\left\lfloor \frac{n}{q^{\ddagger}m} \right\rfloor - 2\right)\right)$
θ	$N_P(d,S) = N_Q(d,S)$	$O\left(\binom{m}{k}(m-k) \lg n\right)$	$\Omega((m-k) \lg n)$
SeqPl	$N(P, \lambda^k) = N(Q, \lambda^k)$	$\Theta\left(\frac{m!}{(m-k)!} \lg\left(1 + \frac{nm!}{(m-k)!}\right) + n \lg\left(1 + \frac{m!}{n(m-k)!}\right)\right)$	
MAV	$sc(a,P) - sc(b,P)$ $= sc(a,Q) - sc(a,Q)$	$\Theta(m \lg n)$	
α-CC	$sc(W,P) - sc(S,P)$ $= sc(W,Q) - sc(S,Q)$	$O\left(\binom{m}{k} \lg n\right)$	$\Omega\left(\lfloor \frac{m}{k} \rfloor \lg n\right)$
PAV		$O\left(\binom{m}{k}(k \lg k + \log(n \lg n))\right)$	$\Omega\left(\lfloor \frac{m}{k} \rfloor (\lg(n \lg n))\right)$

$^\dagger \hat{k} = \min(k, m-k)$
‡ for some $q > 0$

6 Conclusion and Future Work

We characterise the equivalence classes and compilation complexity bounds of many common ordinal and approval-based multi-winner rules. This helps us understand how they fare against each other regarding the amount of memory needed to store the partial results. Although some results are inspired by ideas used in [4,19] for single-winner voting rules, several of our results, such as those for θ-Winning sets, Sequential Plurality and Thiele's Optimization Method are quite novel. Recall that the reason why the lower and upper bounds for α-CC, β-CC, θ-Winning Sets, and PAV do not match is that it is not easy to count the number of functions from $\mathcal{S}_k(A)$ to $\mathbb{N}$ that correspond to $sc(.,P)$ for some profile P for each of these rules. These gaps could be reduced in future. Another open problem is the extension of our results to the setting where the number of additional voters is already known. Finally, equivalence classes and compilation complexity bounds for Single Transferrable Vote, Phragmén, and the Method of Equal Shares, are still open. We present below a summary of our results.

Acknowledgements. We are grateful to the anonymous reviewers for their careful reading and helpful comments. This work was supported in part by project ANR-22-CE26-0019 "Citizens", funded by the French Agence Nationale de la Recherche.

References

1. Aziz, H., Elkind, E., Faliszewski, P., Lackner, M., Skowron, P.: The Condorcet principle for multiwinner elections: from shortlisting to proportionality. In: Proceedings of IJCAI 2017, pp. 84–90 (2017)

2. Barberà, S., Coelho, D.: How to choose a non-controversial list with k names. Soc. Choice Welf. **31**(1), 79–96 (2008)
3. Chevaleyre, Y., Lang, J., Maudet, N., Monnot, J.: Compilation and communication protocols for voting rules with a dynamic set of candidates. In: Proceedings of TARK 2011, pp. 153–160 (2011)
4. Chevaleyre, Y., Lang, J., Maudet, N., Ravilly-Abadie, G.: Compiling the votes of a subelectorate. In: Proceedings of IJCAI 2009, pp. 97–102 (2009)
5. Coelho, D.: Understanding, evaluating and selecting voting rules through games and axioms. Ph.D. thesis (2004)
6. Conitzer, V., Sandholm, T.: Vote elicitation: complexity and strategy-proofness. In: Proceedings of AAAI-2001, pp. 392–397 (2002)
7. Dey, P., Bhattacharyya, A.: Sample complexity for winner prediction in elections. In: Proceedings of AAMAS 2015, pp. 1421–1430 (2015)
8. Dey, P., Talmon, N., Van Handel, O.: Proportional representation in vote streams. arXiv preprint arXiv:1702.08862 (2017)
9. Elkind, E., Faliszewski, P., Skowron, P., Slinko, A.: Properties of multiwinner voting rules. Soc. Choice Welf. **48**(3), 599–632 (2017)
10. Elkind, E., Lang, J., Saffidine, A.: Condorcet winning sets. Soc. Choice Welf. **44**(3), 493–517 (2015)
11. Faliszewski, P., Skowron, P., Slinko, A., Talmon, N.: Multiwinner voting: a new challenge for social choice theory. Trends Comput. Soc. Choice **74**, 27–47 (2017)
12. Gehrlein, W.: The condorcet criterion and committee selection. Math. Soc. Sci. **10**(3), 199–209 (1985)
13. Hazon, N., Aumann, Y., Kraus, S., Wooldridge, M.: On the evaluation of election outcomes under uncertainty. Artif. Intell. **189**, 1–18 (2012)
14. Janson, S.: Phragmén's and thiele's election methods. Technical report (2016)
15. Konczak, K., Lang, J.: Voting procedures with incomplete preferences. In: Proceedings of the IJCAI-2005 Multidisciplinary Workshop on Advances in Preference Handling, vol. 20. Citeseer (2005)
16. Lackner, M., Skowron, P.: Multi-Winner Voting with Approval Preferences - Artificial Intelligence, Multiagent Systems, and Cognitive Robotics. Springer Briefs in Intelligent Systems, Springer, Cham (2023). https://doi.org/10.1007/978-3-031-09016-5
17. Peters, D., Pierczyński, G., Skowron, P.: Proportional participatory budgeting with additive utilities. In: Proceedings of NeurIPS 2021, pp. 12726–12737 (2021)
18. Skowron, P., Faliszewski, P., Slinko, A.: Axiomatic characterization of committee scoring rules. J. Econ. Theory **180**, 244–273 (2019)
19. Xia, L., Conitzer, V.: Compilation complexity of common voting rules. In: Proceedings of AAAI 2010, pp. 915–920 (2010)

On Approximately Strategy-Proof Tournament Rules for Collusions of Size at Least Three

David Mikšaník[1], Ariel Schvartzman[2(✉)], and Jan Soukup[1]

[1] Computer Science Institute, Charles University, Prague, Czechia
soukup@kam.mff.cuni.cz
[2] Google Research, Mountain View, CA, USA
aschvartzman@google.com

Abstract. A tournament organizer must select one of n possible teams as the winner of a competition after observing all $\binom{n}{2}$ matches between them. The organizer would like to find a tournament rule that simultaneously satisfies the following desiderata. It must be *Condorcet-consistent* (henceforth, CC), meaning it selects as the winner the unique team that beats all other teams (if one exists). It must also be *strongly non-manipulable* for groups of size k at probability α (henceforth, k-SNM-α), meaning that no subset of $\leq k$ teams can fix the matches among themselves in order to increase the chances any of it's members being selected by more than α. Our contributions are threefold. First, wee consider a natural generalization of the Randomized Single Elimination Bracket rule from [18] to d-ary trees and provide upper bounds to its manipulability. Then, we propose a novel tournament rule that is CC and 3-SNM-1/2, a strict improvement upon the recent work of [7] who proposed a CC and 3-SNM-31/60 rule. Finally, we initiate the study of reductions among tournament rules.

Keywords: Tournament design · Strategy-proof rules · Computational Social Choice

1 Introduction

Consider the problem a tournament organizer faces when, after observing all pairwise matches between n teams, they must select one as the winner of the tournament. We model the tournament T as a complete, directed graph on the n teams. A tournament rule r is a (possibly randomized) mapping from the set of tournaments on n teams $\mathcal{T}_n$ to a probability vector in Δ^n, $r : \mathcal{T}_n \to \Delta^n$. The tournament organizer is thus tasked with designing a tournament rule r and would like the rule to satisfy the following natural properties:

1. If there is a team who beats all other teams, termed a *Condorcet-winner*, they should be picked as the winners of the tournament with probability 1. We call such rules *Condorcet-consistent* (or CC).
2. No team should be incentivized to unilaterally throw their own games in order to obtain a better outcome. We call such rules *monotone*.

3. No subset of $\leq k$ teams should have incentives to fix the matches among themselves in order to improve the chances of any of its members being selected as the winner of the tournament. We call such rules *k-strongly non-manipulable* (or k-SNM).

These properties are motivated by real-world sports competitions. It would be unimaginable to violate Property 1 and not award the top prize to an undefeated team. Violations to Properties 2, 3 have been observed in high-stakes competitions such as the Olympic Games and the FIFA World Cup. An infamous scandal in the Women's Doubles Badminton tournament at the 2012 Olympics saw multiple teams purposefully losing the last games of their group stage matches in order to avoid a difficult match-up in the following single-elimination bracket. This clear violation of monotonicity (and sportsmanship) resulted in the disqualification of 4 teams, including many of the likely medalists. A less investigated but equally egregious scandal occurred during the 1982 FIFA World Cup. West Germany and Austria disputed the last match of their group stage with full knowledge of the outcomes of all other games. It is suspected that they colluded in order to produce an outcome that would see both teams advance to the next stage of the tournament, at the expense of Algeria, who unexpectedly defeated West Germany in their opening match. As a result of this possible violation of strong non-manipulability, the last game of every group in every FIFA World Cup since has been played simultaneously.

Observe that if one wanted to simply satisfy Properties 2, 3, there are numerous simple rules that do so. For example, picking a winner uniformly at random, picking a fixed team as the winner (i.e., a dictatorship) or picking the winner proportional to the number of wins in the tournament all satisfy Properties 2, 3 but not Property 1. Similarly, it is easy to satisfy Properties 1, 2. If there is a Condorcet-winner, pick that team. Otherwise, pick a team uniformly at random. Unfortunately, it is known that Properties 1, 3 are directly at odds with each other: [1] showed that there exists no randomized tournament rule that can satisfy both of these properties at the same time, even for $k = 2$. One way to overcome this impossibility result is to relax Property 3 as follows:

4. No subset of $\leq k$ teams should be able to fix the matches among themselves in order to improve the chances of any of its members being selected as the winner by more than α. We call such rules k-SNM-α.

A growing body of work has asked what is the smallest α for which there exists a rule that satisfies Properties 1, 2 and 4 (for some fixed value of k). First, the work of [18] proves that the Random Single-Elimination Bracket (henceforth RSEB) rule is CC and 2-SNM-1/3, and that no other CC rule can do better. Later, [19] show that a rule termed Randomized King-of-the-Hill (henceforth, RKOTH) matches the performance of RSEB and satisfies a condition even stronger than CC. These works completely settle the question of finding CC and minimally manipulable (or *optimal*) strategy-proof rules for collusions of size $k = 2$. On the other hand, very little is known about the case when $k > 2$, even $k = 3$. [18] prove a simple lower bound of $\alpha \geq \frac{k-1}{2k-1}$. [19] proved that there exists an LP-based rule that is CC and k-SNM-2/3 *for all* k simultaneously. Unfortunately, this rule is neither explicit nor monotone. More recently, in concurrent work [7] gave the first explicit, CC, monotone, 3-SNM-α rule for $\alpha = 31/60$ (and this value of α is tight for their rule).

As hinted in the previous paragraph, there are two approaches to proving the existence of CC and approximately strategy-proof tournament rules. One approach takes simple rules (such as RSEB, RKOTH), in the hopes that they are not too manipulable, and provides tight analysis for them. This does not always work: many simple rules, such as picking the team with the most wins, are extremely manipulable (i.e., have $\alpha = 1 - O(1/n)$, see [18]). The other approach provides rules which are not explicitly implementable. For example, the LP-based rule of [19] arises from fixing the tournament rule for tournaments with a Condorcet-winner, relaxing the manipulability constraints for tournaments that are close to having a Condorcet-winner and proving that the resulting polytope is non-empty for some value of $\alpha < 1$. Our results make use of both of these approaches and introduce another.

Our first contribution, inspired by the positive results of [18], introduces a natural generalization of RSEB. The RSEB rule randomly seeds teams on the leaf nodes of a binary tree and recursively labels inner nodes as the winner of the match between its children. The winner of the bracket is the team whose label appears in the root node of the tree. We define the Random d-ary Single Elimination Bracket (henceforth RdSEB) rule similarly with one key difference: instead of using binary trees, we use d-ary trees. If the sub-tournament induced by an inner node's children has a Condorcet-winner, then the inner node will carry the label of that child. Otherwise, the inner node picks a child uniformly at random to advance[1]. We provide an upper bound $\alpha_{d,k} < 1$ on the manipulability of RdSEB on tournaments with n teams and collusions of size up to $k \leq d$.

Theorem 1. *Let* $2 \leq k \leq d$. *The* RdSEB *rule is Condorcet-consistent, monotone and* k-SNM-$\alpha_{d,k}$ *for*

$$\alpha_{d,k} \leq 1 - \left(\frac{2 \cdot (d)_k}{d^{k+1}}\right),$$

where $(d)_k = \prod_{i=0}^{k-1}(d-i)$ *is the falling factorial of* d *with* k *terms.*

For $k = d = 3$, we obtain that $\alpha_{3,3} = .8519$. As a consequence of Theorem 1, we get the first explicit family of CC, monotone rules whose manipulability for any (fixed) k is bounded away from 1. This stands in contrast to the LP-based rule of [19] which was neither monotone nor explicit and to several of the rules analyzed in [18] which had $\alpha \to 1$ as $n \to \infty$, even for $k = 2$. In other words, the bound from Theorem 1 is independent of n, the number of competing teams.

Our second contribution, inspired more by the second approach to finding approximately optimal tournament rules, is a new explicit tournament rule which strictly improves upon the results of [7].

Theorem 2. *The* SIGNIFICANTONLY *rule is Condorcet-consistent, monotone,* 2-SNM-$1/3$ *and* 3-SNM-$1/2$, *and this is tight.*[2]

---

[1] This decision is inspired by the observation that tournaments on three teams either have a Condorcet-winner or have three teams that beat each other cyclically.

[2] The best lower bound for this problem, due to [18], is $2/5$.

The SIGNIFICANTONLY rule, while explicitly describable, arises from an approach similar to the LP-based rule of [19]. We identify the tournaments which are close to having a Condorcet-winner as those where teams have more incentives to manipulate outcomes. Given such a close-to-Condorcet tournament T, our rule deems a small number of teams as *significant* and distributes most of the probability mass on these teams according to additional properties of the tournament itself (and the rest uniformly across the remaining teams).

Finally, our last contribution introduces a new way of designing Condorcet-consistent and *asymptotically* optimal tournament rules. If one had substantial computational power, one could compute a top-cycle consistent,[3] optimal rule for fixed values of n, k. How could we use such a rule r_n to construct a rule $r_{n'}$ that works for $n' > n$? First pad the tournament with dummy teams that lose to all real teams until the number of teams $n' := n \cdot M$ is a multiple of n. Partition the teams into n groups of equal size. Within each group, pick a team from the top-cycle uniformly at random as a finalist. The number of finalists will be exactly n. Finally, run r_n on the n finalists and declare its winner as the overall winner. We prove that this simple idea suffices to transform top-cycle consistent, k-SNM-α rules for n teams to CC, k-SNM-α' rules for $n' > n$ teams where α' is close to α.

Theorem 3. *If there exists a top-cycle consistent, and k-SNM-α rule r for n teams, then there exists a top-cycle consistent and k-SNM-α' rule r' for $n' > n$ teams where*

$$\alpha' \leq \alpha\left(1 - \frac{(k-1)^2}{n}\right) + \frac{(k-1)^2}{n}.$$

Theorem 3 can be thought of as reducing the problem of finding approximately optimal tournament rules for large n to the same problem for small n. As an example, if we could verify the existence of a top-cycle consistent 3-SNM-2/5 rule for $n = 25$, then this would imply top-cycle consistent 3-SNM-α rules for $n \geq 25$ and $\alpha < 1/2$, directly improving on Theorem 2.

1.1 Related Work

Most of the related work has been mentioned already. There exist two other results that are directly related to our problem. Whereas in this paper (and all previously mentioned papers) we evaluate the manipulability of tournament rules based on their worst-case performance, the work of [8] instead studies this question under the lens of average-case analysis. More recently, the work of [6] expanded the model to include prize vectors $v = (v_1, \ldots, v_n)$. Tournament rules with prizes output a complete, linear ranking over the teams where the i-th team earns reward v_i, rather than giving reward 1 to the winner and 0 to everyone else.

Another related line of work involves the Tournament Fixing Problem, where the organizer of the tournament is colluding with a team in order to produce a seeding that

---

[3] Top-cycle consistency is stronger than Condorcet-consistency. We defer its formal definition but informally, the top-cycle is the smallest non-empty set S of teams such that no team in S loses to a team outside of S. A top-cycle consistent rule would only pick teams from the top-cycle.

selects them as the winner (see, e.g. [2, 12, 13, 20, 22, 23]). The central questions here are computational (i.e., can the organizer efficiently decide if there is a winning bracket for their favorite team) and structural (i.e., under what conditions does there exist a bracket that selects the organizer's favorite team, see, e.g. [15]). Due to its connections with voting theory and social choice, there is a long history of analyzing properties of particular tournament rules ([4, 5, 9–11, 16, 17], to name a few). We refer the reader to the survey by [21] on recent developments in tournaments and computational social choice, or other books on computational social choice [3, 14].

2 Notation

In this section, we introduce key concepts to contextualize our results. Recall a tournament graph T is a complete, directed graph $G = ([n], E)$ on n labelled vertices. We refer to a tournament graph's vertices as teams (and the number of teams as the size of the tournament), its undirected edges as matches and its directed edges as outcomes (where if $(i, j) \in E$, we say i beats j in T). For a fixed team i, tournament T we let $\delta^+(i, T) = \{j | j \in [n], (i, j) \in E)\}$ be the set of teams that i beats under T, and let $\delta^-(i, T) = \{j | j \in [n], (j, i) \in E)\}$ be the set of teams that i loses to under T. Let $\mathcal{T}_n$ be the set of all tournaments on n teams. Recall a tournament rule is a mapping $r : \mathcal{T}_n \to \Delta^n$. That is, for every tournament T, $r(T)$ denotes the distribution over teams according to which the organizer will select a winner. We use notation $r_i(T)$ to denote team i's probability of being selected by r as the winner under tournament T. We use the shorthand notation $r_S(T) := \sum_{i \in S} r_i(T)$ to denote the probability that a team in S is selected by r in tournament T. The next definitions formalize Properties 1, 2, 3 and 4.

Definition 1. *Team i is the* Condorcet-winner *of tournament T if i beats every other team under T. A rule r is* Condorcet-consistent *if $r_i(T) = 1$ when i is T's Condorcet-winner.*

Definition 2. *A tournament rule r is* monotone *if for all teams i and all tournaments T, T' where all matches not involving team i are identical and $\delta^+(i, T) \supseteq \delta^+(i, T')$, it holds that $r_i(T) \geq r_i(T')$.*

A tournament rule is monotone if it is not in a team's best interest to unilaterally lose matches it would otherwise win. We present the ways in which manipulations are modelled.

Definition 3. *We say tournaments T, T' are S-adjacent if the only outcomes where T, T' differ on are those matches that involve two teams in S.*

If T, T' are S-adjacent, outcomes involving at least one team outside of S are identical. Motivated by the results from [1], the following relaxation was introduced by [18].

Definition 4. *A tournament rule is k strongly non-manipulable at probability α (k-SNM-α) if for all $S \subseteq [n]$ of size at most k, for all tournaments T, T' that are S-adjacent we have $r_S(T') \leq r_S(T) + \alpha$. For $\alpha = 0$, we simply say the rule is k strongly non-manipulable (k-SNM).*

3 Analysis for the RdSEB Rule

In this section we study the manipulability of the Randomized d-ary Single-Elimination Bracket (RdSEB) rule, a generalization of the Randomized Single-Elimination Bracket (RSEB) rule from [18], against collusions of size $k \leq d$.

Definition 5. *Given a tournament T, a* sub-tournament *on S is the sub-graph induced by T on vertex set S.*

We are now ready to formally define the RdSEB rule.

Definition 6. *The* Randomized d-ary Single-Elimination Bracket *rule operates as follows. Add dummy teams[4] until the number of teams $n = d^{\lceil \log_d(n) \rceil}$ is a power of d. Randomly place teams at the leaf nodes of a complete d-ary tree of height $\lceil \log_d(n) \rceil$. Recursively label a parent node with the label of the Condorcet-winner of the sub-tournament induced by the labels of its children, if there is one. Otherwise, choose one of its children uniformly at random and use that label instead. The winner of the tournament is the team whose label appears at the root of the tree.*

In terms of Definition 6, RSEB is the rule that results from setting $d = 2$. The family of RdSEB rules operates in the same way as the RSEB rule except that if there is no Condorcet-winner in the sub-tournament induced at a node, the rule advances a team uniformly at random. This choice is motivated by the following simple observation when $d = 3$. There are only two non-isomorphic sub-tournament graphs on three teams: one where there is a Condorcet-winner and one where the teams beat each other cyclically. In the former case, it is obvious which team to advance. In the latter case, we argue choosing a team uniformly at random is reasonable.

Theorem 1. *Let $2 \leq k \leq d$. The* RdSEB *rule is Condorcet-consistent, monotone and k-SNM-$\alpha_{d,k}$ for*

$$\alpha_{d,k} \leq 1 - \left(\frac{2 \cdot (d)_k}{d^{k+1}} \right),$$

where $(d)_k = \prod_{i=0}^{k-1}(d - i)$ is the falling factorial of d with k terms.

The main idea is that if some (at least two) colluding teams meet only in the final round, then they can increase the joint probability that one of them will be winner by at most $(1 - 2/d)$ (which happens when the colluding teams create a Condorcet winner). Obviously, if the colluding teams do not meet in a bracket at all or there exists a Condorcet winner outside the colluding teams, then they cannot increase the chance to win the tournament. In the remaining cases, we simply assume that they can increase the chance to win the tournament by 1.

From Theorem 1, given a fixed k, the RkSEB rule is monotone, CC and its manipulability is bounded away from 1 for all n. This is the first explicit family of rules to exhibit this property (since the LP-based rule of [19] is neither monotone nor explicit). We suspect the bound from Theorem 1 is not tight. A finer argument like the one in [18] might yield a better analysis.

---

[4] Dummy teams are teams that lose to all non-dummy teams. The outcome of a match between two dummy teams is arbitrary.

Proof. It is more convenient to consider the following equivalent variation of the RdSEB rule. Instead of choosing one of the children of A uniformly at random, chose a number j_A from $[d]$ uniformly at random. If there is no Condorcet-winner in the sub-tournament, label A by the label of j_A-th child of A. In both cases, mark the node A with j_A (even if there is a Condorcet-winner in the sub-tournament).[5] For the sake of the proof, let us denote r^d as this equivalent variation of the RdSEB rule. Observe that the labels of inner nodes of the complete d-ary tree can be deduced from the labels of leaves and marks of inner nodes: run r^d but all random choices are made accordingly to these labels and marks.

For any non negative number t, let D_t be a reserved set containing t dummy teams. Moreover, let T be a tournament on N' (disjoint from any D_t) of n' teams, $h := \lceil \log_d(n') \rceil$, and $n := d^h$. We assume that the rule r^d initially adds $n - n'$ dummy teams from $D_{n-n'}$ into T.[6] Given $N := N' \cup D_{n-n'}$, a *d-bracket* $G(\pi, m)$ for N is a complete d-ary tree G of height h endowed with a pair (π, m), where

- π is a bijection from leaves of G to N (i.e., labels of leaves),
- m is a mapping from inner nodes of G to $[d]$ (i.e., marks of inner nodes).

Let $\mathcal{B}_N$ be the set of all d-brackets for N. Now we precisely describe how the labels of nodes of G can be deduced from π and m. Given a d-bracket $G(\pi, m)$ for N, the *outcome* of $G(\pi, m)$ under T[7] is a labeling ω_T of nodes of G such that $\omega_T(A) = \pi(A)$, for every leaf A, and if A is a node with children $A_1, \ldots, A_d$, then

$$\omega_T(A) := \begin{cases} x & \text{if } x \text{ is the Condorcet-winner in} \\ & \text{the sub-tournament induced by} \\ & \{\omega_T(A_1), \ldots, \omega_T(A_d)\} \text{ under } T, \\ \omega_T(A_{m(A)}) & \text{otherwise.} \end{cases}$$

Observe that ω_T is a one-to-one correspondence between the set of all outcomes of d-ary brackets on N under T and the set of all runs of r^d on T.

Fix a d-bracket $G(\pi, m)$ for N. The *winner* of $G(\pi, m)$ under T is $\omega_T(R)$, where R is the root of G. Given a team $x \in N$, $G(\pi, m)$ is *winning for* x under T if x is the winner of $G(\pi, m)$ under T. Denote by $\mathcal{B}_{N,T}(x) \subseteq \mathcal{B}_N$ the set of all winning d-brackets for x under T. The motivation behind this reframing of RdSEB is to simply the argument of the proof. Similar to the original argument of [18], we will bound the manipulability of RdSEB by directly counting the number of brackets where colluding teams could gain and compare it to the total number of brackets. We have

$$r_x^d(T) = |\mathcal{B}_{N,T}(x)|/|\mathcal{B}_N|.$$

Observe that $|\mathcal{B}_N| = n! \cdot d^\ell$, where $\ell := \ell(d, h)$ is the number of inner nodes of a complete d-ary tree of height h.

---

[5] Every inner node has a label and a mark. Note that they can be equal but they have different meaning.
[6] Hence every tournament on n' teams is extended by the same set of dummy teams.
[7] The outcome is well-defined only if the teams in T are the same as the set N.

First, we prove that r^d is monotone. Take an arbitrary team $x \in N'$. It sufficient to show that $r_x^d(T) \geq r_x^d(T')$ for every $\{x,y\}$-adjacent tournaments T' of T such that x beats y under T. Let $G(\pi, m)$ be a d-bracket for N. Observe that if $G(\pi, m)$ is winning for x under T', then $G(\pi, m)$ is also a winning for x under T. Hence $\mathcal{B}_{N,T'}(x) \subseteq \mathcal{B}_{N,T}(x)$, and so $r_x^d(T') \leq r_x^d(T)$ as required.

Second, we prove that r^d is Condorcet-consistent. Suppose that $x \in N'$ is the Condorcet-winner in T. For every d-bracket $G(\pi, \ell)$ for N, consider the unique path P from the leaf in G labeled by x to the root of G. Observe that every node of P is labeled by x by the outcome of $G(\pi, m)$ under T. In particular, the root of G is labeled by x. It follows that every d-bracket $G(\pi, \ell)$ is winning for x under T. Hence $\mathcal{B}_{N,T}(x) = \mathcal{B}_N$, and so $r_x^d(T) = 1$ as required.

Lastly, we prove that r^d is k-SNM-$\alpha_{d,k}$ for some $\alpha_{d,k}$ (to be determined). Suppose that $S = \{s_1, s_2, \ldots, s_k\} \subseteq N'$ is a subset of colluding teams. For any S-adjacent tournaments T' of T, we show that $r_S^d(T') - r_S^d(T) \leq \alpha$. Recall that $\mathcal{B}_{N,T}(x)$ is the set of all winning d-brackets for x under T. Moreover, define $\mathcal{B}_{N,T}(S) := \bigcup_{s \in S} \mathcal{B}_{N,T}(S)$. In this notation, we can write

$$r_S(T') - r_S(T) = \frac{|\mathcal{B}_{N,T'}(S)|}{|\mathcal{B}_N|} - \frac{|\mathcal{B}_{N,T}(S)|}{|\mathcal{B}_N|}.$$

We upper bound this expression using the idea introduced in [18]. For that, let us denote by $\mathcal{B}_N^+(S)$ the set of all d-brackets $G(\pi, m)$ for N such that the least common ancestor of any leaves A and B with $\pi(A), \pi(B) \in S$ is the root of $B(\pi, m)$. In other words, $\mathcal{B}_N^+(S)$ is the set of all d-brackets for N such that the colluding teams can meet possibly only in the final round. Set $\mathcal{B}_N^-(S) := \mathcal{B}_N \setminus \mathcal{B}_N^+(S)$. Moreover, for a tournament T, let $\mathcal{B}_{N,T}^+(S) := \mathcal{B}_N^+(S) \cap \mathcal{B}_{N,T}(S)$ and $\mathcal{B}_{N,T}^-(S) := \mathcal{B}_N^-(S) \cap \mathcal{B}_{N,T}(S)$. Then

$$r_S(T') - r_S(T) = \frac{|\mathcal{B}_{N,T'}(S)|}{|\mathcal{B}_N|} - \frac{|\mathcal{B}_{N,T}(S)|}{|\mathcal{B}_N|}$$

$$= \frac{|\mathcal{B}_{N,T'}^+(S)| + |\mathcal{B}_{N,T'}^-(S)|}{|\mathcal{B}_N|} - \frac{|\mathcal{B}_{N,T}^+(S)| + |\mathcal{B}_{N,T}^-(S)|}{|\mathcal{B}_N|}$$

$$= \frac{|\mathcal{B}_{N,T'}^+(S)| - |\mathcal{B}_{N,T}^+(S)|}{|\mathcal{B}_N|} + \frac{|\mathcal{B}_{N,T'}^-(S)| - |\mathcal{B}_{N,T}^-(S)|}{|\mathcal{B}_N|}$$

$$\leq \frac{|\mathcal{B}_{N,T'}^+(S)| - |\mathcal{B}_{N,T}^+(S)|}{|\mathcal{B}_N|} + \frac{|\mathcal{B}_N^-(S)|}{|\mathcal{B}_N|}.$$

We upper bound the first term in the last expression. Let $G := G(\pi, m)$ be a d-bracket for N in $\mathcal{B}_{N,T'}^+(S)$. Let $\mathbf{x} := (x_1, \ldots, x_d)$ be the d-tuple of finalists in G under T'. More precisely, let R be the root of G with children $A_1, \ldots, A_d$. Then d-tuple of finalists of G under T is $(\omega_{T'}(A_1), \ldots, \omega_{T'}(A_d))$. The crucial observation is that also $\mathbf{x}$ is d-tuple of finalist of G under T. We say that there is a Condorcet-winner in $\mathbf{x}$ under T' if there is a Condorcet-winner in the sub-tournament induced by x under T'. Notice that:

(i) If $|S \cap \{x_1, x_2, \ldots, x_d\}| \leq 1$, then $G \in \mathcal{B}_{N,T}^+(S)$.

(ii) If $|S\cap\{x_1, x_2, \ldots, x_d\}| \geq 2$ and there is no Condorcet-winner in **x** under T', then $G \in \mathcal{B}^+_{N,T}(S)$.

(iii) If $|S\cap\{x_1, x_2, \ldots, x_d\}| \geq 2$ and there is a Condorcet-winner $s_i \in S$ in **x** under T', then $G \notin \mathcal{B}^+_{N,T}(S)$ only if there is no Condorcet-winner $s_j \in S$ in **x** under T and the mark of the root R of G is pointing to a team outside of S (i.e., $\omega_T(A_{m(R)}) \notin S$).

A d-bracket is of type (i) if it satisfies the statement (i). Analogously, for types (ii) and (iii). It follows that, for every d-bracket in $\mathcal{B}^+_{N,T'}(S)$ of type (i) or (ii), there exists at least one d-bracket in $\mathcal{B}^+_{N,T}(S)$ of the same type. Moreover, we claim that, for every d d-bracket in $\mathcal{B}^+_{N,T'}(S)$ of type (iii), there exist at least two d-brackets in $\mathcal{B}^+_{N,T}(S)$ of type (iii). Indeed, take a bracket $G(\pi, m) \in \mathcal{B}^+_{N,T'}(S)$ of type (iii). Then $G(\pi, m') \in \mathcal{B}^+_{N,T'}(S)$ is of type (iii), where only the mark of the root of G can be changed (there are d ways how to change it). On the other hand, let $i \neq j$ be two indices such that $x_i, x_j \in S$. Then $G(\pi, m_1), (\pi, m_2) \in \mathcal{B}^+_{N,T}(S)$ are of type (iii), where m_i and m_j are m but the mark of the root of G is changed to i and j, respectively.

If we denote by p be the number of d-brackets from $\mathcal{B}^+_{N,T'}(S)$ of type (iii), then

$$|\mathcal{B}^+_{N,T'}(S)| - |\mathcal{B}^+_{N,T}(S)| \leq p \cdot \left(1 - \frac{2}{d}\right)$$
$$\leq |\mathcal{B}^+_{N,T'}(S)| \cdot \left(1 - \frac{2}{d}\right) \leq |\mathcal{B}^+_N(S)| \cdot \left(1 - \frac{2}{d}\right).$$

Observe that

$$|\mathcal{B}^+_N(S)| = (d)_k \cdot \left(\frac{n}{d}\right)^k \cdot (n-k)! \cdot d^\ell$$

and hence

$$|\mathcal{B}^-_N(S)| = n! \cdot d^\ell - (d)_k \cdot \left(\frac{n}{d}\right)^k \cdot (n-k)! \cdot d^\ell.$$

Therefore,

$$r_S(T') - r_S(T) \leq \frac{|\mathcal{B}^+_{N,T'}(S)| - |\mathcal{B}^+_{N,T}(S)|}{|\mathcal{B}_N|} + \frac{|\mathcal{B}^-_N(S)|}{|\mathcal{B}_N|}$$
$$\leq \frac{\left((d)_k \cdot \left(\frac{n}{d}\right)^k \cdot (n-k)! \cdot d^\ell\right) \cdot \left(1 - \frac{2}{d}\right)}{n! \cdot d^\ell}$$
$$+ \frac{n! \cdot d^\ell - (d)_k \cdot \left(\frac{n}{d}\right)^k \cdot (n-k)! \cdot d^\ell}{n! \cdot d^\ell}$$
$$= 1 - \frac{1}{n!} \cdot \left(\frac{2}{d} \cdot (d)_k \cdot \left(\frac{n}{d}\right)^k \cdot (n-k)!\right)$$
$$= 1 - \frac{n^k \cdot (n-k)}{n!} \cdot \frac{2 \cdot (d)_k}{d^{k+1}}$$
$$\leq 1 - \frac{2 \cdot (d)_k}{d^{k+1}}.$$

□

We conjecture that the manipulability of RdSEB is bounded away from $1/2$ for all $k \leq d$.

Conjecture 1 (See Table 1). For all $3 \leq k \leq d$, $\alpha_{d,k} \geq 227/420$.

Table 1. Evaluation of $\alpha_{d,k}$ for small values of d and k rounded up to 4 decimals.

d	k				
	3	4	5	6	7
3	0.8519	–	–	–	–
4	0.8125	0.9531	–	–	–
5	0.808	0.9232	0.9846	–	–
6	0.8148	0.9074	0.9691	0.9949	–
7	0.8250	0.9	0.9572	0.9878	0.9983

4 Analysis for the SIGNIFICANTONLY Rule

In this section we formalize the SIGNIFICANTONLY rule and outline the proof of Theorem 2. The motivation behind the SIGNIFICANTONLY rule is captured by the following simple observation. Suppose we are trying to design a rule which is CC and k-SNM-α for some fixed k. Take any tournament graph T. If there is a Condorcet-winner, then the rule is fixed and must declare that team the winner. If T is *far* from having a Condorcet-winner, meaning that the smallest set of teams who could collude and produce a Condorcet-winner among them is larger than k, then pick a team uniformly at random. If T is only a small number of manipulations away from having a Condorcet-winner, then in order to satisfy the SNM constraint (approximately) we must allocate all (resp. most) of the probability mass to the teams who could produce a Condorcet-winner. However, we must be careful in order not to incentivize teams in tournaments that are far from having a Condorcet-winner to manipulate into those which are close to having a Condorcet-winner.

The previous paragraph captures the spirit of the SIGNIFICANTONLY rule. We first partition the set of all tournament graphs $\mathcal{T}_n$ into three groups: those with a Condorcet-winner (*Condorcet* tournaments), those where (maybe multiple) sets of at most k teams can produce a Condorcet-winner (*near-Condorcet* tournaments) and those where no set of k teams can produce a Condorcet-winner (*far-Condorcet* tournaments). The rule is fixed for the first part of the partition, and will select a winner uniformly at random on the last part of the partition. The middle part of the partition is further partitioned into four categories depending on the exact number and size of the groups that could produce a Condorcet-winner. For each of the parts in this sub-partition, we propose a way to distribute the probability mass. We must balance two different sets of incentives here. On the one side, we must give sufficient probability mass to the groups that can produce a Condorcet-winner. On the other, we can't give them *too* much mass, since otherwise teams in far-Condorcet tournaments might be substantially incentivized into manipulating the tournament into a near-Condorcet one, where some of the colluding members are also in groups that can now produce Condorcet-winners. We formalize the above with the following definitions.

Definition 7. *A tournament T is said to be* near-Condorcet *for k if there is no Condorcet-winner in T but there exists at least one team i with $|\delta^-(i,T)| \leq k-1$. Call the set of teams $MW(i,T) := \{i\} \cup \delta^-(i,T)$ a minimal winning group (MW group). We call team i the* leader *of $MW(i,T)$ and any team in $j \in MW(i,T)$ significant. If $|MW(i,T)| = 2, 3$ we call it an MW pair or an MW triple, respectively.*

We first prove simple structural properties of near-Condorcet tournaments.

Lemma 1. *Every minimal winning group has exactly one leader.*

Proof. Suppose there exists some other leader j of $MW(i,T)$. By definition, j must lose to i and vice versa, a contradiction. □

Notice that even though every MW group has a different unique leader, a leader of one group can be a member in a different group. Moreover, it can happen that one MW group is a subset of another MW group.

Lemma 2. *Let T be a near-Condorcet tournament, and $MW(i,T)$ and $MW(j,T)$ be distinct MW groups in T with leaders i and j, respectively. Then either i is in $MW(j,T)$, or j is in $MW(i,T)$. In particular, $MW(i,T)$ and $MW(j,T)$ must have a non-empty intersection.*

Proof. Suppose that both i and j in $MW(i,T) \cap MW(j,T)$, or they are both outside. Then, by the definitions of $MW(i,T), MW(j,T)$, they should beat each other, a contradiction. Thus, exactly one of them must be in the intersection. □

We now prove the main lemma about the structure of near-Condorcet tournaments for $k = 3$, showing that there are not too many significant teams and leaders.

Lemma 3. *If $k = 3$ and T is a near-Condorcet tournament, then the number of significant teams in T is at most 6, and the number of teams in the union of MW pairs is at most 3. Furthermore, there cannot be more than 3 MW pairs. If there is exactly one MW pair, then the maximal number of significant teams is 5, if there are exactly two MW pairs, then the maximal number of significant teams is 4, and if there are exactly three pairs, then the maximal number of significant teams is 3.*

Proof. Assume there are p MW pairs and t MW triples in T. By Lemmas 1 and 2, we know that each such group has a unique leader, and these leaders are pairwise distinct. Furthermore, each leader of an MW pair loses exactly one match, and each leader of an MW triple loses exactly two matches. Therefore, all leaders together lose exactly $p + 2t$ matches. On the other hand, there are exactly $\binom{p+t}{2}$ matches between leaders, and in each of these matches, one leader loses. Hence,

$$\binom{p+t}{2} \leq p + 2t.$$

We can equivalently rewrite is as $p^2 + t^2 + 2pt - 3p - 5t \leq 0$. If $p \geq 4$ we get that $4+3t+t^2 \leq 0$, which does not have a solution for a non-negative integer t. Hence, there are either 3, 2, 1, or no MW pairs. Furthermore, significant teams are either leaders of some groups or they beat some leaders. Hence, there are at most the number of leaders plus the number of matches lost by leaders to some non-leaders many of them. In other words, there at most $(p+t) + p + 2t - \binom{p+t}{2}$ significant teams. If there are three pairs, the maximum of this expression for non-negative integer t is 3. Similarly, if $p = 2$ the maximum is 4, if $p = 1$ the maximum is 5, and if $p = 0$ the maximum is 6. Moreover, there cannot be more than 3 teams in the union of all MW pairs. Otherwise, there would be at least 3 MW pairs (there cannot be only two because they intersect), but we already showed that if we have three or more MW pairs, there can be at most 3 teams in the union of all MW groups. □

This lemma implies that in every near-Condorcet tournament at most 6 teams can be directly part of some manipulating group creating a Condorcet-winner. Hence, we can directly design a rule that is 3-SNM-$\frac{2}{3}$. It is sufficient to assign probability 1 to Condorcet-winners, probability $\frac{1}{6}$ to significant teams in near-Condorcet tournaments, and distribute the remaining probabilities in all tournaments uniformly between the remaining teams. We will not prove this formally since we design a better rule, but we believe this intuition is useful in understanding our construction. We now formalize the partition of tournament graphs $\mathcal{T}_n$.

- Let $\mathcal{CC}_n \subset \mathcal{T}_n$ be the set of tournaments with a Condorcet-winner.
- Let $\mathcal{FC}_n \subset \mathcal{T}_n$ be the set of far-Condorcet tournaments.
- Let $i-\mathcal{NCP}_n \subset \mathcal{T}_n$ be the set of near-Condorcet tournaments with exactly i MW pairs for $i = 0, 1, 2, 3$.

Now we are ready to define our main rule.

Definition 8 (SIGNIFICANTONLY Tournament Rule). *For $n \geq 6$ teams, the* SIGNIFICANTONLY *tournament rule does the following.*

1. *If $T \in \mathcal{CC}_n$ the Condorcet-winner gets 1, and the remaining teams get zero.*
2. *If $T \in \mathcal{FC}_n$, pick a winner uniformly at random.*
3. *If $T \in 3-\mathcal{NCP}_n \cup 2-\mathcal{NCP}_n$, teams in MW pairs get $\frac{1}{3}$, and the remaining teams get zero.*
4. *If $T \in 1-\mathcal{NCP}_n$, teams in the only MW pair get $\frac{1}{3}$, teams in MW triples that are not in any MW pair get $\frac{1}{9}$, and the remaining teams get the remaining probability mass evenly distributed between them.*
5. *If $T \in 0-\mathcal{NCP}_n$, teams in MW triples get $\frac{1}{6}$, and the reaming teams get the remaining probability mass evenly distributed between them.*

We restate the main result of this section.

Theorem 2. *The* SIGNIFICANTONLY *rule is Condorcet-consistent, monotone, 2-SNM-1/3 and 3-SNM-1/2, and this is tight.* (See Footnote 2)

The fact that SIGNIFICANTONLY is Condorcet-consistent follows directly from the definition. The proof of Theorem 2 and other missing proofs can be found in the full version of this paper.

5 Almost Optimal Rules for Large n from Optimal Rules for Small n

For fixed values of n, k, [18] present an LP that can compute the minimally manipulable, Condorcet-consistent and monotone rule. Unfortunately, solving this LP is highly intractable as the number of variables and constraints grows exponentially in the number of teams. For small values of n, k, and with sufficient computational power, one could compute such solutions, in the hope that one could use that rule in order to construct approximately optimal ones for larger n. Consider the following simple procedure to scale a k-SNM-α rule r_n for n teams to a k-SNM-α' rule $r_{n'}$ for $n' > n$. First, increase n' to the nearest multiple of n, $n' := nM$ of n. Partition the teams into n groups of equal size. Within each group, compute the top-cycle on the induced sub-tournament and select a team from it uniformly at random as a finalist. This will reduce the number of teams to exactly n finalists. Run r_n on the n finalists and declare that rule's winner as the overall winner. The result of this section bounds the value of α' for $r_{n'}$.

Theorem 3. *If there exists a top-cycle consistent, and k-SNM-α rule r for n teams, then there exists a top-cycle consistent and k-SNM-α' rule r' for $n' > n$ teams where*

$$\alpha' \leq \alpha\Big(1 - \frac{(k-1)^2}{n}\Big) + \frac{(k-1)^2}{n}.$$

We now outline the proof of Theorem 3 and defer its proof to the full version of this paper. It is easy to show that r' is top-cycle consistent. The derivation of the upper bound on α' is more complicated but it is based on a fairly simple idea. Let S be a set of colluding teams in a tournament T on a set of n' teams. If a permutation π on a set of n' teams is chosen uniformly at random, then either every group in the partition contains at most one team from S or there exists a group containing at least two teams from S. In the former case, we use the fact that fixing matches inside S does not change a team's chances of surviving the group. Then we use the assumption that r is k-SNM-α to conclude that S can increase the probability to win the tournament T by at most α by fixing matches inside S. In the latter case, we simply assume that they can increase the probability by 1. The latter case happens with probability $(k-1)^2/n$, which finishes the proof. An explicit consequence of this result is the following. If there exists a top-cycle consistent 3-SNM-2/5 rule for $n = 25$ teams (which would be the best possible as per [18]), then there exists a 3-SNM-α rule for all $n \geq 25$ for some $\alpha < 1/2$.

6 Conclusion and Future Directions

This paper extends our knowledge of non-manipulable tournament rules in several ways. First, we generalize RSEB from [18] into RdSEB. This rule, at every node, picks a Condorcet-winner if one exists among its children and otherwise chooses a team uniformly at random. We show that for $k \leq d$ this rule is k-SNM-$\alpha_{d,k}$ for some $\alpha_{d,k}$ bounded away from 1, providing the first explicit family of rules that are bounded away from 1 for any n. We suspect that a more careful analysis for small values of d, k might yield better rules but conjecture, however, that $\alpha_{d,k} > 1/2$ for all $k \leq d$.

We present a new rule, the SIGNIFICANTONLYRule, which is monotone, Condorcet-consistent, 3-SNM-1/2 and 2-SNM-1/3 (which is best possible). The rule identifies a small set of teams as significant, awards them substantial probability mass and distributes it uniformly among non-significant teams. The motivation for the rule is that tournaments with a Condorcet-winner and tournaments which are far from having a Condorcet-winner are easy to resolve. We find a way to resolve the intermediate tournaments in a way to avoid substantial gains from manipulation.

Finally we propose a way of reducing the problem of finding good rules for large values of n to the problem of finding good rules for small values of n. Our result implies that if there exists a 3-SNM-2/5 just for $n = 25$, then there exist 3-SNM-α rules for $n \geq 25$ and $\alpha < 1/2$, which would directly improve on the state of the art results.

Acknowledgement. This research is part of a project that has received funding from the European Union's Horizon 2020 research and innovation program under the Marie Skłodowska-Curie grant agreement No. 823748, and while D. M. and J. S. were participants in the DIMACS REU program at Rutgers University, supported by NSF grant CNS-2150186.

References

1. Altman, A., Kleinberg, R.: Nonmanipulable randomized tournament selections. In: Proceedings of the National Conference on Artificial Intelligence, vol. 2, pp. 686–690 (2010)
2. Bartholdi, J.J., Tovey, C.A., Trick, M.A.: How hard is it to control an election? Math. Comput. Model. **16**(8), 27–40 (1992). https://doi.org/10.1016/0895-7177(92)90085-Y. http://www.sciencedirect.com/science/article/pii/089571779290085Y
3. Brandt, F., Conitzer, V., Endriss, U., Lang, J., Procaccia, A.D.: Handbook of Computational Social Choice. Cambridge University Press, Cambridge (2016)
4. Copeland, A.: A 'reasonable' social welfare function. Seminar on Mathematics in Social Sciences (1951)
5. Csato, L.: 2018 FIFA World Cup qualification can be manipulated (2017)
6. Dale, E., Fielding, J., Ramakrishnan, H., Sathyanarayanan, S., Weinberg, S.M.: Approximately strategyproof tournament rules with multiple prizes. In: Proceedings of the 23rd ACM Conference on Economics and Computation, EC 2022, pp. 1082–1100. Association for Computing Machinery, New York (2022). https://doi.org/10.1145/3490486.3538242
7. Dinev, A., Weinberg, S.M.: Tight bounds on 3-team manipulations in randomized death match. In: Hansen, K.A., Liu, T.X., Malekian, A. (eds.) WINE 2022. LNCS, vol. 13778, pp. 273–291. Springer, Cham (2022). https://doi.org/10.1007/978-3-031-22832-2_16
8. Ding, K., Weinberg, S.M.: Approximately strategyproof tournament rules in the probabilistic setting. In: 12th Innovations in Theoretical Computer Science Conference, LIPIcs. Leibniz International Proceedings in Informatics, vol. 185, p. 20. Schloss Dagstuhl. Leibniz-Zent. Inform., Wadern (2021). Art. No. 14
9. Dutta, B.: Covering sets and a new Condorcet choice correspondence. J. Econ. Theory **44**(1), 63–80 (1988). https://doi.org/10.1016/0022-0531(88)90096-8. http://www.sciencedirect.com/science/article/pii/0022053188900968
10. Fishburn, P.C.: Condorcet social choice functions. SIAM J. Appl. Math. **33**(3), 469–489 (1977). https://doi.org/10.1137/0133030
11. Gibbard, A.: Manipulation of voting schemes: a general result. Econometrica **41**(4), 587–601 (1973)

12. Kim, M.P., Suksompong, W., Vassilevska Williams, V.: Who can win a single-elimination tournament? In: Proceedings of the Thirtieth AAAI Conference on Artificial Intelligence, Phoenix, Arizona, USA, 12–17 February 2016, pp. 516–522 (2016). http://www.aaai.org/ocs/index.php/AAAI/AAAI16/paper/view/12194
13. Kim, M.P., Vassilevska Williams, V.: Fixing tournaments for kings, chokers, and more. In: Proceedings of the Twenty-Fourth International Joint Conference on Artificial Intelligence, IJCAI 2015, Buenos Aires, Argentina, 25–31 July 2015, pp. 561–567 (2015). http://ijcai.org/Abstract/15/085
14. Laslier, J.F.: Tournament Solutions and Majority Voting, vol. 7. Springer, Heidelberg (1997)
15. Maurer, S.B.: The king chicken theorems. Math. Mag. **53**(2), 67–80 (1980). http://www.jstor.org/stable/2689952
16. Miller, N.R.: A new solution set for tournaments and majority voting: further graph-theoretical approaches to the theory of voting. Am. J. Polit. Sci. **24**(1), 68–96 (1980). http://www.jstor.org/stable/2110925
17. Moulin, H.: Choosing from a tournament. Soc. Choice Welf. **3**(4), 271–291 (1986). http://www.jstor.org/stable/41105842
18. Schneider, J., Schvartzman, A., Weinberg, S.M.: Condorcet-consistent and approximately strategyproof tournament rules. In: 8th Innovations in Theoretical Computer Science Conference, LIPIcs. Leibniz International Proceedings in Informatics, vol. 67, p. 20. Schloss Dagstuhl. Leibniz-Zent. Inform., Wadern (2017). Art. No. 35
19. Schvartzman, A., Weinberg, S.M., Zlatin, E., Zuo, A.: Approximately strategyproof tournament rules: on large manipulating sets and cover-consistence. In: 11th Innovations in Theoretical Computer Science Conference, LIPIcs. Leibniz International Proceedings in Informatics, vol. 151, p. 25. Schloss Dagstuhl. Leibniz-Zent. Inform., Wadern (2020). Art. No. 3
20. Stanton, I., Vassilevska Williams, V.: Rigging tournament brackets for weaker players. In: IJCAI 2011, Proceedings of the 22nd International Joint Conference on Artificial Intelligence, Barcelona, Catalonia, Spain, 16–22 July 2011, pp. 357–364 (2011). https://doi.org/10.5591/978-1-57735-516-8/IJCAI11-069
21. Suksompong, W.: Tournaments in computational social choice: recent developments. In: Zhou, Z.H. (ed.) Proceedings of the Thirtieth International Joint Conference on Artificial Intelligence, IJCAI-2021, pp. 4611–4618. International Joint Conferences on Artificial Intelligence Organization (2021). https://doi.org/10.24963/ijcai.2021/626. Survey Track
22. Vassilevska Williams, V.: Fixing a tournament. In: Proceedings of the Twenty-Fourth AAAI Conference on Artificial Intelligence, AAAI 2010, Atlanta, Georgia, USA, 11–15 July 2010 (2010). http://www.aaai.org/ocs/index.php/AAAI/AAAI10/paper/view/1726
23. Vu, T., Altman, A., Shoham, Y.: On the complexity of schedule control problems for knockout tournaments. In: 8th International Joint Conference on Autonomous Agents and Multi-agent Systems (AAMAS 2009), Budapest, Hungary, 10–15 May 2009, vol. 1, pp. 225–232 (2009). https://doi.org/10.1145/1558013.1558044

As Time Goes By: Adding a Temporal Dimension to Resolve Delegations in Liquid Democracy

Evangelos Markakis[1,2] and Georgios Papasotiropoulos[3(✉)]

[1] Athens University of Economics and Business, Athens, Greece
markakis@gmail.com
[2] Input Output Global (IOG), Athens, Greece
[3] University of Warsaw, Warsaw, Poland
gpapasotiropoulos@gmail.com

Abstract. In recent years, the study of models and questions related to Liquid Democracy has been of growing interest among the community of Computational Social Choice. A concern that has been raised by practitioners is that the current academic literature focuses solely on static inputs, concealing a key characteristic of Liquid Democracy: the right for a voter to change her mind as time goes by. Our work initiates the incorporation of a time-horizon into decision-making problems in Liquid Democracy systems. Our approach, via a computational complexity analysis, exploits concepts and tools from temporal graph theory.

1 Introduction

Liquid Democracy (LD) is a novel voting framework that aspires to revolutionize the typical voter's perception of civic engagement and ultimately elevate both the quantity and quality of community involvement. At its core, LD is predicated on empowering voters to determine their mode of participation. This can be achieved by either casting a vote directly, as in direct democracy, or by entrusting a proxy to act on their behalf, as in representative democracy. Notably, delegations are transitive, meaning that a delegate's vote can be delegated afresh, and at the end of the day a voter that has decided to cast a ballot, votes with a weight dependent on the number of agents that she represents, herself included. As a result of its flexibility, LD is alleged to reconcile the appeal of direct democracy with the practicality of representative democracy, yielding the best of both worlds. The origin of the "liquid" metaphor remains a matter of debate up to date, with one view being that it stems from the ability of votes to flow along delegation paths, while an alternative view argues that it arises from the ability of voters to revoke delegation approvals and continuously adjust their choices. The second opinion serves as a significant driving force behind our research, providing the essential impetus for our endeavors.

According to [6] there is a number of features that suffice to establish a framework as a Liquid Democracy one. Most of them are related to the transitivity property and to the options given to the voters about casting a ballot or choosing representatives. These are more or less taken into account in all relevant works that come from the field of Computational Social Choice. A further aspect, called *Instant Recall*, encompasses the ability of voters to withdraw their delegation at any time. As a matter of fact, in practice, elections allow for extended (sometimes structured) periods of deliberation, until the votes are finalized, and LD could serve as a means of debate empowerment. A withdrawal of delegation may occur for a variety of reasons: a voter may develop doubts on the integrity of her existing representative, or she may discover a higher alignment of opinion with a different representative, or she may simply obtain a better understanding of the election issue under consideration. A characteristic that is being shared by the works that come from Computational Social Choice is that they seem to ignore the Instant Recall feature, and examine isolated static profiles. This oversight was identified and criticized by the team behind the LiquidFeedback platform [1], the most influential and large scale experiment of LD. In [2], inter alia, they state the following, which serves as the main motivation of our work:

"In a governance system with a continuous stream of decisions, participants observe the actions (and non-actions) of others, in particular the activities of their (direct and indirect) delegates as well as the activities of other participants. Based on their observations, we expect participants to adapt their own behaviour in respect to setting, changing, and removing delegations and their own participation. Based on the track records of the participants, a network of trust or dynamic scheme of representation proves itself to be a responsible power structure. The effects that occur through observation and adaptation over time are an essential prerequisite for a comprehensive understanding of liquid democracy, (which) requires a broader view, namely adding a temporal dimension to delegation models."

Leaving aside the lack of temporal aspects in the literature, there are also additional concerns to address in traditional LD models. A crucial disadvantage is that we may experience delegation cycles or paths towards abstainers, resulting to inevitably lost votes. A way that has been suggested in theory [7,11,13] and has been implemented in practice [12], in order to mitigate such issues is to allow each delegating agent to specify an entire set of agents she approves as potential representatives together with a ranking that indicates her preferences. Nevertheless, even with these efforts, the discussed issues may still arise at the election-day. And here is where the temporal dimension can come into play! The main focus of our work is in proposing a framework that leverages temporal information to address the previously discussed concerns, while also providing a valuable tool for deliberation. In the realm between the algorithmic and axiomatic approaches, our study examines the existence of efficient *delegation rules*, i.e., centralized algorithms that take as input the information from the deliberation phase and prescribe for each non-abstaining participant, a delegation path to a voter who casts a ballot, so as to meet the axioms outlined below (for formal definitions and further elaboration we refer to Sect. 2).

Confluent Delegation Rules. In models incorporating multiple, ranked, delegations, as the one under consideration, an esteemed property is *confluence*, which posits that each voter should have at most one other immediate representative in the final outcome [7]. This desirable attribute guarantees that every voter is instructed to take a single action among the three options: vote, abstain, or delegate her own and all received ballots to a specific voter. On the contrary, a non-confluent rule may prompt a voter to delegate different ballots received from different voters to different representatives, or fractionally distribute her acquired ballots. Such suggestions can be challenging for a voter to follow (even though they may indeed be meaningful in certain occasions). In addition to its intuitive nature, confluence is significant for maintaining transparency and preserving the high level of accountability inherent in classic Liquid Democracy [11].

Time-Conscious Delegation Rules. The necessity for incorporating the temporal dimension in the design of delegation rules becomes evident when considering an election where the ballots at the very end of the deliberation phase produce a cycle or a path to an abstainer, for some voter(s). This, unquestionably problematic, scenario has been a widely recognized issue in the literature, that results in ballot loss. Our main insight proposes that a viable compromise to address such situations is by looking into approvals expressed during the previous time-steps of the deliberation phase. Our work operates under the premise that if a voter v decides to trust a voter u, at a given time t, then v accepts any decisions made by u at time t or earlier (up to a certain number of time-steps prior to t, which could be given as a parameter by voter v). This is because the decision to approve a delegation to u is based on what v observes in the previous time-steps and up until time t. However, voter v still retains the right to revoke her approval to u at a later point in time. If this occurs, then u is permitted to finally represent v only if she chooses an action that she had declared at or before time t and, consequently, u will represent v with a specific opinion that v had indeed approved. We refer to the rules that produce delegation paths respecting in such a way the ordering of the time-instants at which a delegation is made available, as *time-conscious*. We highlight that, in instances where phenomena of ballot loss do not appear at the end of the deliberation phase, the solutions we suggest need not involve delegations made in previous time-steps; for the remaining instances avoiding ballot losses is impossible without utilizing past delegations. Hence, one can view our procedures as unavoidable compromises that strike a balance between voters' participation and satisfiability.

The framework we suggest generalizes the model in [7]. Furthermore, our optimization objective coincides with the one in [10,14]. To our knowledge, our work is the first that incorporates temporal aspects in LD models towards the avoidance of ballot loss. Dynamic aspects of LD have also been explored in [5,10], but from a different angle, namely focusing on the game-theoretic perspective.

Our Contributions

Conceptually, our main contribution lies in explicitly incorporating a temporal dimension into an existing election framework, broadening the solution space to alleviate recognized drawbacks. This is putting a stake in the ground in bridging a significant research gap identified by practitioners. From a technical perspective, our work is making a pioneering contribution to the Computational Social Choice literature, since our study incorporates concepts and techniques from temporal graph theory: a notably novel approach in the field.

We study the compatibility of computational tractability with desirable axioms of delegation rules, and with the objective of reducing the loss of votes. More specifically, we are interested in polynomially computable rules that maximize the total utility of the electorate and are also time-conscious and confluent. Unfortunately, despite the natural appeal of the studied requirements, it turns out that this is too much to ask for: our main result establishes that such a delegation rule does not exist, unless $P = NP$, even for simple variants of the model. Therefore, the best one could hope for is to sacrifice one of the considered axioms or to resort to special cases of the problem. We present results in both directions that effectively circumvent the hardness. First, we show that dropping the requirement of time-consciousness is sufficient for designing an efficient rule. Second, in contrast, we prove that certain instances remain hard even when we give away confluence. Then, in response, we introduce natural restrictions under which positive results emerge in the latter case. Finally, insisting on both axioms, we describe a rule that, despite being non-polynomial, significantly outperforms the brute-force solution in terms of computational complexity.

2   The Framework of Temporal Liquid Democracy

In this section, we provide a detailed description of Temporal Liquid Democracy, the time-aware voting formalism that we study. We consider elections in which a set V of n voters should reach a decision on a certain issue. Apart from voting themselves, the participants are given two additional options: abstaining or delegating to other voters. The voters also have some time available to consider what to do (e.g., to get informed on the issue at hand or to observe other voters' choices) and they are allowed to change their mind, perhaps multiple times, until the actual election-day.

We say that an election is a *Temporal Liquid Democracy Election* (t-LD in short) if it consists of two phases:

- A *deliberation phase* of L rounds, where each voter v tries to decide whether to personally vote or not. If she decides to cast a ballot at some time $t \in [L]$, we consider this as her final decision that will not change in the remaining rounds. Otherwise, at every time-instant $t \in [L]$, and as long as a voter v has not decided to cast a ballot herself, she is asked to specify the following:
    - A set of approved voters $S_v^t \subseteq V \setminus \{v\}$ (which may be the empty set, if v wants to abstain at round t), indicating the voters that she trusts to cast a ballot on her behalf, possibly with different levels of confidence.

- A (weak) preference ranking over the voters in S_v^t, which induces a partition of S_v^t into preference groups, according to v. This is accompanied by a positive integer score $sc_v^t(i)$, indicating the utility or happiness level that v experiences if a voter from her i-th most preferred group at time t, will ultimately be selected as her immediate representative.
- A trust-horizon parameter $\delta_v^t \leq t-1$, indicating approval for the views held by any voter in S_v^t up to $\delta_v^t \in \mathbb{N}_{\geq 0}$ time-steps prior to t.

- A *casting phase*, in which all the voters that, during the deliberation phase, expressed willingness to vote (and only these), cast a ballot on the issue under consideration. Every voter who did not declare an intention to abstain at $t = L$, or to vote, is assigned a representative through a prespecified delegation rule that takes into account the entire deliberation phase. The winner(s) of the election then are elected using some weighted voting rule, with each voter's weight derived from the number of participants she represents.

For an illustrative exposition of our model we refer to Example 1. At what follows we elaborate on the input that is required from the voters during the deliberation phase and the assumptions made. Customization is a key to the proposed decision-making model, which offers a range of possibilities to enhance its practicality and reproducibility, without affecting the theoretical guarantees we present. From what follows, it can be deduced that our findings, whether pertaining to tractability or intractability, hold consistently across multiple natural variations of the framework.

The preference ranking, which facilitates voters to express different levels of confidence towards potential representatives, mainly offers notational convenience, since solely requesting scores (and excluding rankings) from voters is sufficient. However, to streamline voters' ballots towards improving the model's practicality, we could even permit voters to submit only rankings, (omitting scores), and then, scores could be automatically generated (the same applies to trust-horizon parameters, if necessary). The scoring function allows to capture cases where a voter is willing to either increase her scores over rounds due to becoming gradually more informed about others' opinions or, in the opposite direction, decrease scores due to becoming more hesitant about who represents her. Therefore, scores are implicitly assumed to be comparable not only between voters but also across different time-instants for a given voter. Realistically, we expect voters to have just a few preference groups and our results hold for such cases as well. In summary, our presentation choice regarding voters' preferences aims to facilitate adaptation to model variants.

Numerous meaningful variants of the described model could be readily devised by making slight adjustments to the suggested model, as evident for example from the preceding paragraph. Another noteworthy case concerns instances where voters prioritize being represented by someone approved by them as late as possible. We will now underscore particular assumptions made which, crucially, do not impact our findings as well. For technical convenience, we have assumed that (a) each voter provides the same time-horizon parameter for all her approved representatives, and (b) once a voter declares the intention to cast a ballot, she cannot change her stance (voters' decision to abstain could be also finalized earlier than $t = L$, if preferred). Furthermore, we note that the described model holds particular relevance in situations where voters prioritize being represented by a previously approved voter over forfeiting their ballot (e.g., by delegating to an abstainer). Hence, essentially, it captures scenarios where voters make adjustments rather than completely reversing their preferences; nonetheless, our results also hold for the model variant that allows for full revocation of vote.

The intuition behind the trust-horizon parameters comes from the fact that the decision of a voter v to approve a delegation to u at time t, can be based on looking only at the behavior of u in the previous rounds and up until time t of the deliberation phase. Since a voter v may not agree with u in all previous time-steps, the parameter δ_v^t specifies that v agrees with the choices made by u (regarding voting herself or further delegate to others with her own trust-horizon) at any preceding time, that is no more than δ_v^t time-instants before t. The simplest case appears when $\delta_v^t = t - 1$ (i.e., v trusts whatever u has chosen at any time in the past). If this holds for every voter v and for any $t \in [L]$, we say that the election profile is of *retrospective trust*.

In the elections we consider, we have three types of participants. We refer to the voters that declared intention to vote as *casting voters*. The non-casting voters that will abstain from the election are precisely those who do not approve anyone at the final time-step, e.g., a voter v such that $S_v^L = \emptyset$. We refer to those as *abstaining voters*. The rest of the voters will be called *delegating voters*. As evident from Sect. 1, and as will be further illustrated by Example 1, the temporal dimension could be considered valuable, via the notion of time-consciousness, when the examination of the isolated instance at $t = L$ cannot produce a feasible solution (i.e., delegation cycles or paths towards abstainers are unavoidable). A delegation rule is a mechanism that "resolves delegations" to address such problematic cases, i.e., a procedure that ultimately assigns to each delegating voter, a casting voter, possibly via following some path of trust relationships. More formally, it is a function that takes as input the voters' preferences, as reported during the deliberation phase, and outputs a path to a casting voter, for every delegating voter. A delegation rule should ask casting voters to vote, abstaining voters to abstain and should not suggest any delegation path towards an abstainer or introduce delegation cycles.

Temporal Graphs. The driving force in our work is to model and analyze t-LD elections using principles from temporal graph theory. We start with a basic overview of the concept and the terminology and following this, we will introduce some notation that we will use in the remainder. At a high level, a temporal graph is nothing more than a simple, called *static*, graph in which a temporal dimension is being added, i.e., a graph that may change over time. Frequently, a temporal (multi)graph is being expressed as a time-based sequence of static graphs. For convenience, we will use an equivalent definition, under which, a (directed) temporal (multi)graph $G(V, E, \tau, L)$ is determined by a set of vertices

V, a (multi)set of directed, temporal edges E, a discrete time-labelling function τ that maps every edge of E to a non-empty subinterval of $[1, L]$, and a lifespan $L \in \mathbb{N}$. If the edges of E are weighted according to a function $w : E \to \mathbb{N}$, then we say that G is weighted. The interval $\tau(e) = [s_e, t_e]$, that is assigned to an edge e, indicates that e is available at the time-instants that belong to $\tau(e)$ (it is possible also that, $s_e = t_e$). By allowing G to be a multigraph[1] it is permitted for an edge to be present in multiple (disjoint) time-intervals. Unless otherwise stated, henceforth, by the term *graph*, we denote a weighted directed temporal multigraph. For more details on temporal graphs we refer to a relevant survey [16]. The *static variant* of a temporal graph is the static (multi)graph that emerges if we ignore the time-labels of its edges. We call a (temporal) graph *temporal directed tree rooted at vertex r* if its static variant contains a directed path towards r from every other vertex and its undirected variant is a tree. A crucial concept for our work, in the context of temporal graphs, is the notion of time-conscious paths, that satisfy a monotonicity property regarding the temporal dimension of their edges. Consider a temporal graph $G(V, E, \tau, L)$, coupled with a tuple $\delta_v = (\delta_v^t)_{t \in [L]} \in \mathbb{N}^{[L]}$ for every vertex v of V. Let also $\delta = (\delta_v)_{v \in V}$ and for notational convenience we will occasionally include δ in the description of a temporal graph, denoted then by $G(V, E, \tau, L, \delta)$. We say that a path in G from v_1 to v_{k+1} is δ-*time-conscious* if it can be expressed as an alternating sequence of vertices and temporal edges $(v_i, (e_i, t_i), v_{i+1})_{i \in [k]}$, such that for every $i \in [k]$ it holds that $e_i = (v_i, v_{i+1}) \in E$, $t_i \in \tau(e_i)$ and for every $i \in [k-1]$ it holds that $t_i \geq t_{i+1} \geq t_i - \delta_{v_i}^{t_i}$. Hence, time-consciousness specifies a traversal of an edge e at a time that is no later than the previous edge in the path, and while e is present.

In the remainder of Sect. 2, it will become more clear how this notion fits in our framework. We also call δ-time-conscious, a temporal directed tree, rooted at a vertex r, if all its paths towards r are δ-time-conscious.

Modelling t-LD Elections as Temporal Graphs. The deliberation phase of a t-LD election, as defined in the beginning of this section, can be modeled as a weighted directed temporal multigraph $G(V \cup \{\triangledown\}, E, \tau, L, w, \delta)$ formed by:

- a vertex in V for every voter, as well as a special vertex $\triangledown$ representing the voters' commitment to cast a ballot,
- a multiset E of temporal edges that contains the following: (i) edges that represent the approvals for delegation per round via a function τ that assigns a time-label to every edge, (ii) an edge $(v, \triangledown)$, for which $L \in \tau((v, \triangledown))$, for every vertex v that corresponds to a casting voter,
- a lifespan L that represents the duration of the deliberation phase,
- a function w that assigns a weight to every edge (v, u) of E, according to sc_v^t, provided that $t \in \tau((v, u))$,
- a vector δ that, for every voter v, contains a tuple $(\delta_v^t)_{t \in [L]}$, as declared by v during the deliberation phase. For convenience, we allow δ to have empty

---

[1] We are using multigraphs instead of (simple) graphs for technical convenience. Alternatively, one could work with graphs by letting τ be a function that maps edges to a set of subintervals of $[1, L]$.

entries, corresponding to casting voters or to time-steps during which the corresponding voter abstained.

If C is the set of casting voters, then if any such voter had indicated preferences for potential representatives before deciding to cast a ballot, these preferences, and their corresponding edges, can be safely disregarded. More precisely, only the following two types of edges may exist: directed edges of the form $e = (v, u)$ for $v \in V \backslash C$ and $u \in V$ with $\tau(e) = [s_e, t_e]$, indicating that at any time-instant $t \in [s_e, t_e]$, voter u belongs to S_v^t, and directed edges $e = (v, \triangledown)$ for $v \in C$ with $\tau(e) = [s_e, L]$, indicating that from time s_e and onwards, voter v agrees to cast a ballot. As already explained before, the deliberation phase implicitly partitions the set of voters V into three sets: the set of casting voters C, the set of abstaining voters A and the set of delegating voters D. More formally, $C = \{v \in V : (v, \triangledown) \in E\}$, $A = \{v \in V \backslash C : L \notin \tau((v, u)),$ for any $(v, u) \in E\}$ and $D = V \backslash (C \cup A)$. The weight function w indicates the cardinal preferences of a voter, as implied by the scores that accompany her preference rankings during the deliberation phase. Additionally, for convenience, we set to zero the weights of edges (v, u) such that v corresponds to a casting or an abstaining voter. This choice can be justified by the upcoming discussion of the optimization objective in the "Electorate's Satisfaction" paragraph. Given a graph $G(V \cup \{\triangledown\}, E, \tau, L, w, \delta)$ that models a t-LD election, a delegation rule returns, for every delegating voter v, a weighted directed temporal path from v to $\triangledown$, inferring which infers an assignment of every delegating voter to a casting one. We call a delegation rule *efficient* if its output can be computed in polynomial time in the input size.

Our Target Axioms Under the Temporal Framework. We proceed to define, within the framework of temporal graphs, the axioms that constitute the core focus of our work, previously discussed at a high level in Sect. 1.

Definition 1. *Let $G(V \cup \{\triangledown\}, E, \tau, L, w, \delta)$ be a graph modelling a t-LD election.*

1. *A delegation rule is time-conscious, if for every delegating voter v, the delegation path output for v is a δ-time-conscious directed temporal path.*
2. *A delegation rule is confluent, if the union of the paths output for all the delegating voters is a directed temporal tree, rooted at vertex $\triangledown$, that spans the vertices of $V \backslash A$.*

The definition of time-consciousness guarantees that all paths suggested by the delegation rule satisfy the constraints imposed by the voters, regarding their trust-horizon parameters. Hence, for any edge (v, u) in an output path, u must perform an action (i.e., choose an edge corresponding to a further delegation or vote directly) that she had declared at a time that was approved by v. The definition of confluence guarantees that for every delegating voter v, there is a unique path to a casting voter, that is intended to serve both v and all voters who delegated to v.

We make the usual assumptions for Liquid Democracy models that (a) voters completely trust their representatives and (b) trust between voters is transitive.

This implies that if voter v accepts voter u as her potential representative, she concurs with any subsequent choice made by u and also extends trust to any voter w who may be entrusted by u. Hence, we note that the utility experienced by a delegating voter from a delegation rule can be considered as a local one, being contingent solely on the voter's immediate representative and not influenced by further choices made by the chosen representative. Therefore, the utility of a delegating voter can be determined by the score that she declared for her immediate representative, specified by the delegation rule. Note that two different time-instants t, t' may exist such that $u \in S_v^t \cap S_v^{t'}$. In these cases, given that the output of a delegation rule is a set of temporal paths, if the rule suggests a delegation from v to u, it also explicitly specifies the time-instant at which the delegation will occur, say e.g., at time t' and, thereby the utility of v is equal to $sc_v^{t'}(i)$, if u belongs to the i-th most preferred group of v, at time t'. Regarding now the casting voters, we do not take into account their utility since their will to cast a ballot has been realized; we do the same for abstaining voters. We consider as infeasible every solution that asks a casting (resp. abstaining) voter to delegate her ballot or abstain (resp. vote), and hence, our focus will be on the welfare of the delegating voters. Finally, the quality of a rule is assessed by the total satisfaction it elicits from the electorate which is expressed as the sum of utilities of all delegating voters.

As extensively discussed in Sect. 1, our goal is to consider the entire deliberation phase so as to address instances where it is unattainable to achieve feasibility by looking only at the final time-step. Nevertheless, the proposed framework also facilitates exploration of a broader setting, wherein allowing past consultation can improve a solution's quality (with respect to the electorate's welfare) even when feasible solutions at the final time-step do exist. Our optimization objective is to maximize the electorate's satisfaction. The algorithmic problem our work focuses on will be called RESOLVE-DELEGATION and is formally defined as follows: Given a graph $G(V \cup \{\nabla\}, E, \tau, L, w, \delta)$ representing the deliberation phase of a t-LD election, compute a weighted directed temporal path from each delegating voter to ∇, with the aim of maximizing the total utility derived from the delegating voters, defined as the sum of the weights of the paths' first edges.

Example 1. Consider the following instance of a t-LD election with 5 rounds and 6 voters, namely Alice, Bob, Charlie, Daisy, Elsa, and Fred. Their preferences are outlined below:

- Alice initially intended to delegate to Charlie. In the second round, she decided to get informed about the considered issue and vote.
- Bob did not participate in the deliberation during the first round, but approved Alice in the second round. In the third round, he revoked his approval of Alice and instead approved Charlie and Elsa. Bob's approval of Elsa remained until the end of the deliberation.
- Charlie approved Alice only in the beginning of the election. He also approved Bob in the first and third round, but removed his approval (and abstained) in the second round. In the fourth round, Charlie approved both Daisy's and

Fred's perspectives on the topic at hand, but he removed his approval of Daisy in the final round.
- Daisy expressed interest in being a casting voter from the beginning until the end of the deliberation phase.
- Although Elsa intended to delegate her ballot to Fred at rounds 2 and 3, ultimately both refrained from participating.

The described instance can be visualized using the graph shown in Fig. 1. We assume that $\delta_B^t = 1$ and $\delta_C^t = t-1$ for every $t \in \{1, 2, \ldots, 5\}$. The scores assigned by the voters to their approved representatives are encoded by the form of the edges, where curly edges have weight 1, straight edges have weight 2, and double-lined edges have weight 3. Dotted edges indicate the casting voters. The labels of the edges represent the time-intervals of their presence. In this instance, Alice and Daisy form the set of casting voters, while Elsa and Fred abstain. Therefore, edge (A, C) can be removed, since Alice will definitely cast a ballot. In a δ-time conscious solution, Bob would not delegate to Charlie, since no δ-time-conscious path to $\triangledown$ using the edge (B, C) exists, for instance, edge (C, A) violates the time-horizon declared by Bob. Similarly, Charlie would not delegate to Alice or to Bob at time 1. Since we do not allow Bob to delegate to an abstainer, he must delegate to Alice, whom he trusted at time 2. Then, there are two possible outcomes for the delegation rule, depending on the choice made for Charlie. The edge that maximizes Charlie's utility is (C, D). Therefore, the optimal delegation rule that is both time-conscious and confluent, would suggest the paths $\{((C, D), (D, \triangledown)), ((B, A), (A, \triangledown))\}$, achieving a total satisfaction score of 4.

It is plainly evident how the temporal dimension comes to the rescue: if one were to focus solely on the snapshot taken at time 5, disregarding the information garnered from the deliberation phase, the only option would be to ask Bob and Charlie to delegate to abstaining voters since at time 5 they only approve Elsa and Fred respectively. Instead, our framework utilizes the information obtained throughout the deliberation phase to propose an outcome that avoids paths towards abstainers and also avoids delegating cycles.

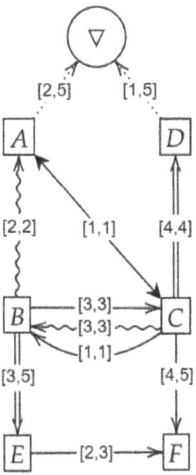

Fig. 1. Illustration of the associated graph.

3 On Efficient Utility Maximizing Delegation Rules

We will now explore the compatibility of the axioms we have put forward from Sects. 1 and 2, with efficient computation. Due to page limitations, we only present intuitive insights and sketches; complete proofs appear in the full version.

Our first result shows that it is impossible to have polynomially computable utility maximizing delegation rules that satisfy simultaneously the axioms of time-consciousness and confluence, unless P = NP, even under simple, natural restrictions. Before stating the result, we discuss the types of instances for which we establish hardness. It is expected that in real-life elections, voters tend to exhibit a relatively stable opinion over time, and do not revise their preferences numerous times during the deliberation phase, due to the required effort to process new information. Similarly, it is reasonable to expect that due to limited cognitive capacity, the voters are only able to partition their accepted representatives into a few disjoint preference groups. The theorem that follows demonstrates that the computational intractability of RESOLVE-DELEGATION persists even when we limit the voters to changing their mind at most once during the deliberation phase and partitioning their accepted representatives into at most two groups, at each round. Furthermore, it holds even for instances of retrospective trust, and for simple scoring functions. Therefore, the primary takeaway is that incorporating temporal aspects in conjunction with natural requirements does come at a computational cost, even in restricted profiles.

Theorem 1. RESOLVE-DELEGATION *in a time-conscious and confluent manner is NP-hard, even for profiles of retrospective trust.*

We now explore roads to circumvent this impossibility result. Our proposal is to relinquish either the necessity for efficiency or one of the axioms of time-consciousness and confluence, in hopes of solving RESOLVE-DELEGATION. Our findings show that this strategy proves successful for some of the problems that emerge, which highlights that Theorem 1 is not devastating. Notably, most of the suggested procedures are simple enough and therefore are confirmed as strong contenders for practical applications.

We begin with studying the easiest variant of RESOLVE-DELEGATION in which the requirement of time-consciousness is being disregarded. This is mainly done for the sake of completeness since studying it requires overlooking the temporal dimension of the instance, which is the defining characteristic of our work. In order to efficiently solve RESOLVE-DELEGATION in a confluent but not necessarily time-conscious manner, the delegation rule can treat any input submitted by a voter at any time as if it was not subject to time-related constraints. Since confluence implies that the output should be a directed tree, and since the utility of each delegating voter is determined by its outgoing edge, then all edges of the tree with non-zero weight will contribute exactly once to the total satisfaction, and therefore, the objective is to find a (static) directed tree of maximum total weight, that is rooted at $\triangledown$ and spans the vertices of $V \setminus A$. To solve this problem we leverage the well-known algorithm by Edmonds [9] for the directed analog of the classic MINIMUM SPANNING TREE problem. In this problem, given a weighted directed static graph $G(V, E, w)$ and a designated vertex $r \in V$, we are asked for a subgraph T of G, the undirected variant of which is a tree, of minimum total cost, such that every vertex of G is reachable from r by a directed path in T. We note that, in our case, the paths we need to compute are towards a fixed vertex, rather than originate from it. To apply Edmonds's algorithm, an adjustment of the graph G is needed, as described in the proof of the following theorem.

Theorem 2. RESOLVE-DELEGATION *in a confluent manner is solvable in polynomial time.*

We now shift our focus to efficient utility maximizing delegation rules that satisfy time-consciousness but are not necessarily confluent. Despite not necessarily resulting in a tree structure, such a rule should still suggest a precise path to a casting voter for every delegating voter v. Then, the utility of v will be derived from her immediate representative (i.e., the weight of the first edge) in that path, regardless of whether other paths going through v may exist for serving other voters who have delegated to v.

The question of why non-confluent delegation rules merit investigation is discussed in [7]. It was discovered that, among a large family of delegation rules, only non-confluent rules possessed the potential to satisfy the axiom of *copy-robustness*, an axiom that is also motivated by practical considerations [3] which, at a conceptual level, guarantees that neither a delegating voter nor her representative can be better off if the former casts an identical ballot to the latter instead of delegating. Moreover, there are non-confluent rules with desirable properties that have been previously studied, such as the Depth-First-Delegation rule that precludes the possibility of Pareto-dominated delegations [13]. Hence, it is not unprecedented to sacrifice confluence on the altar of attaining other attributes.

Quite surprisingly, even in the absence of a requirement for a confluent rule, RESOLVE-DELEGATION remains NP-hard, as shown by the following theorem. Notably, this holds even for simple scenarios that involve only a brief deliberation phase, uniform trust-horizon parameters across all voters and a lone delegating voter. Importantly, it is orthogonal to the result of Theorem 1, since it explicitly uses the fact that the considered elections are not of retrospective trust.

Theorem 3. RESOLVE-DELEGATION *in a time-conscious manner is NP-hard, even for profiles with only a single delegating voter.*

Continuing with our study of efficient utility-maximizing delegation rules that are time-conscious but not necessarily confluent, we now turn to exploring potential workarounds to the impossibility result of Theorem 3. To overcome the intractability, we restrict ourselves to the still hard variant where the voters share the same time-horizon parameter and propose the following relaxations:

- *Assuming retrospective trust profiles*, i.e., $\delta_v^t = t - 1$, for every voter v and every time-step t, in which v delegates to others. These profiles are motivated by the fact that in real life, as long as voters do not change their opinion

arbitrarily, it is likely that a delegating voter will trust another voter for all the past steps instead of taking the effort to further refine her trust interval.
– *Permitting walks instead of only paths*, or in other words allowing for revisits to vertices, along a path from a delegating voter to a casting one. This enlarges the solution space and can be helpful towards achieving time-consciousness in certain instances, as it may be necessary to go through a cycle before being able to satisfy the time constraints.

We highlight that the NP-hardness reduction of Theorem 3 does not use instances of retrospective trust and does not allow for walks. The approach of neglecting confluence, enables the development of local delegation rules, likewise the rules studied in [8], that make a decision for every voter completely independent of the choices made for the rest of the electorate. For the two relaxations suggested in the previous discussion, we propose a procedure that, at a high level, visits every vertex v, corresponding to a delegating voter v, in a sequential manner, and for each such vertex, it detects a feasible, i.e., δ-time-conscious, way to reach $\triangledown$, that uses the out-going edge of v of maximum possible weight. The aforementioned way of reaching a casting voter can be computed by a suitable modification of the temporal analog of the Breadth-First search algorithm from [15], in the case where the input profile is of retrospective trust and by using the polynomial procedure based on Dijkstra's algorithm from [4], in the case where walks are allowed and all voters share the same trust-horizon parameter.

Concerning the first relaxation, in [15], a polynomial-time algorithm was suggested to solve a (more general than what we need in our setting) problem, called FOREMOST PATH. In this, we are given a (unweighted) directed temporal graph $G(V, E, \tau, L)$, a source vertex $v \in V$, a sink vertex $u \in V$, and a time-instant $t_{start} \in [L]$, and we are asked to compute[2] a time-respecting path[3] from v to u, that starts no sooner than t_{start} (or report non-existence).

For the second relaxation, we begin by the following definition: Given, a temporal graph $G(V, E, \tau, L, \delta)$ in which all entries of the vector δ coincide with a fixed value Δ, a temporal walk p of G of length ℓ, say $p = (v_{i-1}, (e_i, t_i), v_i)_{i \in [\ell]}$ such that v_i's are not necessarily all pairwise distinct, is called Δ-restless if for every $i \in [\ell]$ it holds that $e_i = (v_{i-1}, v_i)$ and that $t_i \in \tau(e_i)$ and for $i \in [\ell-1]$ it holds that $t_i \leq t_{i+1} \leq t_i + \Delta$. To efficiently solve the relaxation of RESOLVE-DELEGATION in which walks are allowed, in a time-conscious manner, we utilize the procedure from [4], that outputs a Δ-restless temporal walk between two vertices, for any fixed Δ.

Theorem 4. RESOLVE-DELEGATION *in a time-conscious manner is polynomially solvable for profiles of retrospective trust. Moreover, the same holds for the variant of the problem where walks are allowed, for profiles in which there is a common, fixed trust-horizon parameter, for all voters and all time-steps.*

---

[2] To be more precise, the goal is to select the path that minimizes the arrival time but for our purposes, this objective is superfluous (but harmless).

[3] The term *time-respecting* is being used in the literature to describe a temporal path $(v_i, (e_i, t_i), v_{i+1})_{i \in [\ell]}$ such that for every $i \in [\ell]$ it holds that $e_i = (v_i, v_{i+1})$, $t_i \in \tau(e_i)$, and $1 \leq t_1 \leq t_2 \leq \cdots \leq t_\ell \leq L$ (also called "journey" or simply "temporal").

Finally, we conclude with studying RESOLVE-DELEGATION in a time-conscious and confluent manner, but now without the requirement of computational efficiency. Clearly, if polynomial solvability is no longer a worry, a straightforward brute-force procedure can examine all possible trees to maximize the voters' satisfaction. However, our objective goes beyond this. Such a procedure is exponential in the number of edges (and hence in n) and also in L. In contrast, we aim at developing a procedure that could be well-suited for scenarios where the deliberation phase is prolonged, by avoiding exponentiality in L. On top of that, we focus on designing an algorithm that is superior to the brute-force procedure also with respect to the exponential dependence in n. Specifically, we aim for a running time dependent only on the number of delegating voters $|D|$ (upper bounded by n), which would be suitable in any relatively small community. Yet, this is not possible without further assumptions, given the negative result of Theorem 3 that holds even for $|D| = 1$. Therefore, as before, we resort to instances of t-LD elections of retrospective trust.

Theorem 5. RESOLVE-DELEGATION *in a time-conscious and confluent manner is solvable in time exponential in $|D|$ and polynomial in the remaining input parameters, for profiles of retrospective trust.*

Summarizing, while fulfilling all the desired attributes concurrently seems implausible, positive results can indeed be attained in various natural directions.

4 Takeaways and Directions for Future Works

The main features of LD (cf. [6]) are the transitivity of delegations together with the voters' ability to cast a ballot, to delegate voting rights and to modify or recall a delegation. Our work is the first in the literature of Computational Social Choice, that presents a model satisfying every each of the above and suggesting rules designed to prevent ballot loss. Inspired by the proposal of [2] to add a temporal dimension in the algorithmic considerations of LD, and building upon [7], we studied a framework from a viewpoint that lies in the middle ground between algorithmic and axiomatic approaches. By leveraging concepts and tools from temporal graph theory (a fresh perspective for Computational Social Choice), we unveiled the impossibility of simultaneously fulfilling a natural pair of desired axioms. Despite this, we achieved positive results in various directions.

We intentionally gave significant emphasis on developing a general model for incorporating temporal aspects and we feel it opens up the way for several promising avenues for future research. One such is to examine whether time-consciousness (or other time-related axioms) is compatible with other established axioms, beyond confluence. Also an intriguing topic is to extend our positive results to further generalizations of t-LD elections, e.g., when the voters are able to use a more powerful language to express complex preferences. Furthermore, studying an egalitarian objective instead of the social welfare or imposing restrictions on the maximum in-degree or on the maximum path-length in the output of a delegation rule, are perfect candidates for future works on the area. It is also interesting to incorporate the preferences of the delegating voters over their (final) casting voter representative, instead of their immediate representative in the output graph. Finally, we also identify as an important future work topic the use of experimental or empirical evaluations of LD frameworks that take into account temporal considerations.

Acknowledgments. This research was supported by the framework of H.F.R.I call "Basic research Financing (Horizontal support of all Sciences)" under the National Recovery and Resilience Plan "Greece 2.0" funded by the European Union – NextGenerationEU (H.F.R.I. Project Number: 15877). Georgios Papasotiropoulos was supported by the Hellenic Foundation for Research and Innovation (H.F.R.I.) under the "3rd Call for H.F.R.I. PhD Fellowships" (Fellowship Number: 5163), as well as by the European Union (ERC, PRO-DEMOCRATIC, 101076570). Views and opinions expressed are however those of the author only and do not necessarily reflect those of the European Union or the European Research Council. Neither the European Union nor the granting authority can be held responsible for them.

Disclosure of Interests. The authors have no competing interests.

References

1. Behrens, J., Kistner, A., Nitsche, A., Swierczek, B.: The Principles of LiquidFeedback. Interaktive Demokratie (2014)
2. Behrens, J., Kistner, A., Nitsche, A., Swierczek, B.: The temporal dimension in the analysis of liquid democracy delegation graphs. Liq. Democracy J. **7** (2021). https://liquid-democracy-journal.org/issue/7/The_Liquid_Democracy_Journal-Issue007-04-The_Temporal_Dimension_in_the_Analysis_of_Liquid_Democracy_Delegation_Graphs.html
3. Behrens, J., Swierczek, B.: Preferential delegation and the problem of negative voting weight. Liq. Democracy J. **3**, 6–34 (2015)
4. Bentert, M., Himmel, A.S., Nichterlein, A., Niedermeier, R.: Efficient computation of optimal temporal walks under waiting-time constraints. Appl. Netw. Sci. **5**(1), 1–26 (2020)
5. Bloembergen, D., Grossi, D., Lackner, M.: On rational delegations in liquid democracy. In: Proceedings of the AAAI Conference on Artificial Intelligence, AAAI, pp. 1796–1803 (2019)
6. Blum, C., Zuber, C.I.: Liquid democracy: potentials, problems, and perspectives. J Polit Philos **24**(2), 162–182 (2016)
7. Brill, M., Delemazure, T., George, A., Lackner, M., Schmidt-Kraepelin, U.: Liquid democracy with ranked delegations. In: Proceedings of the AAAI Conference on Artificial Intelligence, AAAI, pp. 4884–4891 (2022)
8. Caragiannis, I., Micha, E.: A contribution to the critique of liquid democracy. In: Proceedings of the International Joint Conference on Artificial Intelligence, IJCAI, pp. 116–122 (2019)
9. Edmonds, J.: Optimum branchings. Math. Decis. Sci. **1**(335–345), 25 (1968)

10. Escoffier, B., Gilbert, H., Pass-Lanneau, A.: The convergence of iterative delegations in liquid democracy in a social network. In: Fotakis, D., Markakis, E. (eds.) SAGT 2019. LNCS, vol. 11801, pp. 284–297. Springer, Cham (2019). https://doi.org/10.1007/978-3-030-30473-7_19
11. Gölz, P., Kahng, A., Mackenzie, S., Procaccia, A.D.: The fluid mechanics of liquid democracy. ACM Trans. Econ. Comput. **9**(4), 1–39 (2021)
12. Hardt, S., Lopes, L.: Google votes: a liquid democracy experiment on a corporate social network. Technical Disclosure Commons (2015)
13. Kotsialou, G., Riley, L.: Incentivising participation in liquid democracy with breadth-first delegation. In: Proceedings of the International Conference on Autonomous Agents and Multiagent Systems, AAMAS, pp. 638–644 (2020)
14. Markakis, E., Papasotiropoulos, G.: An approval-based model for single-step liquid democracy. In: Caragiannis, I., Hansen, K.A. (eds.) SAGT 2021. LNCS, vol. 12885, pp. 360–375. Springer, Cham (2021). https://doi.org/10.1007/978-3-030-85947-3_24
15. Mertzios, G.B., Michail, O., Spirakis, P.G.: Temporal network optimization subject to connectivity constraints. Algorithmica **81**(4), 1416–1449 (2019)
16. Michail, O.: An introduction to temporal graphs: an algorithmic perspective. Internet Math. **12**(4), 239–280 (2016)

Game Theory, Optimization, and Decision Making

Simple Stochastic Stopping Games: A Generator and Benchmark Library

Avi Rudich, Isaac Rudich(✉)[iD], and Rachel Rue[iD]

Polytechnique Montréal, Montréal, QC, Canada
isaac.rudich@gmail.com

Abstract. Stochastic Games are used for modeling decision-making processes in environments characterized by uncertainty and adversarial interactions. They are particularly relevant for multi-agent systems, where understanding the equilibrium of multiple decision-makers is essential. Simple Stochastic Games (SSGs) were introduced by Anne Condon in 1990, as the simplest version of Stochastic Games for which there is no known polynomial-time algorithm [5]. Condon showed that Stochastic Games are polynomial-time reducible to SSGs, which in turn are polynomial-time reducible to Stopping Games. SSGs are games where all decisions are binary and every move has a random outcome with a known probability distribution. Stopping Games are SSGs that are guaranteed to terminate. There are many algorithms for SSGs, most of which are fast in practice, but they all lack theoretical guarantees for polynomial-time convergence. The pursuit of a polynomial-time algorithm for SSGs is an active area of research. This paper is intended to support such research by making it easier to study the graphical structure of SSGs. Our contributions are: (1) a generating algorithm for Stopping Games, (2) a proof that the algorithm can generate any game, (3) a list of additional polynomial-time reductions that can be made to Stopping Games, (4) an open source generator for generating fully reduced instances of Stopping Games that comes with instructions and is fully documented, (5) a benchmark set of such instances, (6) and an analysis of how two main algorithm types perform on our benchmark set.

Keywords: Stochastic Games · Decision Theory · Algorithmic Game Theory

1 Introduction

Stochastic Games were first introduced in 1953 to model situations where strategic decisions are made under conditions of uncertainty. They provide a framework for modeling situations where multiple decision-makers interact in a stochastic environment, such as optimizing the transmission of data over shared wireless networks [1,15]. A strategy is the set of decisions each respective player makes

---

A. Rudich, I. Rudich and R. Rue—Contributed equally.

about how to move when the game is at a position they control, and the outcome of each move is governed by a known probability distribution [12]. There exist optimal strategies that are both global (all moves are chosen in advance of the game) and deterministic (exactly one move is chosen for each game position) [5].

There is no known polynomial-time algorithm for general stochastic games. In 1990, Condon defined Simple Stochastic Games (SSGs), gave a polynomial-time reduction of Stochastic Games to SSGs, and proved that the decision problem associated with SSGs is in $\mathcal{NP} \cap$ co-$\mathcal{NP}$ [5]. In an SSG, there are two players, and all of the decisions made by the players are binary. Simple Stochastic Games have become a focus of research because they are the simplest version of stochastic games for which there is no known polynomial-time algorithm. They are also among the few combinatorial problems where the associated decision problem is in $\mathcal{NP} \cap$ co-$\mathcal{NP}$ and not known to be in P [5].

SSGs are played on a directed graph composed of four types of nodes: max, min, average, and terminal. The two terminal nodes are labeled 1 and 0, and have out-degree 0. All other nodes have out-degree 2. At max nodes, the max player determines which arc to take; at min nodes, the min player determines which arc to take; and at average nodes, each arc is taken with probability $\frac{1}{2}$. The goal of the max player is to maximize the probability that a random walk on the graph from a given start node reaches terminal-1, and the goal of the min player is to minimize that probability. The value of a node given a pair of max/min strategies is the probability that a random walk starting from that node reaches terminal-1. A stable assignment of values is any assignment compatible with mutually optimal max/min strategies.

While it is an open question whether SSGs admit a polynomial-time algorithm, many algorithms are fast in practice [3,4,10,11]. These algorithms generally fall into two categories: value improvement and strategy improvement. Both types of algorithms start from some initial position and make improvements by updating either node values or player strategies until the constraints that define an SSG are satisfied. Another approach is the permutation-improvement algorithm introduced in [7]. The algorithm starts from an initial permutation of stochastic (generalized average) nodes where a permutation represents an ordering of values, it computes the optimal strategies for that permutation, and then it updates the permutation until optimal strategies are found.

Despite considerable work in refining existing approaches, there remains a large gap between the best known lower and upper bounds. Halman [8] showed that an existing algorithm [13] for LP-type problems has an upper bound of $e^{O(\sqrt{n \log n})}$ on its running time when applied to SSGs. A lower bound can be demonstrated by constructing an example that forces the Hoffman-Karp strategy improvement algorithm, which solves an LP in each iteration, to go through a linear number of iterations [10]. To our knowledge, no one has been able to produce an example that requires more than a linear number of iterations.

Condon also proved that SSGs are polynomial-time reducible to Stopping Games, which terminate with probability 1 and have a unique stable value vector [5]. Additionally, it is possible to make several other simple reductions to remove all subgraphs that are solvable in polynomial time. Improved algorithms have been proposed that leverage the graphical structure of SSGs [2–4,6,7,9,11,16]. Further exploration of the structure might produce a better understanding of why it is hard to prove polynomial-time convergence, or why it is challenging to produce hard instances. For that purpose, it is useful to have a benchmark set of SSGs that are as simple as possible without losing complexity.

Past research on algorithms for solving SSGs used small or non-stopping instances [10,11]. We generate a set of large stopping games. The contributions of this paper are as follows: (1) an algorithm for generating Stopping Game instances, (2) proof that our algorithm can generate any Stopping Game, (3) several simple polynomial-time reductions used by the generator, (4) an open-source, fully documented implementation of our algorithm with instructions, (5) a benchmark set of fully reduced problems, ranging from 32 to 4096 nodes and (6) a study of Hoffman-Karp (a strategy improvement algorithm), and a permutation improvement algorithm, on our benchmark set.

1.1 Definitions

A *Simple Stochastic Game* (SSG) is a directed graph G with four types of nodes: max, min, average, and terminal nodes. Each max, min, and average node v has exactly two out-arcs. There are two terminal nodes, terminal-1 and terminal-0, with no out-arcs. By convention, the nodes of G are numbered $1, \ldots, n$, with the terminals 0 and 1 numbered $n-1$ and n, respectively. Each node $i \in G$ is assigned a value $v_i \in [0,1]$, and the values of the terminal nodes are always set to 0 and 1 ($v_{n-1} = 0$, $v_n = 1$).

A *stable assignment* to G is an assignment of values to nodes such that:

$$v_i = \max(v_j, v_k) \quad \text{for each max node } i \text{ with children } j, k$$
$$v_i = \min(v_j, v_k) \quad \text{for each min node } i \text{ with children } j, k$$
$$v_i = \frac{(v_j + v_k)}{2} \quad \text{for each average node } i \text{ with children } j, k$$

We say that nodes whose values satisfy the equations above are *satisfied*.

A *strategy* σ for the max player is a set of arcs consisting of one arc (i, j) from each max node i. Equivalently, a strategy can be represented by the *strategy subgraph* G_σ produced from G by removing every max node out-arc not contained in σ. Similarly, a strategy τ for the min player is a set of arcs consisting of one arc (i, j) from each min node i. The strategy subgraph $G_{\sigma,\tau}$ is the subgraph of G induced by the max/min strategy pair (σ, τ).

Given a pair of strategies (σ, τ), the *solution* value of each node is the probability that a random walk starting at that node will reach terminal-1 by following the unique out-arc from each max and min node in $G_{\sigma,\tau}$, and by randomly following one of the two out-arcs from each average node with equal probability [5]. The solution to $G_{\sigma,\tau}$ can be found by setting the value of all nodes with no path to a terminal to zero, and then solving the system of equations below for the remaining node values [5].

$$v_i = v_j \qquad \text{for each max node } i \text{ with children } j, k, \text{ where } j \in \sigma$$
$$v_i = v_j \qquad \text{for each min node } i \text{ with children } j, k, \text{ where } j \in \tau$$
$$v_i = \frac{(v_j + v_k)}{2} \qquad \text{for each average node } i \text{ with children } j, k$$

The max player aims to maximize, and the min player aims to minimize, the probability that a random walk ends at terminal-1. A max strategy σ is *optimal* with respect to a min strategy τ if all max nodes are satisfied in G by the solution values for $G_{\sigma,\tau}$, and similarly for the optimality of a min strategy τ with respect to a max strategy σ. It was shown in [5] that there always exists at least one mutually optimal pair of strategies (σ^*, τ^*), and that the vector of solution values is the same for every pair of mutually optimal strategies. Thus, the algorithmic problem for the max player is to find a strategy that maximizes the solution value of every node if the min player plays optimally.

The associated decision problem for SSGs is: Given a simple stochastic game G and a node v, does there exist a max strategy σ such that for any min strategy τ, a token starting at node v in $G_{\sigma,\tau}$ arrives at terminal-1 with probability at least $\frac{1}{2}$? The decision problem is known to be in $\mathcal{NP} \cap \text{co} - \mathcal{NP}$ [5].

A *Stopping Game* is a simple stochastic game G with the property that for every pair of strategies (σ, τ) and every node v in G, there is a path from v to a terminal in $G_{\sigma,\tau}$. Equivalently, a Stopping Game is an SSG where, for any pair of strategies, a random walk starting from any node will reach a terminal with probability 1.

Anne Condon proved in [5]:

1. A Stopping Game has exactly one stable assignment, which is the solution to the game played with optimal max/min strategies.
2. For any SSG G, it is possible to construct a Stopping Game G' such that a solution to G can be recovered in polynomial time from the unique solution to G'. In addition, if the max player has a strategy in G' such that the start node has a value $\geq \frac{1}{2}$, then the max player also has a winning strategy in G.

It follows that if Stopping Games are in $\mathcal{P}$, then Simple Stochastic Games (and Stochastic Games) are in $\mathcal{P}$. For a Stopping Game G, the algorithmic problem is to find the unique stable assignment of values to nodes in G.

2 Stopping Game Generator

In this section, we provide an algorithm that generates Stopping Games, with parameters for the number of max, min and average nodes. We prove that it generates only Stopping Games, and that, given fixed parameters, every Stopping Game has a non-zero probability of being generated. In Sect. 3, we describe simple subgraphs that can be solved quickly and give a modified version of the generator that produces Stopping Games with no such subgraphs.

2.1 Bad Subgraphs

A simple stochastic game is a Stopping Game if and only if it does not contain certain subgraphs, which we call *bad subgraphs*. The generator outputs simple stochastic games that are guaranteed not to contain any bad subgraphs.

Definition 1. *Let G be a simple stochastic game. A subgraph S of G is a bad subgraph if all of the following conditions are met:*

1. *Every max node in S has at least one arc pointing into S.*
2. *Every min node in S has at least one arc pointing into S.*
3. *Every average node in S has both arcs pointing into S.*
4. *S contains no terminal nodes.*
5. *S is strongly connected (any node in S is reachable from any other node in S).*

If there is a subgraph S of G that meets conditions 1–4, then G is not a stopping game (Lemmas 1 and 2). Condition 5 is useful because it allows the representation of the set of bad subgraphs meeting conditions 1–4 as a unique set of disjoint subgraphs.

Lemma 1. *Let G be a simple stochastic game. If S is a subgraph of G satisfying criteria 1–4 in the definition of a bad subgraph, then S contains a bad subgraph.*

Proof: Let G be a simple stochastic game, and let S be a subgraph of G satisfying criteria 1–4 in the definition of a bad subgraph. Let (σ, τ) be a pair of max/min strategies such that the single arc from each max and min node in $S_{\sigma,\tau}$ points back into $S_{\sigma,\tau}$. $S_{\sigma,\tau}$ also satisfies criteria 1–4. Any path in $G_{\sigma,\tau}$ starting from a node in $S_{\sigma,\tau}$ must eventually reach a cycle inside $S_{\sigma,\tau}$. A cycle is a strongly connected component (SCC). Let $C = \{C1, \ldots, C_i\}$ be the set of maximal SCCs in $S_{\sigma,\tau}$. Let C' be a directed graph with one node for each SCC in C, and an arc from C_i to C_j if there is a path in $S_{\sigma,\tau}$ from C_i to C_j. C' is an acyclic graph with at least one node with no out-arcs. Let C_k be an SCC in $S_{\sigma,\tau}$ represented by a node with no out-arcs in C'. No node in C_k can have any arc at all pointing out of C_k, because any path starting with such an arc must eventually cycle inside of $S_{\sigma,\tau}$. By assumption such a cycle cannot be an SCC outside of C_k. Suppose there is a path starting with an arc from a node in C_k to a node outside of C_k, ending in a cycle contained in C_k. The union of C_k and a path leading out of it and back would be an SCC, but this violates the maximality of C_k. So there can be no arc out of C_k. It follows that C_k satisfies all five criteria in the definition of a bad subgraph. □

Lemma 2. *A simple stochastic game G is a stopping game if and only if it contains no bad subgraphs.*

Proof: Let G be a simple stochastic game containing a bad subgraph S. Let σ be a max strategy that chooses arcs pointing into S for all max nodes in S, and let τ be a min strategy that chooses arcs pointing into S for all min arcs in S.

Then in $G_{\sigma,\tau}$, there is no path to a terminal from any node in S, so G is not a stopping game. Conversely, let G be a simple stochastic game with no bad subgraphs. By Lemma 1, G cannot contain any subgraph satisfying criteria 1–4 of the definition of bad subgraphs. Therefore for any subset V of non-terminal nodes of G and any strategy pair (σ, τ), there must exist some node with an arc out of V in $G_{\sigma,\tau}$. It follows that it is possible to construct a path from any node v_0 to a terminal in $G_{\sigma,\tau}$, by starting with $V = v_0$ and in each step i finding a node $v_i \in G \backslash V$ such that there is an arc from a node in V to v_i, then setting $V = V \cup v_i$. The number of nodes is finite, so eventually a terminal node must be reached. □

2.2 A Simple Stopping Game Generator

We introduce a simple Stopping Game generator that has a non-zero probability of producing any stopping game with one minor condition. The generator only generates games where no max or min node points directly to a terminal. These nodes can be solved independently of the rest of the graph in constant time (this is proven in Sect. 3.1).

The simple Stopping Game generator works in two phases. In phase 1, all nodes are numbered and each non-terminal node is randomly assigned an arc to a higher-numbered node. Lemma 3 shows that this does not prevent any stopping games from being generated. In phase 2, all nodes receive second arcs. First, all average nodes receive second arcs uniformly at random. To assign second arcs to max and min nodes, we pick a max or min node m uniformly at random and find the set Q of all nodes q such that adding arc (m, q) to the graph will not create a bad subgraph. Q is non-empty because there is always an average node pointing to a terminal, and that node is in Q. We then pick a node v from Q uniformly at random and add the arc (m, v) to the graph. We prove below that this is sufficient to guarantee that the final constructed graph contains no bad subgraphs.

Lemma 3. *Let G be a Stopping Game with n nodes. Then there is a numbering of nodes in G with the terminals numbered $n - 1$ and n, such that for every non-terminal node v, v has an arc to a higher-numbered node.*

Proof. Fix any strategy pair (σ, τ) in G, and let $S = G_{\sigma,\tau}$ be the strategy subgraph induced by (σ, τ). By definition of a Stopping Game, for each node v, there is a path in S from v to a terminal. Let S' be a shortest path tree on S with all paths ending in a terminal node. Define a partial order P on nodes of S as follows: For any non-terminal node a in S' at distance k from a terminal, let b be the next node in the unique path from a to the terminal (so b is at distance $k - 1$ from the terminal). Let $a < b$ in the partial order P. Let T be any total order on the nodes of S that preserves P, and number the nodes of G according to T. By construction, every non-terminal node in G has an arc to a higher-numbered node. □

Algorithm 1: Stopping Game Generator

1. **node labeling** Number the nodes from 1 to n. Pick integers $a, b, c \geq 1$, such that $n = a + b + c + 2$. Label nodes $n - 1$ and n as terminal-0 and terminal-1, respectively. Label node $n - 2$ as an average node. Assign the remaining numbers to $a - 1$ average nodes, b min nodes, and c max nodes uniformly at random.
2. **foreach** *non-terminal node v* **do**
3.   Pick a higher-numbered node w uniformly at random and add an arc from v to w.
   /* (Assigns an out-arc to each non-terminal node that has none.) */
4. **while** *there are average nodes with exactly one out-arc* **do**
5.   Pick an average node m with exactly one out-arc (m, p) uniformly at random.
6.   Pick any node $q \notin \{m, p\}$ uniformly at random and add arc (m, q).
7. **while** *there are max or min nodes with exactly one out-arc* **do**
8.   Pick a node m with exactly one out-arc (m, p) uniformly at random.
9.   Let Q be the set of nodes such that arc (m, q) can be added to the graph without adding a bad subgraph, and $q \notin \{m, p\}$ (Algorithm 2).
10.   Remove the terminal nodes from Q.
11.   Pick a node $q \in Q$ uniformly at random, and add arc (m, q).
    /* (Assigns a second out-arc to each max, and min node.) */
12. **return** *The constructed graph.*

Proof of Correctness. *Proof that the generator can produce any stopping game:* By Lemma 3, in every stopping game the nodes can be numbered in such a way that every non-terminal node has at least one arc to a higher numbered node. Let G be a stopping game, and let G' be a subgraph of G where every non-terminal node has degree 1 and its single arc goes to a higher-numbered node. Every arc in G' can be generated in Line 3 of the Stopping Game Generator.

All remaining arcs can be generated in Lines 6–11. Second arcs from average nodes are generated first, followed by second arcs from max and min nodes, in any order. The only arcs that cannot be added are those that would create a bad subgraph; bad subgraphs persist as additional arcs from max and min nodes are added to the graph, so if a bad subgraph were created at any point, the final constructed graph would not be a stopping game. Consequently, each arc in G can be generated in Lines 6–11.

Proof that the generator produces only stopping games: Let G_0 be the graph constructed in lines 1–6. In G_0 each max and min arc has out-degree one. Therefore, there is only one possible pair of max/min strategies, and G_0 is also the only possible strategy subgraph. G_0 must be a stopping game, since in G_0, every node has a path through successively higher-numbered nodes to a terminal node. Let G_k be the graph after k second arcs from max and min nodes have been added, with $k \geq 0$. Assume inductively that G_k is a stopping game, and so it contains no bad subgraphs. Let m be the next max or min node to receive a second arc. Since Algorithm 2 generates a set of arc choices that will not create a bad subgraph, G_{k+1} must also be a stopping game for any of these arc choices.

It remains to be shown that the algorithm for finding valid arcs (Algorithm 2) correctly identifies the set of nodes Q such that $\forall q \in Q$, adding the arc (m, q) to G_k does not create a bad subgraph. Suppose adding arc (m, v) to G_k would create a bad subgraph S. By definition, S is a strongly connected component and all nodes in S must be ancestors of m. Algorithm 2 removes all ancestors of m from the list of candidates for a new arc, so the initial list Q is guaranteed to contain no node v such that adding arc (m, v) would create a bad subgraph. Ancestor nodes are then added back to the candidate list one at a time, with a node q restored to the candidate list Q if and only if:

1. q is an average node with an arc to a node in Q or a terminal node (Lines 13–17, 22–26); or
2. q is a max or min node such that all of its arcs point to a node in Q (Lines 26–29).

Let S be any bad subgraph that could be created from a subset of m's ancestors by adding an arc from m to a node in S. Initially, $Q \cap S = \emptyset$. Condition 1 guarantees that any average node restored to the candidate list Q has at least one arc pointing out of S and so could not be contained in S. Condition 2 guarantees that any max or min node restored to Q has no arc pointing into S, and so could not be contained in S. Thus, no arc added back into Q is in S, and it remains the case that $Q \cap S = \emptyset$.

Let v be any node such that adding arc (m, v) would not create a bad subgraph. Suppose $v \notin Q$, the set returned by Algorithm 2. Let v-reachable be the set of nodes reachable from v without going through nodes in Q. All arcs out of nodes in v-reachable must either point back into v-reachable or point into Q. Define the *perimeter* of v-reachable to be the set of nodes in v-reachable with at least one arc into Q. None of the perimeter nodes can be average nodes, because an average node with an arc into Q would be added to Q in Lines 22–26. Any max or min node in the perimeter must have at least one arc into v-reachable, or it would have been added to Q in Lines 26–29. Thus all average nodes in v-reachable have both arcs pointing into v-reachable and all max and min nodes in v-reachable have at least one arc pointing into v-reachable. By Lemma 1, v-reachable must contain a bad subgraph, contradicting the assumption that adding arc (m, v) will not create a bad subgraph. Thus all valid arcs in the ancestor set of m must be added back to Q.

It follows that the final list Q of valid arcs is correct. □

3 Simple Reductions

The reduction of SSGs to Stopping Games is not the only useful reduction for generating instances of theoretical interest. In this section we discuss subgraphs that can be solved in polynomial (linear) time, and further reductions. Any research on the complexity of SSGs can safely assume that each instance is a

Algorithm 2: Find Valid Arcs

Input : a node m that has exactly one out-arc (m, n)

1. Let Q be a list containing all of the nodes in the graph.
2. $Q = Q \backslash \{m, \text{terminal-0}, \text{terminal-1}\}$.
3. Let $P = \{m\}$
4. **while** $P \neq \emptyset$ **do**
5.   Let p be a random node $\in P$.
6.   Set $P = P \backslash \{p\}$.
7.   **foreach** parent p' of p **do**
8.   **if** $p' \in Q$ **then**
9. $Q = Q \backslash \{p'\}$.
10.   Add p' to P.

/* (Removes all ancestors from the candidate list.) */

11. Let $T = \emptyset$
12. **foreach** node $v \notin Q$ **do**
13.   Let (v, n) be v's first out-arc, and (v, p) be v's second out-arc if it exists.
14.   **if** v is an *average node* **then**
15.   **if** $n \in \{Q, \text{terminal-0}, \text{terminal-1}\}$ OR $p \in \{Q, \text{terminal-0}, \text{terminal-1}\}$ if (v, p) exists **then**
16.   Set $T = T \cup \{v\}$.
17.   Set $Q = Q \cup \{v\}$.

/* (Finds initial nodes for final processing phase.) */

18. **while** $T \neq \emptyset$ **do**
19.   Let p be a random node $\in T$.
20.   Set $T = T \backslash \{p\}$.
21.   **foreach** parent p' of p **do**
22.   Let (p', p) be p''s first out-arc, and (p', u) be p''s second out-arc if it exists.
23.   **if** $p' \notin Q$ **then**
24.   **if** p' is an *average node* **then**
25.   Set $T = T \cup \{p'\}$.
26.   Set $Q = Q \cup \{p'\}$.
27.   **else if** $u \in Q$ **then**
28.   Set $T = T \cup \{p'\}$.
29.   Set $Q = Q \cup \{p'\}$.

/* (Add back in nodes that don't cause a bad subgraph.) */

30. **return** Q.

Stopping Game from which all of the following subgraphs have been removed, and to which all of the following reductions have been applied.

3.1 Trivially Removable Subgraphs

Let G be a Stopping Game. We show that we may assume for the purposes of complexity analysis that G contains none of the following subgraphs. The strategy will proceed by defining a subgraph S of a Stopping Game G, and then showing that a solution to G can be recovered in constant time from a solution to $G \backslash S$.

Max or min node with at least one arc to a terminal. Let v be a max node with arcs (v, a), (v, b) and suppose that a is a terminal node. If a is terminal-0, then v can be merged with b without changing the solution value of any other nodes. If a is terminal-1, then v can be merged with a. The reverse is true for min nodes.

Any node with two identical arcs. Let (v, w), (v, w) be the identical arcs. v can be merged with w without changing the solution value of any other nodes.

An average node v with a self-arc. Let v have arcs (v, v), (v, w). Then:

$$\text{value}(v) = \frac{\text{value}(v) + \text{value}(w)}{2} \Rightarrow \text{value}(v) = \text{value}(w)$$

So, v can be merged with w without changing the solution value of any other nodes.

A max, min, or average node with in-degree zero. Let v be a node with in-degree zero. The value of v has no effect on the value of any other node. We can find a solution to the original SSG by removing v, solving the remaining graph, and then solving v.

A terminal with in-degree zero. Suppose that one of the two terminal nodes has in-degree zero. Then the value of all nodes in the graph is trivially equal to the value of the other terminal node.

After removing the above subgraphs, either G contains at least two distinct average nodes, one with an arc to terminal-1 and one with an arc to terminal-0, or there is only one such average node and the entire graph has a value of $\frac{1}{2}$. In our algorithm, we eliminate the second possibility.

3.2 Collapsing Clusters

In this section we provide a linear-time algorithm for solving all nodes with values equal to 1 or 0. It is based on the observation that a max node will always choose the arc that points to the higher-value node. There is no higher value than 1, and no lower value than 0. The reverse is true for the min nodes.

The algorithm for finding 1-valued nodes, detailed in Algorithm 3, starts by defining a cluster set of nodes containing only terminal-0. Any average nodes that point to terminal-0 have a value less than 1 and are added to the cluster. Any min nodes or average nodes that have an arc to the cluster have a value less than 1. Similarly any max nodes with both arcs to the cluster have a value less

Algorithm 3: Find 1-Valued Nodes

1. Label the nodes from 1 to n, with terminal-0 labeled $n-1$.
2. Let q be a list containing only terminal-0.
3. Let bv be a binary string of 1s with length n.
4. Set $bv[n-1]$ to 0.
5. **while** q *is not empty* **do**
6.   Let u be a node popped from q.
7.   **foreach** *parent p of u* **do**
8.   Let i be the index of node p.
9.   **if** $bv[i]$ *is a 1* **then**
10.   **if** p *is a min or average node* **then**
11.   Set $bv[i]$ to 0.
12.   Add p to q.
13.   Let a and b be the indexes of the nodes that p has arcs to.
14.   **if** p *is a max node AND* $bv[a] = 0 = bv[b]$ **then**
15.   Set $bv[i]$ to 0.
16.   Add p to q.

17. **return** *The indexes of the ones in bv.*

---

than 1. Nodes are added to the cluster until no remaining nodes can be added to the cluster. The remaining nodes have a value of 1. The algorithm can be trivially reversed to find 0-valued nodes.

The algorithm is linear because the number of arcs, and therefore parents, is also linear. Each node can be considered only once, and when a node is considered, its parents are iterated over. Each node is a parent to only two nodes and will only appear twice as a parent in the algorithm. Thus, the total number of parents considered is no more than $2n$.

3.3 Strongly Connected Components

A strongly connected component (SCC) in a graph is a set of nodes S such that every node in S is reachable from every other node in S. A graph can be decomposed into SCCs in linear time [14]. Note that each terminal node is its own SCC. Given a decomposition of a graph G into SCCs, there must be an SCC C such that all of C's out-arcs are to terminal nodes. The values of nodes in C are independent of the values of nodes in all other non-terminal SCCs, and so the nodes in C can be solved independently of the rest of the nodes. Once nodes in C are solved and set to constant values, there must be a new SCC C' such that all of C''s out-arcs are to terminals and other constant-valued nodes. The process can be repeated until all nodes are solved. In other words, each SCC can be solved independently and it is safe to assume that your SSG is a single SCC. However, it is not necessarily true that the complexity of solving an SSG with a single non-terminal SSC is the same as that of solving an SSG with multiple non-terminal SSCs. After the first SSC is solved, the remaining SSCs may have arcs to any number of solved nodes with values $\in (0,1)$. By assuming that every SSG is a single SSC, you lose the assumption that terminal-0 and terminal-1 are the only nodes with constant values.

3.4 Useful Assumptions Summary

To summarize this section, if you are pursuing a constructive proof that Stochastic Games are in $\mathcal{P}$, you may assume all of the following about the game you are trying to solve:

1. It is both a Simple Stochastic Game, and a Stopping Game.
2. There are no max or min nodes with arcs to the terminal.
3. There are no nodes with identical arcs or self-arcs.
4. There are no nodes with in-degree zero.
5. There is at least one pair of average nodes where one has an arc to terminal-0, and the other has an arc to terminal-1.
6. There are no nodes with a value of 1 or 0.
7. **EITHER** it is a single SCC with the only out-arcs going to nodes with constant values in $[0,1]$, **OR** there are only two nodes with constant values, terminal-0 and terminal-1.

3.5 Generator Implementation

A modified version of the generator algorithm with all the modifications from Sect. 3 is shown in Algorithm 4. While it can generate instances with in-degree 0 nodes, this occurs infrequently. To target specific sizes, such instances can be discarded, and the generator can be rerun.

4 Benchmark Instances

In this section, we discuss the implementation of our generator algorithm and the benchmark set of problems we generated. The code and benchmark set is publicly available, with documentation explaining how to generate new instances[1].

4.1 Benchmark Set

We generated instances with sizes in powers of two from 2^5 to 2^{12}. For each size, we generated 800 instances, with 100 instances at each ratio of average nodes to max nodes $\in \{1:4, 2:4, ..., 7:4, 8:4\}$. In order to maintain the intended ratio, some instances are slightly larger or slightly smaller than the labeled size.

Each instance in our benchmark set is fully reduced using all of the reductions listed in Sect. 3.4. To do this we generated instances until we found enough in each category that were already fully reduced when they were generated. This means that each instance in our benchmark set has the property of being both

---

[1] github.com/isaacrudich/simplestochasticgamesbenchmark.

Algorithm 4: Modified Stopping Game Generator

1 **node labeling** Number the nodes from 1 to n. Pick integers $a \geq 2$ and $b, c \geq 1$, such that $n = a + b + c + 2$. Label nodes $n - 1$ and n as terminal-0 and terminal-1, respectively. Label node $n - 2$ as an average node, and add an arc from node $n - 2$ to terminal-0. Label node $n - 3$ as an average node, and add an arc from node $n - 3$ to terminal-1. Randomly assign the remaining numbers to $a - 2$ average nodes, b min nodes, and c max nodes.
2 **foreach** *average node v with no out-arcs* **do**
3   Pick a higher-numbered node w uniformly at random and add arc (v, w).
    `/* (Assigns first out-arc to each non-terminal average node.) */`
4 **foreach** *max and min node v* **do**
5   Pick a higher-numbered non-terminal node w uniformly at random and add arc (v, w).
6 Let z be the number of nodes with in-degree zero.
7 Pick a random number r between $\max(z - (b+c), 0)$ and $\min(a, z)$.
8 **if** $r \neq 0$ **then**
9   **foreach** *integer from 1 to r* **do**
10   Select an average node m with exactly one out-arc (m, p) uniformly at random.
11   Pick a node $q \notin \{m, p\}$ with in-degree zero uniformly at random, and add arc (m, q).
12 **while** *there are average nodes with exactly one out-arc* **do**
13   Pick an average node m with exactly one out-arc (m, p) uniformly at random.
14   Pick any node $q \notin \{m, p\}$ uniformly at random and add arc (m, q).
15 **while** *there are max or min nodes with exactly one out-arc* **do**
16   Pick a node m with exactly one out-arc (m, p) uniformly at random.
17   Let Q be the set of nodes such that arc (m, q) can be added to the graph without creating a bad subgraph, and $q \notin \{m, p\}$ (Algorithm 2).
18   **if** *there are nodes with in-degree zero $\in Q$* **then**
19   Randomly pick a node $q \in Q$ with in-degree zero, and add arc (m, q).
20   **else**
21   Randomly pick a node $q \in Q$, and add arc (m, q).
    `/* (Assigns a second out-arc to each max, and min node.) */`
22 Merge 1- and 0-valued nodes into their respective terminals (Algorithm 3).
23 **return** *The constructed graph.*

---

a single SCC and having only two nodes with constant value. However, if you are interested in studying the behavior of instances with several constant-valued nodes, that can be easily accomplished using the same instances. Simply reassign each arc that points to a terminal to a node with a random fixed value $\in [0, 1]$. The instance will still be fully reduced, and will still be a single SCC.

4.2 Experimental Results

We solved each of the 6400 benchmark problems 100 times using random seeds with two different algorithms and recorded the iterations and solve time. The experiments were conducted on a 2023 M2 Pro CPU with 32 GB RAM. We tested Hoffman-Karp, the standard strategy-iteration algorithm, and a permutation improvement algorithm based on Gimbert and Horn [7]. We achieved a substantial speed-up over a naive implementation by generalizing the concept of collapsing clusters to nodes of any value, and then pre-calculating the collapsing clusters for any candidate ordering.

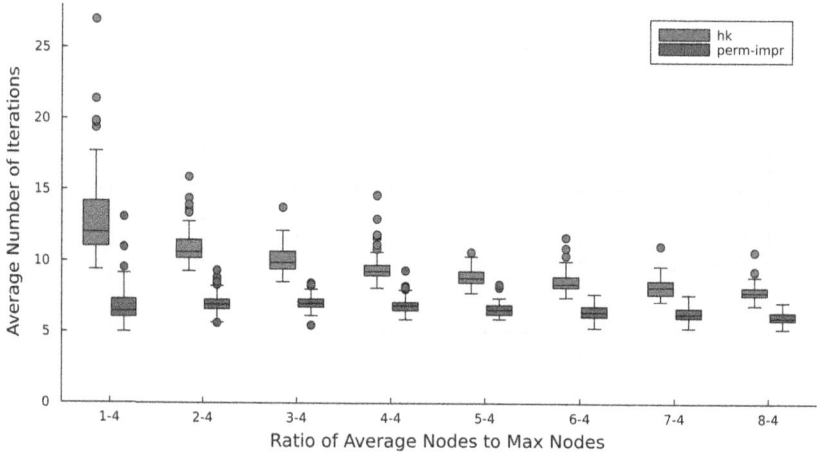

Fig. 1. Average Iterations to Solve for Size 4096 Games

Figure 1 shows data for the two algorithms, grouped by ratio, for the size 4096 SSGs. The graphs for the other sizes look similar but with globally lower numbers of iterations; they are available with the code for the generator. The data shows that the permutation improvement algorithm consistently outperforms Hoffman-Karp in terms of both iterations and time. Hoffman-Karp performs better as the number of decision nodes goes down, while the permutation improvement algorithm shows consistent performance for all of the ratios tested.

Additional summary data showing the average run-time, in both time and iterations, for both algorithms and each category of problem is available in the repository with the code. In terms of average number of iterations, the permutation improvement algorithm out-performed Hoffman-Karp on every category of problem. In terms of average amount of time, the permutation improvement algorithm out-performed Hoffman-Karp on almost every category of problem. The 4 categories (out of 48) where Hoffman-Karp out-performed the permutation improvement algorithm are shown in bold.

5 Conclusion

This paper presented a fast algorithm for generating Simple Stochastic Stopping Games. It also provided several polynomial reductions and assumptions for further simplifying and reducing Simple Stochastic Games. These assumptions are useful for anyone attempting a constructive proof that the complexity of Stochastic Games is in $\mathcal{P}$. They are also useful for trying to understand why hard-to-solve instances are so challenging to generate. Our code for generating instances is open source, as well as the benchmark instances we generated.

References

1. Altman, E., Avratchenkov, K., Bonneau, N., Debbah, M., El-Azouzi, R., Menasché, D.S.: Constrained stochastic games in wireless networks. In: IEEE GLOBECOM 2007-IEEE Global Telecommunications Conference, pp. 315–320. IEEE (2007)
2. Auger, D., Coucheney, P., Strozecki, Y.: Solving simple stochastic games with few random nodes faster using Bland's rule. arXiv preprint arXiv:1901.05316 (2019)
3. Auger, D., de Montjoye, X.B., Strozecki, Y.: A generic strategy iteration method for simple stochastic games. CoRR abs/2102.04922 (2021)
4. Condon, A.: On algorithms for simple stochastic games. In: Advances in Computational Complexity Theory, vol. 13, pp. 51–72 (1990)
5. Condon, A.: The complexity of stochastic games. Inf. Comput. **96**(2), 203–224 (1992)
6. Dai, D., Ge, R.: New results on simple stochastic games. In: Dong, Y., Du, D.-Z., Ibarra, O. (eds.) ISAAC 2009. LNCS, vol. 5878, pp. 1014–1023. Springer, Heidelberg (2009). https://doi.org/10.1007/978-3-642-10631-6_102
7. Gimbert, H., Horn, F.: Simple stochastic games with few random vertices are easy to solve. In: Amadio, R. (ed.) FoSSaCS 2008. LNCS, vol. 4962, pp. 5–19. Springer, Heidelberg (2008). https://doi.org/10.1007/978-3-540-78499-9_2
8. Halman, N.: Simple stochastic games, parity games, mean payoff games and discounted payoff games are all LP-type problems. Algorithmica **49**, 37–50 (2007)
9. Ibsen-Jensen, R., Miltersen, P.B.: Solving simple stochastic games with few coin toss positions. In: Epstein, L., Ferragina, P. (eds.) ESA 2012. LNCS, vol. 7501, pp. 636–647. Springer, Heidelberg (2012). https://doi.org/10.1007/978-3-642-33090-2_55
10. Klingler, C.W.: An empirical analysis of algorithms for simple stochastic games. Graduate theses, dissertations, and problem reports (2023)
11. Křetínský, J., Ramneantu, E., Slivinskiy, A., Weininger, M.: Comparison of algorithms for simple stochastic games. Inf. Comput. **289**, 104885 (2022). Special Issue on 11th Int. Symp. on Games, Automata, Logics and Formal Verification
12. Shapley, L.S.: Stochastic games. Proc. Natl. Acad. Sci. **39**(10), 1095–1100 (1953)
13. Sharir, M., Welzl, E.: A combinatorial bound for linear programming and related problems. In: Finkel, A., Jantzen, M. (eds.) STACS 1992. LNCS, vol. 577, pp. 567–579. Springer, Heidelberg (1992). https://doi.org/10.1007/3-540-55210-3_213
14. Tarjan, R.: Depth-first search and linear graph algorithms. SIAM J. Comput. **1**(2), 146–160 (1972)
15. Tembine, H., Vilanova, P., Assaad, M., Debbah, M.: Mean field stochastic games for SINR-based medium access control. In: Gamecomm2011. pp. 10–p (2011)
16. Tripathi, R., Valkanova, E., Kumar, V.A.: On strategy improvement algorithms for simple stochastic games. J. Discrete Algorithms **9**(3), 263–278 (2011)

Protective and Nonprotective Subset Sum Games: A Parameterized Complexity Analysis

Jaroslav Garvardt[1](✉), Christian Komusiewicz[1], Berthold Blatt Lorke[2], and Jannik Schestag[1]

[1] Friedrich Schiller University Jena, Jena, Germany
{jaroslav.garvardt,c.komusiewicz,j.t.schestag}@uni-jena.de
[2] TU Berlin, Berlin, Germany
lorke@tu-berlin.de

Abstract. In SUBSET SUM GAME as studied by Pieterse and Woeginger [Theory of Computing Systems, 2021], two players alternatingly fill a common knapsack each with items from a private collection. The goal of Player A is to reach a value of at least T_A, whereas Player B may follow different strategies. SUBSET SUM GAME is NP-complete and solvable in pseudopolynomial time if Player B greedily selects the biggest available item in each turn; the game is PSPACE-complete, however, if Player B plays a hostile strategy where the only aim is to avoid that Player A wins. We continue the study of the game with these two strategies for Player B.

First, we provide a faster pseudopolynomial-time algorithm for a greedy Player B and show that the problem with a hostile Player B is fixed-parameter tractable with respect to the knapsack capacity C. Moreover, we study the influence of further parameters such as T_A, the number of rounds in the game, and the number of different numbers in the input on the complexity of the problem. Second, we consider a further variant of the game, called PROTECTIVE SUBSET SUM GAME, where Player A additionally has the goal that Player B reaches a value of at least T_B. In a nutshell, we show that most algorithms for the nonprotective variant can be transferred to PROTECTIVE SUBSET SUM GAME.

1 Introduction

A core problem in decision theory, and in a sense the epitome of profit maximization in the face of limited resources is the KNAPSACK problem. Here, the input is a set of items, each with a value and a weight, and the task is to add items to a knapsack of bounded capacity such that the total value is maximized and the total weight does not exceed the capacity of the knapsack. In some scenarios, weight and value are equal for each item. This special case is one of the many variants of the SUBSET SUM problem. Both KNAPSACK and SUBSET SUM have been studied intensively from an algorithmic point of view over the years.

In many applications, however, several agents compete for the same resources. The SUBSET SUM GAME, introduced by Darmann et al. [3], is a combinatorial problem that models such a situation. In this game, two players A and B have two disjoint item sets and alternatingly add one item from their set to the knapsack. Each item has a weight and at any point, the total weight of the items may not exceed the capacity C of the knapsack. Player A wants to achieve a total weight of at least T_A and, in principle, Player B wants to achieve a total weight of at least T_B. When the game is studied from the perspective of one of the players, say for Player A, then the winning condition naturally only concerns T_A. Any reasoning about a strategy for Player A, however, needs a model for the strategy of Player B who may be considered as the adversary of Player A. Darmann et al. [3] and Pieterse and Woeginger [11] formulated altogether three strategies for Player B: the *greedy* strategy where Player B always selects the largest item that fits in the knapsack, the *selfish* strategy where Player B aims to maximize the total weight of its items in the knapsack, and the *hostile* strategy, where B only wants to prevent that Player A reaches its goal. The greedy strategy and the hostile strategy thus ignore the value of T_B.

The SUBSET SUM GAME as described above was first shown to be NP-hard for any strategy of Player B by a reduction from SUBSET SUM [3]. This NP-hardness holds even if all items of Player B have the same value. Later, it was shown that if Player B plays a selfish or hostile strategy, then SUBSET SUM GAME is even PSPACE-complete [11]. The case with greedy Player B admits a pseudopolynomial-time algorithm with running time $\mathcal{O}(c^2(n_A)^2(n_B)^2)$ [11]. Here, n_A and n_B are the number of items for players A and B, respectively. In addition, Pieterse and Woeginger [11] studied the polynomial-time approximability of SUBSET SUM GAME, showing that for hostile and selfish Player B, constant-factor approximations are unlikely and that for greedy Player B SUBSET SUM GAME admits a polynomial time approximation scheme (PTAS). SUBSET SUM GAME has also been studied in a Stackelberg setting, where each player has only one turn in which it may select a *set* of items [8]. A more general type of games are KNAPSACK games [9,10,13] where each item has a weight and a profit.

We continue the algorithmic study of SUBSET SUM GAME, with a focus on the effect of different structural input properties on the problem difficulty. In addition, we extend the SUBSET SUM GAME to a *protective* variant where the aim of A is not only to reach its own threshold but also to guarantee that Player B reaches its threshold. To illustrate this, one may consider a situation in which a parent packs the trunk of a car with their child. While the parent wants to ensure that sufficiently many important items are packed in the trunk, they also want that the child is happy, in other words, that it can bring sufficiently many toys. Hence, the parent which is Player A wants that both thresholds are met. It is reasonable to assume that the child will follow a greedy strategy for choosing its items. Hence, this situation corresponds to a protective SUBSET SUM GAME with greedy Player B.

Our Results. We study SUBSET SUM GAME in the protective and the standard non-protective variant with two adversarial strategies: a greedy and a hostile

Player B. This leads to altogether four problems P-Hostile-SSG, P-Greedy-SSG, Greedy-SSG, and Hostile-SSG, where the P indicates the protective variant. Note that the protective variant is a strict generalization of the standard variant: when we set $T_B = 0$ the winning condition only depends on the value of Player A. For an overview of our results, refer to Tables 1 and 2.

The first parameters that we consider are the knapsack capacity C and the threshold T_A for Player A. We show in particular that P-Greedy-SSG can be solved in $\mathcal{O}(T_A \cdot n^7)$ time, where n denotes the total number of items. This result thus makes progress in two directions: First, it extends pseudopolynomial-time solvability from the standard problem to the protective problem. Second, since the pseudopolynomial part depends only on T_A and not on C, we also obtain efficient algorithms when the items for Player A are small compared to the items for Player B. Next, we consider these parameters with a hostile adversary and show that the protective and nonprotective problems are FPT with respect to C and presumably not FPT with respect to T_A.

We then follow the "number of numbers" paradigm of parameterization [5] where one considers the case that the input numbers may be large but that there are only few different numbers in the input. We show that with a greedy Player B the problems are FPT with respect to var, the total number of *different* items for both players. If we consider only the total number of different items for Player A, then we obtain polynomial-time solvability for constant parameter values. This gives a strong contrast to the parameterization by var_B, since Greedy-SSG is NP-hard even if all items in B have the same weight.

Finally, we consider the case that the total number of rounds R for the game is bounded. We show that while the problem can be solved in polynomial time for constant R, it is unlikely to be FPT with respect to R. In other words, it seems that the degree of the polynomial necessarily depends on R.

Due to lack of space, proofs of statements marked with a (*) are deferred to a long version of this work.

Table 1. Results for Subset Sum Game.

Parameter	Greedy-SSG		Hostile-SSG	
Capacity C	$\mathcal{O}(C \cdot n^5)$	Corollary 2	$\mathcal{O}^*(2^C \cdot C)$	Theorem 2
Threshold T_A of A	$\mathcal{O}(T_A \cdot n^5)$	Corollary 2	not in FPT, XP	Proposition 4, Corollary 3
Variety var of $\mathcal{A} \cup \mathcal{B}$	FPT	Theorem 3	open	
Variety var_A of $\mathcal{A}$	XP (FPT is *open*)	Proposition 1	para-NP-h	Proposition 4
Variety var_B of $\mathcal{B}$	para-NP-h	[3]	para-NP-h	[3]
Rounds R	W[1]-h, XP	Propositions 2, 3	W[1]-h, XP	Propositions 2, 3

2 Preliminaries

Notation and Problem Definition. For $a, b \in \mathbb{N}$ with $a \leq b$ we write $[a, b]$ for the set $\{i \in \mathbb{N} \mid a \leq i \leq b\}$ and define $[n] := [1, n]$. We consider multisets of

Table 2. Results for PROTECTIVE-SUBSET SUM GAME

Parameter	P-GREEDY-SSG		P-HOSTILE-SSG	
Capacity C	$\mathcal{O}(C \cdot n^7)$	Theorem 1	$\mathcal{O}^*(2^C \cdot C)$	Theorem 2
Threshold $T_\mathtt{A}$ of A	$\mathcal{O}(T_\mathtt{A} \cdot n^7)$	Theorem 1	not in FPT	Proposition 4
Threshold $T_\mathtt{B}$ of B	para-NP-h	[3]	para-NP-h	[3]
Variety var of $\mathcal{A} \cup \mathcal{B}$	FPT	Theorem 3	open	
Variety var_A of $\mathcal{A}$	XP (FPT is *open*)	Proposition 1	para-NP-h	Proposition 4
Variety var_B of $\mathcal{B}$	para-NP-h	[3]	para-NP-h	[3]
Rounds R	W[1]-h, XP	Propositions 2, 3	W[1]-h, XP	Propositions 2, 3

integers of the form $S = \{s_1, s_2, \ldots, s_{|S|}\}$, where distinct elements s_i and s_j with $i \neq j$ may be the same integer. For a multiset $S = \{s_1, s_2, \ldots, s_k\}$ we write $\sum S = \sum_{i=1}^{k} s_i$.

Definition 1. *Let $\mathcal{A} = \{a_1, a_2, \ldots, a_n\}$ be a multiset. A picking order for $\mathcal{A}$ is a sequence $A = (a_{i_1}, a_{i_2}, \ldots, a_{i_r})$ of distinct elements in $\mathcal{A}$. We say A is nonincreasing if $a_{i_j} \geq a_{i_{j+1}}$ for each $j \in [r-1]$. We associate A with the multiset $A' = \{a_{i_1}, a_{i_2}, \ldots, a_{i_r}\}$ of the elements of $\mathcal{A}$ that are present in A and write $\sum A := \sum A' = \sum_{j=1}^{r} a_{i_j}$. For $\ell \leq r$, we define the partial picking order $A_\ell = (a_{i_1}, a_{i_2}, \ldots, a_{i_\ell})$.*

An instance of SUBSET SUM GAME is given by a tuple $\mathcal{I} = (\mathcal{A}, \mathcal{B}, C, T_\mathtt{A}, T_\mathtt{B})$, where $\mathcal{A}$ is the multiset Player A picks from, $\mathcal{B}$ is the multiset Player B picks from, $T_\mathtt{A}$ is the target threshold for A, $T_\mathtt{B}$ is the target threshold for B, and C is the capacity of the shared knapsack. A picking order A for $\mathcal{A}$ and a picking order B for $\mathcal{B}$ are *valid* if $\sum A + \sum B \leq C$. The outcome of a game is represented by valid picking orders $A = (a_{i_1}, a_{i_2}, \ldots, a_{i_r})$ and $B = (b_{i_1}, b_{i_2}, \ldots, b_{i_r})$ for Player A and Player B respectively, where Player A picks item a_{i_1} in the first round, then Player B picks item b_{i_1} in the first round, then Player A picks item a_{i_2} in the second round and so on. From the perspective of Player A, we define the following winning conditions:

- *Nonprotective*: Player A wins if $\sum A \geq T_\mathtt{A}$.
- *Protective*: Player A wins if $\sum A \geq T_\mathtt{A}$ and $\sum B \geq T_\mathtt{B}$.

In the nonprotective case we will omit $T_\mathtt{B}$ from the input. We define the following strategies for the adversary Player B.

- *Greedy*: Player B picks the largest item $b \in \mathcal{B}$ that is allowed in every round.
- *Hostile*: Player B wants to prevent that Player A wins.

Note that after each choice of Player A, the next item chosen by a greedy Player B is uniquely determined.

Observation 1. *Let A be a given picking order for Player A such that there is a picking order B for Player B where A and B are valid. If Player B plays greedily, then B is unique and can be determined in linear time.*

In the case of a greedy adversary B we can thus say that a picking order A is valid if for the unique greedy response B of Player B we have $\sum A + \sum B \leq C$. We now define the following decision problem.

τ-ϕ-SUBSET SUM GAME
Input: Two multisets of non-negative integers $\mathcal{A} = \{a_1, \ldots, a_{|\mathcal{A}|}\}$ and $\mathcal{B} = \{b_1, \ldots, b_{|\mathcal{B}|}\}$, a capacity C, two thresholds $T_\mathtt{A}$ and $T_\mathtt{B}$.
Question: Does Player A have a strategy to achieve winning condition τ when Player B plays according to strategy ϕ?

In this work we consider the following variants of τ-ϕ-SUBSET SUM GAME:

- GREEDY-SSG: Nonprotective winning condition, greedy adversary
- HOSTILE-SSG: Nonprotective winning condition, hostile adversary
- P-GREEDY-SSG: Protective winning condition, greedy adversary
- P-HOSTILE-SSG: Protective winning condition, hostile adversary

Parameterized Complexity. An instance (x, k) of a parameterized problem L consists of the input x of a decision problem and a parameter k. A parameterized problem L is called *fixed-parameter tractable* if there is a computable function f such that for every instance (x, k) it can be decided in $f(k) \cdot |x|^{\mathcal{O}(1)}$ time whether (x, k) is a yes-instance of L. The complexity class FPT contains all parameterized problems that are fixed-parameter tractable. The class W[1] is a complexity class of presumed parameterized intractability. The complexity class XP contains all parameterized problems that can be solved in $|x|^{f(k)}$ time for a computable function f. We say an NP-hard problem is *para-NP-hard* for a parameter k if the parameterized version of the problem remains NP-hard even if the parameter k is constant. A *parameterized reduction* is an algorithm that takes an instance $I_1 = (x_1, k_1)$ of a problem L_1 and transforms it in time $f(k_1) \cdot |x_1|^{\mathcal{O}(1)}$ into an instance $I_2 = (x_2, k_2)$ of a problem L_2 such that $I_1 \in L_1$ if and only if $I_2 \in L_2$ and $k_2 \leq g(k_1)$ for some computable functions f and g. Parameterized reductions can be used to show that a parameterized problem L is not likely to be fixed-parameter tractable by reducing some W[1]-hard parameterized problem to L. For further background on parameterized complexity theory, we refer to the standard monographs [2,4].

3 Parameterization by Knapsack Capacity or Target Threshold

In this section, we show that all considered problem variants are FPT when parameterized by the capacity C or the target threshold for Player A. First, we show that P-GREEDY-SSG and GREEDY-SSG can be solved in $\mathcal{O}(T_\mathtt{A} \cdot n^7)$ time and $\mathcal{O}(T_\mathtt{A} \cdot n^5)$ time, respectively. This implies a running time of $\mathcal{O}(C \cdot n^7)$ and $\mathcal{O}(C \cdot n^5)$, respectively for P-GREEDY-SSG and GREEDY-SSG. The latter result improves over the previous pseudopolynomial-time algorithm in terms of the running time dependence on C. Afterwards, we show that the PSPACE-complete problems P-HOSTILE-SSG and HOSTILE-SSG can be solved in $\mathcal{O}(2^C \cdot C \cdot n^2)$ time.

In the case of the P-GREEDY-SSG variant, we may in the following assume for convenience that the multiset $\mathcal{B}$ of Player B contains a sufficiently large number of zeros, so that the number of elements chosen by Player B is always at least the number of elements chosen by Player A. Since Player B always greedily chooses the largest available element, adding these zeros to $\mathcal{B}$ clearly results in an equivalent instance.

The following lemma was observed already by Darmann et al. [3], we extend it to the protective case.

Lemma 1 (*). *Let $\mathcal{I} = (\mathcal{A}, \mathcal{B}, C, T_A, T_B)$ be a* yes*-instance of* P-GREEDY-SSG*. Then, there is a winning strategy for Player* A *with a nonincreasing picking order* A*.*

We now make a somewhat surprising observation for nonincreasing picking orders: if they are both valid and the sum of the chosen items is the same, then the greedy response will be exactly the same.

Lemma 2. *Let $\mathcal{I} = (\mathcal{A}, \mathcal{B}, C, T_A, T_B)$ be a* yes*-instance of* P-GREEDY-SSG*. Let $A^* = (a^*_{i_1}, a^*_{i_2}, \ldots, a^*_{i_r})$ and $A' = (a'_{i_1}, a'_{i_2}, \ldots, a'_{i_r})$ be different nonincreasing picking orders for Player* A *such that $\sum A^* = \sum A' =: t$ and for the respective greedy responses $B^* = (b^*_{i_1}, b^*_{i_2}, \ldots, b^*_{i_r})$ and $B' = (b'_{i_1}, b'_{i_2}, \ldots, b'_{i_r})$ of Player* B *the picking orders are valid. Then $B^* = B'$.*

Proof. Assume towards a contradiction that $B^* \neq B'$. Let $\ell < r$ be the first round such that Player B chooses differently for the partial picking orders A^*_ℓ and A'_ℓ. In other words, $b^*_{i_j} = b'_{i_j}$ for $j \in [\ell - 1]$ and $b^*_{i_\ell} \neq b'_{i_\ell}$. Note that this implies $\sum A^*_\ell \neq \sum A'_\ell$. Without loss of generality, assume $\sum A^*_\ell > \sum A'_\ell$. This implies that $\sum A^*_\ell + \sum B^*_{\ell-1} + b'_{i_\ell} > C$, since otherwise Player B would also pick b'_{i_ℓ} in response to A^*_ℓ instead of $b^*_{i_\ell}$. However, $\sum A^*_\ell + \sum B^*_{\ell-1} + b'_{i_\ell} = \sum A'_\ell + \sum B'_{\ell-1} + b'_{i_\ell} < t + \sum B'_\ell < C$, a contradiction. The last inequality holds, since A' with $\sum A' = t$ is a (valid) picking order and B' is the corresponding greedy response. □

We can now show our algorithm for P-GREEDY-SSG with parameter T_A.

Theorem 1. P-GREEDY-SSG *can be solved in $\mathcal{O}(T_A \cdot n^7)$ time.*

The algorithm behind this theorem solves in a subroutine the following auxiliary problem: In STRICT-B-FIRST we are given two sets of integers $\mathcal{A}$ and $\mathcal{B}$ from which players Player A and B can choose, a capacity C, targets T_A and T_B and a *ghost-item* $Z \in \mathbb{N}$, $Z \geq \max \mathcal{A}$. The question is whether Player A can win the game in a protective greedy scenario if we require that Player B starts, Player A picks the items in nondecreasing order and, if Player A can not choose another item because of the capacity constraint, then also Z would be too big to choose.

That is, $(\mathcal{A}, \mathcal{B}, C, T_A, T_B, Z)$ is a yes-instance of STRICT-B-FIRST if and only if there are sets $A \subseteq \mathcal{A}$ and $B \subseteq \mathcal{B}$ such that these four conditions hold:

C1 both players reach their target: $\sum A \geq T_\mathtt{A}$ and $\sum B \geq T_\mathtt{B}$,
C2 the capacity is not exceeded: $\sum A \cup B \leq C$,
C3 Player B starts and plays greedily, thus picking B, and Player A picks the items in A in nondecreasing order, and
C4 if $|A| < |B| + 1$ then $\sum A + \sum B_{|A|+1} + Z > C$. Here, B_q denotes the set containing the q biggest items in B.

We will solve STRICT-B-FIRST exactly in those situations when Player A has already achieved its threshold $T_\mathtt{A}$. Consequently, to show Theorem 1, we are only interested in the special case of STRICT-B-FIRST when $T_\mathtt{A} = 0$. In the following, we show that in this case, an instance of STRICT-B-FIRST is a yes-instance if and only if Player A continuously picking the smallest item gives a solution.

Lemma 3. *An instance $\mathcal{I} = (\mathcal{A}, \mathcal{B}, C, T_\mathtt{A} = 0, T_\mathtt{B}, Z)$ of STRICT-B-FIRST is a yes-instance if and only if there are sets $A \subseteq \mathcal{A}$ and $B \subseteq \mathcal{B}$ that fulfill Conditions C1 to C4 and $a \leq a'$ for each $a \in A$ and $a' \in \mathcal{A} \setminus A$.*

Proof. By definition, if there are sets $A \subseteq \mathcal{A}$ and $B \subseteq \mathcal{B}$ that fulfill Conditions C1 to C4, then $\mathcal{I}$ is a yes-instance of STRICT-B-FIRST.

Let $\mathcal{I}$ be a yes-instance, then there are sets $A \subseteq \mathcal{A}$ and $B \subseteq \mathcal{B}$ that fulfill Conditions C1 to C4. Because Condition C3 is fulfilled, the items of A are selected in nondecreasing order and the items in B are selected in nonincreasing order. Let $a_{i_1} \leq \cdots \leq a_{i_{|A|}}$ be the items of A and let $b_{j_1} \geq \cdots \geq b_{j_{|B|}}$ be the items of B. Assume that there is an $a_{i_q} \in A$ with $a_{i_q} > \min \mathcal{A} \setminus A =: a'$ and assume further that $a_{i_p} \leq a'$ for each $p \in [q-1]$. We show that if Player A picks a' instead of a_{i_q} in their qth turn, then the items of B remain the greedy choices of Player B or the new set of items chosen by Player B also fulfills Conditions C1 to C4. Applying this argument inductively to increasing positions in A then yields the desired result.

Hence, assume that after replacing a_{i_q} with a', for some b_{j_p} with $p \in \{q+1, \ldots, |B|\}$ the greedy choice becomes b instead of b_{j_p} while $b_{j_{q+1}}, \ldots, b_{j_{p-1}}$ remain the respective greedy choice. Because $a_{i_q} > a'$, we have $b > b_{j_p}$. Then, $\sum_{r=1}^{p-1}(b_{j_r} + a_{i_r}) + b > C$ and so $b > \sum_{r=p}^{|A|} a_{i_r} + \sum_{r=p}^{|B|} b_{j_r} > \sum_{r=p}^{|B|} b_{j_r}$. Therefore, $\sum_{r=1}^{p-1} b_{j_r} + b > \sum B \geq T_\mathtt{B}$. Moreover, the threshold $T_\mathtt{A}$ is met since $T_\mathtt{A} = 0$. Consequently, the sets $A := \{a_{i_1}, \ldots, a_{i_{q-1}}, a', a_{i_{q+1}}, \ldots, a_{i_{\min(p-1, |A|)}}\}$ and $B := \{b_{j_1}, \ldots, b_{j_{p-1}}, b\}$ fulfill the Conditions C1 to C4 and $a \leq a'$ for each $a \in A$ and $a' \in \mathcal{A} \setminus A$. □

With Lemma 3, we can solve the special case of STRICT-B-FIRST with $T_\mathtt{A} = 0$ by iterating over the smallest items of $\mathcal{A}$ and in parallel over $\mathcal{B}$ to find the next greedy choice. If $\mathcal{A}$ and $\mathcal{B}$ are already ordered, this needs only linear time.

Corollary 1. *STRICT-B-FIRST can be solved in linear time if $T_\mathtt{A} = 0$ and the items of $\mathcal{A}$ and $\mathcal{B}$ are ordered.*

Now we are ready to prove the correctness of Theorem 1.

Proof (of Theorem 1). Order $a_0 \geq a_1 \geq \cdots \geq a_{|\mathcal{A}|}$ and $b_1 \geq \cdots \geq b_{|\mathcal{B}|}$. For technical reasons, we assume $a_0 = C + 1$. We fill a dynamic programming table DP with entries of the type $\mathrm{DP}[h,t,i,j,z]$, where $h,i \in [|\mathcal{A}|]$, $t \in [T_\mathtt{A}]$, $j \in [|\mathcal{B}|]$, and $z \in [i-1]_0$. An entry $\mathrm{DP}[h,t,i,j,z]$ stores the value $\sum B$ of any multiset $B \subseteq \mathcal{B}$ such that $j = \max\{q \mid b_q \in B\}$ and B is the greedy response to a size-h multiset $A \subseteq \mathcal{A}$ picked in nonincreasing order where $\sum A = t$, a_i is the element with the largest index in A, $a_z \notin A$, and $\{a_{z+1}, \ldots, a_i\} \subseteq A$. Note that by Lemma 2, there is at most one set with these properties and hence this value is uniquely determined. If there is no such set B, then $\mathrm{DP}[h,t,i,j,z] := -\infty$. For the sake of readability, in the rest of the proof we violate notation a bit and write $a_i = \min A$ and $b_j = \min B$ when we actually mean $i = \max\{p \mid a_p \in A\}$ and $j = \max\{q \mid b_q \in B\}$, respectively.

Algorithm. The base case for the reccurence for computing DP are singleton sets A. We set the values as follows:

$$\mathrm{DP}[1,t,i,j,z] := \begin{cases} b_j & z = i-1 \wedge t = a_i \wedge a_i + b_j \leq C \wedge j = 1 \\ b_j & z = i-1 \wedge t = a_i \wedge a_i + b_j \leq C < a_i + b_{j-1} \\ -\infty & \text{otherwise.} \end{cases}$$

Moreover, for $h > i$ we set $\mathrm{DP}[h,t,i,j,0] := -\infty$.

To state the recurrence, we define an auxiliary function Ψ as follows:

$$\Psi(Q,t,q,j) := \begin{cases} Q & Q + t \leq C \wedge q = j - 1 \\ Q & Q + t \leq C \wedge Q + t + (b_{q-1} - b_q) > C \\ -\infty & \text{otherwise.} \end{cases}$$

Intuitively, this function checks whether in the recurrence, a previous set can be combined with a current choice of a_i and b_q without violating either the validity of the picks or the greedy strategy for b.

To compute values $\mathrm{DP}[h,t,i,j,z]$ with $z = i-1$ and $t > a_i$ we use the recurrence

$$\mathrm{DP}[h,t,i,j,i-1] = \max_{p < i-1,\, z' < p,\, q < j} \Psi(\mathrm{DP}[h-1,t-a_i,p,q,z'] + b_j, t, q, j). \quad (1)$$

And to compute values $\mathrm{DP}[h,t,i,j,z]$ with $z < i-1$ and $t > a_i$ we use the recurrence

$$\mathrm{DP}[h,t,i,j,z] = \max_{q < j} \Psi(\mathrm{DP}[h-1,t-a_i,i-1,q,z] + b_j, t, q, j). \quad (2)$$

Once all entries of the table DP are computed, we return **yes** if there are $t_a \in [T_\mathtt{A}]$, $h,i \in [|\mathcal{A}|]$, $j \in [|\mathcal{B}|]$, $z \in [i-1]_0$ and $p \in \{i+1,\ldots,|\mathcal{A}|\}$ such that $\mathrm{DP}[h,t_a,i,j,z] = t_b$ and $t_a + a_p > T_\mathtt{A}$ and $t_a + a_p + t_b < C$ and $(\mathcal{A}' := \{a_{p+1},\ldots,a_{|\mathcal{A}|}\}, \mathcal{B}' := \{b_{j+1},\ldots,b_{|\mathcal{B}|}\}, C - t_a - t_b - a_p, T'_\mathtt{A} := 0, T_\mathtt{B} - t_b, a_z)$ is a **yes**-instance of STRICT-B-FIRST. Otherwise, if no such h, t_a, i, j, z and p exist, then we return **no**.

Correctness. If $\mathcal{A}$ is a singleton $\{a_i\}$, we only need to check that b_j is the greedy response. If $a_i > t$ then we can not find a set A with $a_i = \min A$ and $\sum A = t$. Therefore, the base cases are correct.

As induction hypothesis, for fixed $h, i \in [|\mathcal{A}|]$ let the table DP store the correct value in $\mathrm{DP}[h-1, t, i', j, z]$ for any $i' \in [i-1]$, $t \in [T_\mathsf{A}]$, $j \in [|\mathcal{B}|]$ and $z \in [i'-1]_0$. We show first that if $\mathrm{DP}[h, t, i, j, z] = Q$ then there are multisets $A \subseteq \mathcal{A}$ and $B \subseteq \mathcal{B}$ with $b_j = \min B$, $a_i = \min A$, $h = |A|$, $\sum A = t$, $\sum B = Q$, $a_z \notin A$ and $\{a_{z+1}, \ldots, a_i\} \subseteq A$, and B is the greedy response to Player A picking the elements in A in nonincreasing order. Then, we show that if there are $A \subseteq \mathcal{A}$ and $B \subseteq \mathcal{B}$ with $b_j = \min B$, $a_i = \min A$, $h = |A|$, $\sum A = t$, $a_z \notin A$ and $\{a_{z+1}, \ldots, a_i\} \subseteq A$ and B is the greedy response to Player A picking the elements in A in nonincreasing order then $\mathrm{DP}[h, t, i, j, z] \geq \sum B$.

Let $\mathrm{DP}[h, t, i, j, z] = Q \in \mathbb{N}$. Then, we know with Recurrence (1) or (2) that there are $p < i$, $z' < p$, and $q < j$ such that ($z = i-1$ and $Q = \mathrm{DP}[h-1, t-a_i, p, q, z'] + b_j$ or $z < i-1$ and $Q = \mathrm{DP}[h-1, t-a_i, i_1, q, z] + b_j$) and $Q + t < C$ and ($q = j - 1$ or $Q + t + (b_{q-1} - b_q) > C$). By the induction hypothesis we know that there are $A \subseteq \mathcal{A}$ and $B \subseteq \mathcal{B}$ with $b_q = \min B$, $a_p = \min A$, $h = |A|$, $\sum A = t - a_i$ and B is the greedy response to Player A picking the elements in A in nonincreasing order. Define $A' := A \cup \{a_i\}$. We conclude $\sum A' = a_i + \sum A = t$ and $|A'| = |A| + 1 = h$. Further, the greedy response of Player B after $A \cup B \cup \{a_i\}$ have been selected is b_j. Thus, $\mathrm{DP}[h, t, i, j, z] = \sum B \cup \{b_j\}$ and $B \cup \{b_j\}$ also fulfills all other necessary conditions.

Now let $A \subseteq \mathcal{A}$ and $B \subseteq \mathcal{B}$ be sets with $b_j = \min B$, $a_i = \min A$, $|A| = h$, $\sum A = t$, $a_z \notin A$ and $\{a_{z+1}, \ldots, a_i\} \subseteq A$, and B is the greedy response to Player A picking the elements in A in nonincreasing order. We define $A' := A \setminus \{a_i\}$ and $B' := B \setminus \{b_j\}$. Let $a_p = \min A'$ and $b_q = \min B'$ and further let z' be the index with $a_{z'} \notin A'$ and $\{a_{z'+1}, \ldots, a_p\} \subseteq A$. Clearly $p < i$ and $q < j$ so that by the induction hypothesis $\mathrm{DP}[h-1, t-a_i, p, q, z'] \geq \sum B'$. Because B is the greedy response to A, we know that $\sum A + \sum B < C$ and further $q = j - 1$ or $\sum A + \sum B + (b_{q-1} - b_q) > C$. Therefore $\Psi(\sum B, \sum A, q, j) = \sum B$ and we conclude $\mathrm{DP}[h, t, i, j, z] \geq \Psi(\mathrm{DP}[h-1, t-a_i, p, q, z'] + b_j, t, q, j) \geq \Psi(\sum B, \sum A, q, j) = \sum B$.

It remains to show that the algorithm returns yes if and only if the instance $\mathcal{I} := (\mathcal{A}, \mathcal{B}, C, T_\mathsf{A}, T_\mathsf{B})$ of P-GREEDY-SSG is a yes-instance. Let $\mathcal{I}$ be a yes-instance of P-GREEDY-SSG. Then, there are $A \subseteq \mathcal{A}$ and $B \subseteq \mathcal{B}$ such that $\sum A \geq T_\mathsf{A}$, $\sum B \geq T_\mathsf{B}$, $\sum A + \sum B < C$, and B is the greedy response to A. Let $a_{i_1} \geq a_{i_2} \geq \cdots \geq a_{i_{|A|}}$ be the items of A and analogous let $b_{j_1} \geq \cdots \geq b_{j_{|B|}}$ be the items of B. By Lemma 1 and the greedy strategy for Player B, we may assume that these items are picked in the order of their indices. Without loss of generality let $p \in [|A|]$ be the index with $\sum_{q=1}^{p-1} a_{i_q} < T_\mathsf{A} \leq \sum_{q=1}^{p} a_{i_q}$. Further, let $z < i_{p-1}$ be the index such that $a_z \notin A$ and $\{a_z, \ldots, a_{i_{p-1}}\} \subseteq A$. Then, $\mathrm{DP}[p-1, \sum_{q=1}^{p-1} a_{i_q}, i_{p-1}, j_{p-1}, z] = \sum_{q=1}^{p-1} b_{j_q}$ because by Lemma 2, the response of Player B to A is unique. Further, $a_{i_p+1} \geq \cdots \geq a_{i_{|A|}}$ and $b_{j_p} \geq \cdots \geq b_{j_{|B|}}$ are a solution for the instance ($\mathcal{A}' := \{a_{i_p+1}, \ldots, a_{|\mathcal{A}|}\}$, $\mathcal{B}' := \{b_{j_{p-1}+1}, \ldots, b_{|\mathcal{B}|}\}$, $C-$

$\sum_{q=1}^{p-1} a_{i_q} - \sum_{q=1}^{p-1} b_{j_q} - a_{i_p}, T'_\mathsf{A} := 0, T_\mathsf{B} - \sum_{q=1}^{p-1} b_{j_q}, a_z)$ of STRICT-B-FIRST. Thus, the algorithm returns yes.

Conversely, if the algorithm returns yes, then there are $t_a \in [T_\mathsf{A}]$, $h, i \in [|\mathcal{A}|]$, $j \in [|\mathcal{B}|]$, $z \in [i-1]_0$ and $p \in \{i+1, \dots, |\mathcal{A}|\}$ such that $\mathrm{DP}[h, t_a, i, j, z] = t_b$ and $t_a + a_p > T_\mathsf{A}$ and $t_a + a_p + t_b < C$ and $(\mathcal{A}' := \{a_{p+1}, \dots, a_{|\mathcal{A}|}\}, \mathcal{B}' := \{b_{j+1}, \dots, b_{|\mathcal{B}|}\}, C - t_a - t_b - a_p, T'_\mathsf{A} := 0, T_\mathsf{B} - t_b, a_z)$ is a yes-instance of STRICT-B-FIRST. Then there are sets $A \subseteq \mathcal{A}$ and $B \subseteq \mathcal{B}$ with $b_j = \min B$, $a_i = \min A$, $\sum A = t_a$, $\sum B = t_b$, $a_z \notin A$ and $\{a_{z+1}, \dots, a_i\} \subseteq A$, and B is the greedy response to Player A picking the elements in A in nonincreasing order. Further, there are $A' \subseteq \mathcal{A}'$ and $B' \subseteq \mathcal{B}'$ such that $\sum B' \geq T_\mathsf{B} - t_b$, and $\sum A' \cup B' \leq C - t_a - t_b - a_p$, and B is the greedy choice on A where Player B also starts. Altogether $A \cup \{a_p\} \cup A'$ and $B \cup B'$ are solutions for the instance $\mathcal{I}$ of P-GREEDY-SSG.

Running Time. The table DP has $\mathcal{O}(T_\mathsf{A} \cdot |\mathcal{A}|^3 \cdot |\mathcal{B}|)$ entries. It can be checked in $\mathcal{O}(\log T_\mathsf{A})$ time if $t \leq a_i$ for any $t \in [T_\mathsf{A}]$ and $a_i \in \mathcal{A}$. Each entry $\mathrm{DP}[h, t, i, j, z]$ with $t \leq a_i$ can be computed in linear time. Computing an entry $\mathrm{DP}[h, t, i, j, z]$ with Recurrence (1) is done in $\mathcal{O}(|\mathcal{A}|^3 \cdot |\mathcal{B}|)$ time and with Recurrence (2) in $\mathcal{O}(|\mathcal{B}|)$ time. Therefore we can compute the entire table DP in time $\mathcal{O}(T_\mathsf{A} \cdot |\mathcal{A}|^5 \cdot |\mathcal{B}|^2) = \mathcal{O}(T_\mathsf{A} \cdot n^7)$.

Once all entries of the table DP are computed, we iterate over $t_a \in [T_\mathsf{A}]$, $i \in [|\mathcal{A}|]$, $j \in [|\mathcal{B}|]$, $z \in [i-1]_0$ and $p \in \{i+1, \dots, |\mathcal{A}|\}$ which takes $\mathcal{O}(T_\mathsf{A} \cdot n^4)$ time and do checks that by Corollary 1 can be done in linear time. Therefore the overall running time is $\mathcal{O}(T_\mathsf{A} \cdot n^7)$. □

Observe that in the algorithm presented in Theorem 1, we need the dimension z of the table only to ensure that after Player A collected enough items to meet their target T_A, we can ensure that in the instance of STRICT-B-FIRST we are not dependent on the item a_z. As we do not need this additional check in GREEDY-SSG, the running time improves by a quadratic factor.

Corollary 2. GREEDY-SSG *can be solved in* $\mathcal{O}(T_\mathsf{A} \cdot n^5)$ *time.*

In addition, we obtain the following for hostile adversaries.

Theorem 2 (*). P-HOSTILE-SSG *can be solved in* $\mathcal{O}(2^C \cdot C \cdot n^2)$ *time.*

4 Further Parameterizations

4.1 Parameterization by the Number of Numbers

We now consider parameterization by the total number of different numbers in both item sets, which we denote with var.

Theorem 3. P-GREEDY-SSG *is FPT with respect to var.*

Proof. In the following, we reduce an instance of P-GREEDY-SSG to an equivalent instance of ILP-FEASIBILITY that has $\mathcal{O}(\text{var})$ variables. It is well-known that ILP-FEASIBILITY is FPT with respect to the number m of different variables [6,7] and can be solved in $\log(m)^{\mathcal{O}(m)} \cdot |I|^{\mathcal{O}(1)}$ time [12] with total input size $|I|$.

Let $z_1, \ldots, z_{\text{var}}$ be the unique numbers in $\mathcal{A}$ and $\mathcal{B}$, where $z_i > z_{i+1}$ for each $i \in [\text{var} - 1]$. Since we can assume that both players pick their items in nonincreasing order due to Lemma 1, for any solution, there are rounds, in which one of the players picks an item z_i for the last time and only chooses items z_j with $j > i$ afterwards. We call each such round where a player picks some item z_i for the last time, a *switching event* for the respective player, and item z_i.

The following definitions and observations for Player A are analogously applied for Player B. If a switching event A_i for A and z_i occurs, we define A_i^{Start} to be the first round where A chooses an item z_i and A_i^{End} to be the last round where A chooses an item z_i. The total number of items z_i that player A chooses is then given by $A_i^{\text{End}} - A_i^{\text{Start}} + 1$. Note that between A_i^{End} and A_i^{Start} several switching events for Player B may occur. Moreover, a round can be a switching event for both players. Also, note that Player A may take none of the elements z_j for some $j \in [\text{var}]$. In that case, there is no switching event A_j and we say that Player A *skips* item z_j.

The idea of the algorithm is as follows. We first use an initial branching to determine what switching events occur and in which order. Observe that for a fixed ordering of switching events the complete progression of the game is determined by the number of rounds between each two consecutive switching events. Thus, after the initial branching to fix the order of switching events, it is sufficient to determine the lengths of the intervals between consecutive switching events. For this, we construct an instance of ILP-FEASIBILITY to compute whether there is a feasible choice of interval lengths for the given fixed switching event order that corresponds to valid picking orders such that Player A wins.

In the following, we assume that we are given a fixed order $X_1, X_2, \ldots, X_k$ with $k \in [2\,\text{var}]$ of all switching events and denote with $r(X_i)$ the round in which the switching event X_i occurs. For technical reasons, we also consider a switching event X_0 at round zero. Note that for two consecutive switching events X_i and X_{i-1} we can have $r(X_i) - r(X_{i-1}) = 0$, which means that both switching events happen in the same round and thus X_{i-1} is a switching event for A and X_i is a switching event for B. For each switching event X_i, let $a(i)$ and $b(i)$ be the item that the respective player is picking at round $r(X_i)$, and let S_i^A and S_i^B be the sum of all current items chosen by A and B after round $r(X_i)$, respectively. For two consecutive switching events X_i and X_{i-1} we thus have

$$S_i^A = S_{i-1}^A + (r(X_i) - r(X_{i-1})) \cdot a(i) \text{ and}$$

$$S_i^B = S_{i-1}^B + (r(X_i) - r(X_{i-1})) \cdot b(i).$$

For each $j \in [\text{var}]$ such that $b(i) = z_j$ for some $i \in [k]$ let $\sigma(j)$ be the largest index i such that $b(i) = z_j$. Moreover, observe that we only have to consider an item z_j if there is some switching event X_i with $a(i) = z_j$ or $b(i) = z_j$, since otherwise that item is not chosen by any player.

Description of the ILP-FEASIBILITY *Instance.* Let α_i and β_i be the number of occurrences of z_i in $\mathcal{A}$ and $\mathcal{B}$, respectively. The variables x_i and y_i describe how many times Player A and Player B select z_i, respectively. For each switching event X_i, we add a variable L_i for the length of the interval $[r(X_{i-1})+1, r(X_i)]$ as well as variables S_i^A and S_i^B for the sum of chosen elements after round $r(X_i)$ for the respective player and set $S_0^A = S_0^B = 0$. We now describe the constraints and argue for each why it is needed.

$$\sum_{i=1}^{\text{var}} (x_i + y_i) z_i \leq C \qquad (3)$$

$$\sum_{i=1}^{\text{var}} x_i z_i \geq T_A \qquad (4)$$

$$\sum_{i=1}^{\text{var}} y_i z_i \geq T_B \qquad (5)$$

Constraint (3) ensures that the sum of all items chosen does not exceed the capacity C. If Constraints (4) and (5) are satisfied, then the threshold is achieved for both players.

$$x_i \leq \alpha_i \qquad \forall i \in [\text{var}] \qquad (6)$$
$$y_i \leq \beta_i \qquad \forall i \in [\text{var}] \qquad (7)$$

Constraints (6) and (7) guarantee that no item is chosen more often by a player than it is available in the multiset of that player.

The next constraints establish the connection between variables x_j, y_j and L_i

$$\sum_{i \in [k]: a(i) = z_j} L_i = x_j \qquad \forall j \in [\text{var}] \qquad (8)$$

$$\sum_{i \in [k]: b(i) = z_j} L_i = y_j \qquad \forall j \in [\text{var}] \qquad (9)$$

Clearly, a player picks an item z_j for the last time in the round corresponding to the switching event for that player and item z_j and starts picking z_j after the previous switching event for that player. Thus, the number of items z_j a player picks in total is given by the sum of the lengths of intervals between (consecutive) switching events in which the respective player has z_j as current item. This is ensured by Constraints (8) and (9).

For Player A, the sum of all items currently chosen by A at a switching event X_i is given by the sum of items chosen by A at the previous switching event X_{i-1} plus the current item $a(i)$ for each round between the two switching events. The same holds for Player B. Constraints (10) and (11) therefore ensure that the current sums at each switching event are correct.

$$S_i^A = S_{i-1}^A + L_i \cdot a(i) \qquad \forall i \in [k] \qquad (10)$$
$$S_i^B = S_{i-1}^B + L_i \cdot b(i) \qquad \forall i \in [k] \qquad (11)$$

Constraints (12) ensure that Player B chooses the items greedily.

$$\max\{1 + y_j - \beta_j; 0\} + \max\{(S_{\sigma(j)}^A + S_{\sigma(j)}^B + z_j - C); 0\} > 0 \quad \forall j \in [\text{var}] \quad (12)$$

For each item z_j Player B must pick all available items z_j or after the round where B is choosing z_j for the last time, which is given by the switching event $X_{\sigma(j)}$, picking another item z_j would exceed the capacity C. If at least one of these conditions is fulfilled, one or both of the max terms is nonzero and the constraint is thus fulfilled. Otherwise, both max terms evaluate to zero and the constraint is not fulfilled. Note that these constraints make use of the max operator, which is not a valid expression in standard ILP formulations. However, the max operator can be expressed by a constant number of additional variables and constraints [1].

Finally, Constraints (13) and (14) ensure that all variables take integer values.

$$x_i, y_i \in \mathbb{N}_0 \qquad \forall i \in [\text{var}] \qquad (13)$$
$$L_i, S_i^A, S_i^B \in \mathbb{N}_0 \qquad \forall i \in [k] \qquad (14)$$

Note that it is possible that Player A already reached the target T_A and cannot pick any more items without exceeding the capacity, but Player B still has to pick more items to reach the target T_B. To handle this case let X_ℓ be the last switching event for Player A and thus let $a(i) = 0$ for each $i \in [\ell+1, k]$. Note that we cannot just assume that Player A has sufficiently many zeros to pick from. However, after Player A picked the last item, which happens at round $r(X_\ell)$, if for each item z_j either Player A already picked all items z_j available in $\mathcal{A}$ or picking z_j after round $r(X_\ell)$ would exceed the capacity C, then Player A cannot pick any nonzero items anymore and setting $a(i) = 0$ for each $i \in [\ell + 1, k]$ is valid. To ensure this we add the following constraints.

$$\max\{1 + x_j - \alpha_j; 0\} + \max\left\{\left(S_\ell^A + S_\ell^B + z_j - C\right); 0\right\} > 0 \qquad \forall j \in [\text{var}] \quad (15)$$

Running Time. In the initial branching for the switching event order we need to consider sequences of up to $2 \cdot \text{var}$ possible switching events. Since a player may not pick some item z_j at all, for each of the $2 \cdot \text{var}$ positions we have four options: A switches to the next item, A skips the next item entirely, B switches to the next item or B skips the next item entirely. Thus, there are $\mathcal{O}(4^{2\text{var}})$ possible sequences for this initial branching. For each branch, an ILP-FEASIBILITY instance I with $\mathcal{O}(\text{var})$ variables has to be solved. This can be done in $\log(\text{var})^{\mathcal{O}(\text{var})} \cdot |I|^{\mathcal{O}(1)}$ time [12] and thus the total running time is $\log(\text{var})^{\mathcal{O}(\text{var})} \cdot n^{\mathcal{O}(1)}$. □

If we instead consider as parameter the number of different items in $\mathcal{A}$ only, denoted with var_A, then we obtain an XP-algorithm.

Proposition 1 (*). P-GREEDY-SSG *can be solved in* $\mathcal{O}(|\mathcal{A}|^{var_A} \cdot n)$ *time.*

4.2 Parameterization by the Number of Rounds

In this section, we consider parameterization by the number of rounds. To incorporate this parameter, we consider a problem variant where an additional parameter R is given and we ask whether Player A has a winning strategy such that A wins after at most R rounds.

Proposition 2. *It is W[1]-hard to decide for an instance of* GREEDY-SSG, *whether Player* A *has a winning strategy where the winning condition is fulfilled after at most R rounds, even if the underlying set of the multiset $\mathcal{B}$ of Player* B *only contains the element 1.*

Proof. We reduce from SUBSET SUM, where the input consists of a multiset $\mathcal{Z}$ and two integers G, k and the question is whether there is a subset S of at most k items of $\mathcal{Z}$ that sums up to a value of exactly G. Let $\mathcal{I} := (\mathcal{Z}, G, k)$ be an instance of SUBSET SUM. Further, let $M = 2k$ be a constant greater than k. We construct from $\mathcal{I}$ an instance $\mathcal{I}' := (\mathcal{A}, \mathcal{B}, C, T_\text{A})$ of GREEDY-SSG as follows. We define the multiset $\mathcal{B}$ of Player B containing k times the item 1 and the multiset $\mathcal{A}$ of Player A containing the item $M \cdot z$ for each $z \in \mathcal{Z}$. We set $C := M \cdot G + k - 1$ and $T_\text{A} := M \cdot G$.

We show that $\mathcal{I}$ is a yes-instance of SUBSET SUM if and only if $\mathcal{I}'$ is a yes-instance of GREEDY-SSG. First, assume that $\mathcal{I}$ is a yes-instance and let $S \subseteq \mathcal{Z}$ be a solution of $\mathcal{I}$. Then, $|S| \leq k$ and $\sum S = G$. Define $A := \{M \cdot z \mid z \in S\}$. Then, $\sum A = M \cdot \sum S = M \cdot G$. Because $|A| = |S| \leq k$, Player B has less than k moves before Player A selects the items of A and so the overall capacity needed is at most $M \cdot G + (k-1) = C$. Therefore $\mathcal{I}'$ is a yes-instance of GREEDY-SSG.

Conversely, let $\mathcal{I}'$ be a yes-instance of GREEDY-SSG. Then, Players A and B have valid picking orders $A := (a_{i_1}, \ldots, a_{i_\ell})$ and $B := (1, \ldots, 1)$, respectively. Let $\ell' := |B| = \sum B$ be the number of rounds Player B has played. Define $S := \{a_{i_1}/M, \ldots, a_{i_\ell}/M\}$. Note that per construction each item in S is an integer in $\mathcal{Z}$. Moreover, since $M > k - 1$ we have $C < M \cdot (G+1)$. Since each item in $\mathcal{A}$ is a multiple of M and $\mathcal{I}'$ is a yes-instance, this implies that A reaches the target $T_\text{A} = M \cdot G$ exactly. We conclude that $\sum A = M \cdot G$ and so $\sum S = G$. Furthermore, since $C - \sum A = k - 1$ we have $k - 1 \geq \ell' \geq \ell - 1$, which implies $|S| \leq k$. Thus, S is a solution for $\mathcal{I}$. □

By the above, it is unlikely that any of the four problems is FPT with respect to R. On the positive side, all four problems are in XP with respect to R.

Proposition 3 (*). *For a given instance of* P-HOSTILE-SSG *it can be decided in $\mathcal{O}(n^{2R+1})$ time whether Player* A *has a winning strategy where the winning condition is fulfilled after at most R rounds.*

4.3 Parameterization by Target Threshold for Hostile Adversaries

In Proposition 3 we saw that P-HOSTILE-SSG and thus also HOSTILE-SSG are in XP when parameterized by the number of rounds. This result can be used to show that HOSTILE-SSG is in XP with respect to T_A.

Corollary 3 (*). HOSTILE-SSG *can be solved in* $\mathcal{O}(n^{2T_A+1} \cdot T_A)$ *time.*

As we now show, it is unlikely that this XP algorithm can be improved to an FPT algorithm.

Proposition 4 (*). HOSTILE-SSG *is not* FPT *for* T_A *unless* FPT = W[1].

The above gives a strong contrast to the greedy case which, by Theorem 1, admits an FPT algorithm with only a pseudopolynomial dependence on T_A.

5 Discussion

We have studied the parameterized complexity of different variants of SUBSET SUM GAME, showing that positive results for the standard variants can be extended to the new protective variant. Moreover, we obtained FPT algorithms for smaller parameters such as the target threshold for Player A or the number of different numbers in the input instance. For the future, it would be interesting to study the Stackelberg variants of the games under similar parameterizations. Moreover, one could consider KNAPSACK GAMES. Can any of the presented algorithms be extended to these more general games?

Acknowledgments. Jaroslav Garvardt is supported by the Carl Zeiss Foundation, Germany, within the project "Interactive Inference".

References

1. Burks, T.M., Sakallah, K.A.: Min-max linear programming and the timing analysis of digital circuits. In: Proceedings of the 1993 IEEE/ACM International Conference on Computer-Aided Design, pp. 152–155. IEEE Computer Society/ACM (1993)
2. Cygan, M., et al.: Parameterized Algorithms. Springer, Cham (2015). https://doi.org/10.1007/978-3-319-21275-3
3. Darmann, A., Nicosia, G., Pferschy, U., Schauer, J.: The subset sum game. Eur. J. Oper. Res. **233**(3), 539–549 (2014)
4. Downey, R.G., Fellows, M.R.: Fundamentals of Parameterized Complexity. Springer, London (2013). https://doi.org/10.1007/978-1-4471-5559-1
5. Fellows, M.R., Gaspers, S., Rosamond, F.A.: Parameterizing by the number of numbers. Theory Comput. Syst. **50**(4), 675–693 (2012). https://doi.org/10.1007/s00224-011-9367-y
6. Frank, A., Tardos, É.: An application of simultaneous diophantine approximation in combinatorial optimization. Combinatorica **7**(1), 49–65 (1987). https://doi.org/10.1007/BF02579200
7. Lenstra, H.W., Jr.: Integer programming with a fixed number of variables. Math. Oper. Res. **8**(4), 538–548 (1983)
8. Pferschy, U., Nicosia, G., Pacifici, A.: On a Stackelberg subset sum game. Electron Notes Discrete Math. **69**, 133–140 (2018)
9. Pferschy, U., Nicosia, G., Pacifici, A.: A Stackelberg knapsack game with weight control. Theor. Comput. Sci. **799**, 149–159 (2019)

10. Pferschy, U., Nicosia, G., Pacifici, A., Schauer, J.: On the Stackelberg knapsack game. Eur. J. Oper. Res. **291**(1), 18–31 (2021)
11. Pieterse, A., Woeginger, G.J.: The subset sum game revisited. Theory Comput. Syst. **65**(5), 884–900 (2021)
12. Reis, V., Rothvoss, T.: The subspace flatness conjecture and faster integer programming. In: Proceedings of the 64th IEEE Annual Symposium on Foundations of Computer Science (FOCS 2023), pp. 974–988. IEEE (2023)
13. Wang, Z., Xing, W., Fang, S.C.: Two-group knapsack game. Theor. Comput. Sci. **411**(7–9), 1094–1103 (2010)

Algorithmic Decision Analysis for Multi-stage Games with Incomplete Information

J. M. Camacho[1(✉)], Roi Naveiro[2], and David Ríos Insua[1]

[1] ICMAT-CSIC, Madrid, Spain
josemanuel.camacho@icmat.es
[2] CUNEF Universidad, Madrid, Spain

Abstract. Adversarial risk analysis (ARA) provides decision-theoretic arguments to manage uncertainty in competitive decision-making environments. This paper introduces efficient algorithmic approaches to approximate ARA solutions in multi-stage games, covering both sequential and simultaneous settings, through augmented probability simulation. Two examples concerning international piracy and air combat illustrate the proposed methodology.

Keywords: Adversarial Risk Analysis · Multi-stage games · Augmented probability simulation

1 Introduction

Adversarial Risk Analysis (ARA) introduces a decision-theoretic framework to solve games supporting just one decision-maker, referred to as the defender (D, she), instead of focusing on all agents simultaneously, the traditional approach in standard game theory (GT). Specifically, in two agent games, ARA aims to maximize the defender's expected utility, transforming the problem into a decision-analytic one, therefore implementing a Bayesian approach to games in the sense of [1,2], providing procedures to forecast the attacker's (A, he) actions. Its major strength lies in mitigating standard GT common knowledge assumptions among participants, including the common prior assumption in Harsanyi's approach, critically reviewed in [3-5] to name but a few. In contrast to GT, ARA does not assume that the agents, in particular the supported one, know the payoffs, preferences, and possible actions of their opponents, but instead, it assigns probability distributions to these elements. A comprehensive overview of ARA along with a comparison with standard GT techniques is available in [6]. It has been applied in various contexts, including counter-terrorism online surveillance [7], combat modeling enhancement [8], and adversarial machine learning [9].

A significant issue in ARA derives from the computational challenges it entails, mainly arising from its two-stage process: the attacker's problem is first simulated to generate a probabilistic forecast of his actions which is then exploited to find the defender's optimal strategy. Ekin et al. [10] proposed using the method of augmented probability simulation (APS, [11]) to solve simple single-stage defend-attack games from an ARA viewpoint, outperforming traditional Monte Carlo methods in scenarios in which the cardinality of the agents' decision sets is large, even continuous. APS transforms a decision analysis problem into a grand simulation within the combined space of decisions and random variables. It constructs an auxiliary augmented distribution proportional to the product of the utility function and the distributions modeling relevant uncertainties. The optimal decision alternative then coincides with the mode of the marginal of the augmented distribution over the decisions. Consequently, expected utility maximization can be achieved through simulation from the augmented distribution.

This paper studies how to apply APS to solve multi-stage sequential and simultaneous games, supporting the defender in her decision-making, and examines the proposed algorithmic decision analytic (ADA) methodology through security problems related to international piracy and air-combat modeling. Code to reproduce the results provided is available at https://datalab-icmat.github.io/software.html.

The ARA framework accommodates various assumptions about the attacker and defender rationality [6]. Here we limit our analysis to a Defender behaving as a level-2 thinker in the Stahl and Wilson [12] sense: she models A as a strategic player (level-1 thinker), who analyzes D's decision problem before making his own decision. However, A considers D to be non-strategic (level-0 thinker), meaning that D does not account for A's problem before making a decision. This choice relies on empirical studies showing that people seldom perform beyond level-2 thinking in complex competitive decision-making [13].

2 Multi-stage Sequential Defend-Attack Games

The approach outlined in [10] for single-stage sequential Defend-Attack games is extended to multiple stages. Of the many possible extensions, we adopt the following n-stage sequential game with incomplete information. The defender goes first by selecting a decision $d_1 \in \mathcal{D}_1$; after observing d_1, the attacker chooses $a_1 \in \mathcal{A}_1$; the defender observes a_1 and makes a decision $d_2 \in \mathcal{D}_2$. This pattern of alternating decisions continues until the n-th stage. After each pair of actions, an uncertainty modeled as a random variable $\theta_i \in \Theta_i$ influenced by both actions is resolved and becomes known to both players. As the game concludes, each player's utility (modeling preferences and risk attitudes over consequences) is determined by all the actions taken and the uncertainties throughout the game. Figure 1 portrays a bi-agent influence diagram (BAID) for a 2-stage version of the problem, with circles, squares, and hexagons, respectively representing uncertainties, decisions, and utilities. White nodes refer solely to the defender, grey ones to the attacker, and striped ones are shared by both agents. Dashed arcs pointing to decision nodes indicate that those decisions are made knowing the values of the preceding nodes. Solid arcs pointing to chance and utility nodes indicate, respectively, that probabilities and utilities depend on their predecessor values.

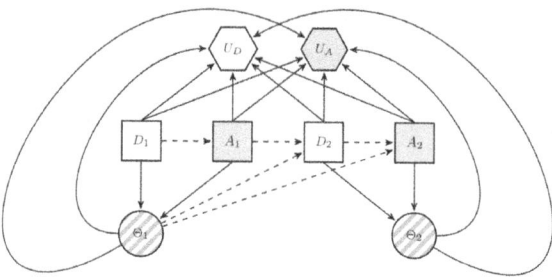

Fig. 1. Template for basic 2-stage sequential defend-attack game.

Let $\theta_{:i}$, $a_{:i}$, and $d_{:i}$ respectively designate the sequence of uncertainties, attacks, and defenses up to stage i. Similarly, let $\theta_{i:}$, $a_{i:}$, and $d_{i:}$ denote the corresponding sequences from stage i onwards. Furthermore, $h_{:i}$ and $h_{i:}$ represent the complete history of uncertainties and actions taken up to and from stage i, respectively. When any of these sequences are unknown, they are considered sequences of random variables and are represented with uppercase letters; as an example, $\theta_{:i}$ would be an instantiation of $\Theta_{:i}$. The Defender's beliefs about θ_i, based on the known sequences $d_{:i}$, $a_{:i}$, and $\theta_{:(i-1)}$, are captured through the probability distribution $p_D(\theta_i | d_i, a_i)$. The utilities of the Defender and Attacker at the final stage are respectively denoted $u_D(d_{:n}, a_{:n}, \theta_{:n})$ and $u_A(d_{:n}, a_{:n}, \theta_{:n})$, reflecting the impact of both agents' actions and the uncertainties throughout the game.

Let us demonstrate how to solve such n-stage sequential games using APS. Start with the simpler case of $n = 2$. Given the history h_1, which includes the actions and uncertainties from the first stage, the last one can be treated as a typical sequential one-stage defend-attack game, similar to those described in [10]. These games can be solved using what we shall designate a *single-stage APS*, which requires nesting two APS (one for the Defender, one for the Attacker) as demonstrated in [10] and illustrated below. With this algorithm, we can determine the optimal defense strategy for the second stage, denoted as $d_2^*(h_1)$. Additionally, APS provides a mechanism for generating samples from the distribution $p_D(a_2 | d_2, h_1)$, which predicts the attacker's actions a_2 in the second stage based on the defensive decisions d_2 and history h_1.

At the initial stage of the game, the Defender selects a decision $d_1 \in \mathcal{D}_1$ by maximizing expected utility. This stage is complex because future decisions, including a_1, d_2, a_2, and uncertainties θ_1 and θ_2, are not yet known and, therefore, are modeled as random variables. The Defender's maximum expected utility problem is expressed as

$$d_1^* = \arg\max_{d_1 \in \mathcal{D}_1} \iiint u_D(d_1, a_1, \theta_1, h_2) p_D(h_2 | d_1, a_1, \theta_1) p_D(\theta_1 | d_1, a_1) \\ p_D(a_1 | d_1) dh_2 d\theta_1 da_1. \quad (1)$$

Assuming we can sample from $p_D(h_2|d_1, a_1, \theta_1)$, and that utilities are non-negative, which can be achieved, in general, via a positive affine transformation, let us define the augmented distribution

$$\pi_{D_1}(d_1, a_1, \theta_1, h_2) \propto u_D(d_1, a_1, \theta_1, h_2) p_D(h_2|d_1, a_1, \theta_1) p_D(\theta_1|d_1, a_1) p_D(a_1|d_1).$$

Then, if we are able to generate samples $(d_1, a_1, \theta_1, h_2) \sim \pi_{D_1}(d_1, a_1, \theta_1, h_2)$, using the marginal samples of d_1, a consistent estimator [14] for the marginal mode of d_1 can be built which, by construction, will coincide with d_1^*.

This sampling can be conducted using a Metropolis-Hastings (MH) algorithm [15]. Let $(d_1, a_1, \theta_1, h_2)$ be the current state of the chain. We iterate through the following three steps until convergence, where $g_D(\cdot|\cdot)$ is a proposal generating distribution

1. Propose $\tilde{d}_1 \sim g_D(\cdot|d_1)$.
2. Sample $\tilde{a}_1 \sim p_D(a_1|\tilde{d}_1)$, $\tilde{\theta}_1 \sim p_D(\theta_1|\tilde{d}_1, \tilde{a}_1)$, $\tilde{h}_2 \sim p_D(h_2|\tilde{d}_1, \tilde{a}_1, \tilde{\theta}_1)$.
3. With probability $\min\left(1, \frac{u_D(\tilde{d}_1, \tilde{a}_1, \tilde{\theta}_1, \tilde{h}_2) g_D(d_1|\tilde{d}_1)}{u_D(d_1, a_1, \theta_1, h_2) g_D(\tilde{d}_1|d_1)}\right)$, set the new state of the chain to $(\tilde{d}_1, \tilde{a}_1, \tilde{\theta}_1, \tilde{h}_2)$.

Standard MCMC convergence results [16] show that the stationary distribution of this chain is $\pi_{D_1}(d_1, a_1, \theta_1, h_2)$ under the condition that the proposal generating distribution g_D has support on D_1.

To sample from $p_D(a_1|d_1)$ in step 2 of the algorithm, it is necessary to address the attacker's first-stage problem from the defender's perspective while accounting for all relevant uncertainties. Given d_1, the attacker seeks to solve

$$\arg\max_{a_1 \in \mathcal{A}_1} \iint u_A(d_1, a_1, \theta_1, h_2) p_A(h_2|d_1, a_1, \theta_1) p_A(\theta_1|d_1, a_1) \mathrm{d}h_2 \mathrm{d}\theta_1.$$

Uncertainty about the attacker's probabilities and utilities is modeled with random utilities U_A and probabilities P_A [6]. This induces uncertainty in the attacker's expected utility; maximizing the corresponding random expected utility yields samples from $p_D(a_1|d_1)$. This problem can be solved using the random augmented probability distribution (assuming U_A is almost surely non-negative)

$$\Pi_A(a_1, \theta_1, h_2|d_1) \propto U_A(d_1, a_1, \theta_1, h_2) P_A(h_2|d_1, a_1, \theta_1) P_A(\theta_1|d_1, a_1).$$

It is simple to see that samples from the mode of the random marginal distribution $\Pi_A(a_1|d_1)$ are distributed according to $p_D(a_1|d_1)$. This sampling procedure involves first generating random utilities and probabilities and, then, applying APS, say using another MH algorithm. Additionally, we need to produce samples from $p_D(h_2|d_1, a_1, \theta_1)$ based on

$$p_D(h_2|d_1, a_1, \theta_1) = p_D(d_2, a_2, \theta_2|h_1) = p_D(\theta_2|d_2, a_2) p_D(a_2|d_2, h_1) p_D(d_2|h_1).$$

Sampling from $p_D(\theta_2|d_2, a_2)$ is standard. Moreover, given h_1, it is possible to generate samples from $p_D(a_2|d_2, h_1)$ and $p_D(d_2|h_1)$ by implementing a single-stage APS for the sequential game constituted by D_2 and A_2, conditioned on

h_1. Importantly, at this point, observe that $p_D(d_2|h_1)$ is essentially a point mass at $d_2^*(h_1)$, the optimal decision at stage two based on h_1. Thus, the proposed single-stage APS provides the required samples from $p_D(h_2|h_1)$, constituted by the APS for D_1 and A_1, enabling us to solve the sequential game.

In conclusion, solving the first stage of a two-stage sequential game involves a nested structure where two single-stage APS algorithms are utilized. During each iteration of the first-stage APS, it is necessary to run the second-stage APS, conditioning it on the information gathered from the first one. The result actually generalizes to the case with n stages by induction.

Proposition 1. *The n-stage sequential game under incomplete information can be solved using n nested single-stage APS algorithms.*

Proof. We prove the result using induction. The case when $n = 1$ has already been established. Assume now that we can solve an $(n-1)$-stage sequential game using APS, which provides us with the means to produce samples from $p_D(h_{2:n}|h_1)$ using $(n-1)$ nested APS algorithms.

Consider the first stage of the n-stage game. The Defender needs to select $d_1 \in \mathcal{D}_1$ to maximize expected utility. At this stage, the future history $(h_{2:n})$ is unknown and, therefore, treated as a random variable. D must solve the optimization problem

$$\arg\max_{d_1 \in \mathcal{D}_1} \iiint u_D(d_1, a_1, \theta_1, h_{2:n}) p_D(h_{2:n}|d_1, a_1, \theta_1)$$

$$p_D(\theta_1|d_1, a_1) p_D(a_1|d_1) \mathrm{d}h_{2:n} \mathrm{d}\theta_1 \mathrm{d}a_1,$$

addressed using APS. For this, define the first-stage augmented distribution

$$\pi_{D_1}(d_1, a_1, \theta_1, h_{2:n}) \propto u_D(d_1, a_1, \theta_1, h_{2:n})$$
$$p_D(h_{2:n}|d_1, a_1, \theta_1) p_D(\theta_1|d_1, a_1) p_D(a_1|d_1).$$

Clearly, the mode of the marginal $\pi_{D_1}(d_1)$ coincides with the optimal first-stage decision. By generating samples $(d_1, a_1, \theta_1, h_{2:n}) \sim \pi_{D_1}(d_1, a_1, \theta_1, h_{2:n})$, we can construct a consistent estimator of the marginal mode using these d_1 samples. This sampling process involves iterating through a MH scheme, similar to the one presented above. Crucially, each iteration of this sampler will require generating samples from $p_D(h_{2:n}|d_1, a_1, \theta_1)$, obtained nesting $(n-1)$ single-stage APS algorithms, as per our induction hypothesis. Therefore, effectively solving the first stage of an n-stage sequential problem necessitates nesting n single-stage APS algorithms. △

3 Multi-stage Simultaneous Defend-Attack Games

Consider now multi-stage simultaneous defend-attack games with incomplete information. In the setup considered, at the first stage, the Defender chooses $d_1 \in \mathcal{D}_1$, while the Attacker simultaneously selects $a_1 \in \mathcal{A}_1$. Following these actions, the uncertainty $\theta_1 \in \Theta_1$, influenced by both a_1 and d_1, is resolved and subsequently affects the uncertainty $\theta_2 \in \Theta_2$, which also depends on subsequent actions $d_2 \in \mathcal{D}_2$ and $a_2 \in \mathcal{A}_2$. This pattern of parallel decisions continues until the n-th stage. Finally, the utility of D is contingent upon all her decisions and the uncertainties $\{\theta_i\}_{i=1}^n$, and, similarly, for A. This setup can be applied to numerous models, such as the air-combat scenario in [17], or adapted to multi-agent reinforcement learning scenarios, e.g., [18]. Figure 2 specifies a BAID for a 2-stage simultaneous game. Note that, in contrast to Fig. 1, decision D_i does not point to decision A_i, as both decisions are made simultaneously at the i-th stage.

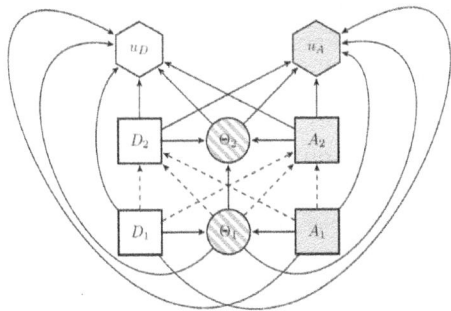

Fig. 2. Template for basic 2-stage simultaneous defend-attack game BAID.

The methodology to solve these games through APS closely mirrors that used for sequential games. We demonstrate it using the two-stage game. Initially, the Defender aims to maximize expected utility by selecting a decision $d_1 \in \mathcal{D}_1$. At this stage, neither the action a_1 nor the future decisions and uncertainties are known; hence, they are modeled as random variables. The Defender's optimization problem is expressed as

$$d_1^* = \arg\max_{d_1 \in \mathcal{D}_1} \iiint u_D(d_1, a_1, \theta_1, h_2) p_D(h_2|d_1, a_1, \theta_1) p_D(\theta_1|d_1, a_1) \\ p_D(a_1) \mathrm{d}h_2 \mathrm{d}\theta_1 \mathrm{d}a_1. \quad (2)$$

Notably, the only difference between this formulation and its sequential counterpart (Eq. 1) is the distribution over the first-stage attack, which is now independent of d_1 due to the simultaneous decision-making feature (i.e., $p_D(a_1)$ vs $p_D(a_1|d_1)$). Consequently, a similar MH APS algorithm is employed, adjusting only step 2 to sample from $p_D(a_1)$ instead of $p_D(a_1|d_1)$. To sample from $p_D(a_1)$, consider A's problem at the first stage, which mirrors D's problem and is stated as

$$a_1^* = \arg\max_{a_1 \in \mathcal{A}_1} \iiint u_A(d_1, a_1, \theta_1, h_2) p_A(h_2|d_1, a_1, \theta_1) p_A(\theta_1|d_1, a_1) \\ p_A(d_1) \mathrm{d}h_2 \mathrm{d}\theta_1 \mathrm{d}d_1.$$

As before, uncertainty about A's utilities and probabilities is accounted for using random utilities U_A and probabilities P_A, generating a random augmented probability distribution that now, as opposed to the sequential case, is unconditional

$$\Pi_A(d_1, a_1, \theta_1, h_2) \propto U_A(d_1, a_1, \theta_1, h_2) P_A(h_2|d_1, a_1, \theta_1) P_A(\theta_1|d_1, a_1) P_A(d_1).$$

Given the minor adjustments required from the sequential to the simultaneous formulation, it becomes apparent that solving the n-stage simultaneous defend-attack game can be resolved by handling n single-stage APS algorithms.

Proposition 2. *The n-stage simultaneous game under incomplete information can be solved using n nested single-stage APS algorithms.*

The proof of Proposition 2 follows a similar approach to that of Proposition 1 and is, therefore, omitted.

4 Computational Issues

The algorithms previously discussed entailed nesting several MH APS schemes, making them computationally intensive. This complexity limits their applicability primarily to games with relatively few stages. To extend these methods to games with a large number of stages, alternative approaches must be considered.

A practical solution for games where decision and uncertainty spaces are discrete and exhibit low cardinality is the implementation of backward induction. Take, for example, the final stage of a two-stage sequential game. Given the history h_1, it is possible to compute the optimal defense $d_2^*(h_1)$ and sample from $p_D(a_2|d_2, h_1)$ using APS, as explained in Sect. 2. This process facilitates creating an empirical estimate of $p_D(a_2|d_2, h_1)$. If the cardinality of the decision and uncertainty spaces is sufficiently small, it becomes practical to calculate these quantities for every possible h_1 scenario and store the results. Moving to the first-stage in the APS, rather than dynamically invoking the second-stage APS to sample future decisions based on the current h_1 sample, we can simply access and utilize the pre-computed values. This method significantly reduces the computational load.

However, the previous approach has its drawbacks, as it can lead to a combinatorial explosion as the cardinality of decision spaces and/or the number of stages increase. Additionally, while the general approach in Sects. 2 and 3 is theoretically applicable to games with continuous decision spaces, the backward induction approach is not. In such cases, it is necessary to develop and employ statistical models that can effectively approximate future optimal decisions and predict attack distributions.

Another potential issue that can arise when applying APS occurs when the expected utility surfaces are very flat around the mode, in which case, the number of required APS samples to approximate the mode could be very high. To address this, we replace the marginal augmented distributions $\pi_D(h)$ in the proposed methodology, being h their corresponding arguments, with their power transformations $\pi_D^J(h)$ since these transfomations are more peaked around the mode. The power J, denoted as augmentation parameter, is treated as a hyperparameter. This adjustment facilitates mode identification, see e.g. [19].

5 Numerical Examples

We illustrate the methodology through two security problems.

5.1 International Piracy

We illustrate practical issues associated with the approach in Sect. 2 for sequential problems through an international piracy example adapted from [20]. Assume we support the owner of a ship in managing piracy risks near the Somali coast. The problem is an asymmetric 2-stage sequential game as the one depicted in the BAID in Fig. 3, being a simplified version of the more general 2-stage sequential problem in Sect. 2, Fig. 1: the second stage attacker decision is omitted, and only one uncertainty is considered. This simplified version allows us to compare the solutions obtained using the proposed scheme with the actual ones provided in [20].

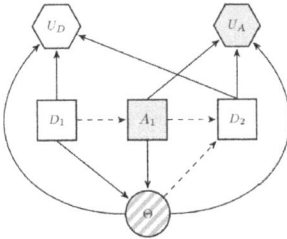

Fig. 3. BAID representing the international piracy example.

In the initial decision-making stage, the ship's owner proactively selects a defensive strategy (D_1) to mitigate piracy risks which includes as options deploying varying levels of armed security or opting for a significantly longer alternative route that circumvents the incumbent high-risk area. Subsequently, the Attacker, in response to the Defender's strategy, decides whether to initiate an attack aimed at hijacking the ship for ransom purposes (A_1). Should the pirates' operation prove successful (Θ), the ship's owner is then faced with the decision (D_2): pay the ransom, refuse it, or ask the armed forces to recover the ship. The utility of the Defender depends on her strategic choices and the results of pirate attacks on the vessel, while the Attacker's utility is influenced by his initial decision, the attack's outcome, and the subsequent decision made by the Defender.

Parametric Setup. A brief summary is provided here. For full explanations, see [20].

Agents. The Defender is the ship owner trying to manage piracy risks. The Attacker are the pirates deciding on eventually seizing the ship for a ransom.

Decisions. The alternatives available to D and A are:

- D_1: Do not implement any protection for the boat (d_1^1), use private protection with one armed person (d_1^2), use private protection with a team of two armed persons (d_1^3), take a more costly alternative route for the trip (d_1^4).
- A_1: The pirates may choose not to attack the owner's ship (a^0), attack it (a^1), or attack another vessel (a^i). For our problem, consider that there are three additional boats in the region, $i \in \{2, 3, 4\}$.
- D_2: Do not respond to the pirates' demands, assuming all associated costs (d_2^1), pay the ransom demanded by the pirates (d_2^2), or request the Navy support to release the boat and crew (d_2^3).

Probabilities. The Defender's beliefs about a successful attack ($\Theta = 1$), conditional on $d_1 \in D_1$, are modeled with probabilities $p_D(\Theta = 1|a^1, d_1^1) = 0.4$, $p_D(\Theta = 1|a^1, d_1^2) = 0.1$, and $p_D(\Theta = 1|a^1, d_1^3) = 0.05$. Besides, the Defender assumes that $p_D(\Theta = 1|a^i) = 0.5, i \in \{2, 3, 4\}$.

Defender Modeling of Attacker's Beliefs. D assesses A's beliefs about an attack being successful as $P_A(\Theta = 1|a^1, d_1^1) \sim Be(40, 60)$, $P_A(\Theta = 1|a^1, d_1^2) \sim Be(10, 90)$, $P_A(\Theta = 1|a^1, d_1^3) \sim Be(50, 950)$, $P_A(\Theta = 1|a^i) \sim Be(1, 1), i \in \{2, 3, 4\}$; and D's actions after a successful attack as $P_A(D_2|d_1^1, A = a^1, \Theta = 1) \sim Dir(1, 1, 1)$, $P_A(D_2|d_1^2, A = a^1, \Theta = 1) \sim Dir(0.1, 4, 6)$, $P_A(D_2|d_1^3, A = a^1, \Theta = 1) \sim Dir(0.1, 1, 10)$, $P_A(D_2|a^i, \Theta = 1) \sim Dir(1, 1, 1)$ for $i = \{2, 3, 4\}$.

Utilities.

- Defender: We employ a constant risk-averse utility function $u_D(d_1, \theta_1, d_2) = -\exp(\rho_D \times c_D(d_1, \theta_1, d_2))$, being ρ_D the risk-aversion coefficient. $c_D(\cdot)$ represents the monetary costs in M €associated with d_1, θ_1 and d_2 (Table 1a). We use $\rho_D = 0.1$ and scale u_D to maintain it positive.
- Attacker: The Defender assumes that the pirates are risk-prone with utility function $u_A(a_1, \theta_1, d_2) = \exp(\rho_A \times c_A(a_1, \theta_1, d_2))$, where $c_A(\cdot)$ represents the costs associated with A_1, Θ_1, and D_2 in M €(Table 1b) and ρ_A is modeled as $\rho_A \sim \mathcal{U}(0, 20)$. In Table 1b, a^j represents an attack on the Defender when $j = 1$, or on the other boats when $j \in \{2, 3, 4\}$.

Table 1. Monetary costs for Defender (c_D) and Attacker (c_A).

(a) $c_D(d_1, \theta_1, d_2)$

D_1	d_1^1	d_1^1	d_1^1	d_1^1	d_1^2	d_1^2	d_1^2	d_1^2	d_1^3	d_1^3	d_1^3	d_1^3	d_1^4
Θ_1	1	1	1	0	1	1	1	0	1	1	1	0	
D_2	d_2^1	d_2^2	d_2^3		d_2^1	d_2^2	d_2^3		d_2^1	d_2^2	d_2^3		
c_D	15.16	2.3	4.28	0	17.25	4.39	6.37	0.05	19.39	6.53	8.51	0.15	0.5

(b) $c_A(a_1, \theta_1, d_2)$

A_1	a^0	a^j	a^j	a^j	a^j
Θ_1		1	1	1	0
D_2		1	1	1	
c_A	0	0.97	2.27	-1.28	-0.53

Results. Using the proposed methodology, we are able to find the optimal first-stage decision d_1^* and the optimal second-stage decision $d_2^*(d_1^*, a_1, \theta)$ given the first one, the eventual attack and its outcome. The results obtained using APS are analogous to those in [20] using Monte Carlo simulation, as we identify the same optimal decision for D_1, demonstrating that both methods yield equivalent solutions. In particular, we find that $d_1^* = d_1^2$, indicating that the optimal proactive measure for the boat owner is to employ private protection with one armed individual. In the second stage, the best course of action, in the event of a successful attack, is to pay the ransom demanded by the pirates ($d_2^* = d_2^2$), as it incurs lower costs for the boat owner compared to either not accepting the pirates' demands (d_2^1) or requesting navy support (d_2^3).

Additionally, we provide the distribution of decisions made by the Attacker in response to the first Defender's decision d_1 illustrated in Table 2. We observe that as the level of proactive measures increases, the probability of the Defender being targeted decreases, indicating that the defense acts as a deterrent. Specifically, when the Defender decides not to implement any protection (d_1^1), the probability of being attacked is very similar to that of the other boats. However, when the Defender employs private protection with a team of two armed personnel (d_1^3), the likelihood of a pirate attack becomes almost negligible. Furthermore, as expected, we observe that when the probability of an attack on the Defender decreases, due to more protective measures or an alternative route, the likelihood of attack is distributed evenly among the other boats. Again, these results are analogous to those in [20], since we obtain a similar forecast of A's distribution of optimal decisions on A_1.

Table 2. Distribution of optimal decisions for A_1.

	a_1^0	a_1^1	a_1^2	a_1^3	a_1^4
d_1^1	0.001	0.221	0.283	0.254	0.241
d_1^2	0.009	0.018	0.336	0.297	0.340
d_1^3	0.008	0.001	0.308	0.355	0.328
d_1^4	0.008	0.000	0.327	0.331	0.334

5.2 Air-Combat Game

Consider a case in which a Defender operates an aerial defense system to protect a strategic infrastructure from an Attacker trying to destroy it. We support D

in allocating infrastructure protection resources as structured in Fig. 2 therefore illustrating the model in Sect. 3. In its initial decision (D_1), the Defender will choose the number of anti-aircraft missiles to allocate; in parallel, the Attacker will decide (A_1) how many drones to deploy, targeting the anti-air defenses of the infrastructure to leave it unprotected against future attacks. Θ_1 represents whether the drones succeed in eliminating the Defender's anti-aircraft missiles. At the second stage, the Defender selects the number of drones to surveil the aerial space of the infrastructure in anticipation of a direct attack. Conversely, the Attacker decides (A_2) how many drones to send for a bombing targeting the infrastructure. Both decisions, along with the success in neutralizing the anti-air defenses, will determine if the infrastructure is destroyed (Θ_2). The utilities of both the agents will depend on the decisions (D_1, D_2, A_1, A_2) and the uncertainties (Θ_1, Θ_2) as next detailed.

Parametric Setup

Agents. The Defender (D) allocates resources to protect its infrastructure against aerial attacks. The Attacker (A) aims to select the optimal strategy to destroy such infrastructure.

Decisions. The alternatives available to D and A are:

- D_1: Install minimal anti-air systems (d_1^1), build a moderate anti-air defense (d_1^2), or a substantial one (d_1^3).
- A_1: Do not attack the aerial defenses (a_1^1), target the anti-missiles with a few drones (a_1^2), or a high number of drones (a_1^3).
- D_2: No surveillance (d_2^1), minimal surveillance (d_2^2), or substantial surveillance (d_2^3) by drones.
- A_2: No direct attack on infrastructure (a_2^1), low-intensity (a_2^2), or high-intensity (a_2^3) attack.

Probabilities. The Defender's belief about the success of an attack (Θ_1) destroying her anti-air defenses, conditional on d_1 and a_1, is modeled through the probabilities in Table 3a. Additionally, the probabilities that her infrastructure is destroyed, conditional on a_2, d_2 and whether the attack on the anti-air missiles was successful, are outlined in Tables 3b and 3c.

Defender Modeling of Attacker's Beliefs. D models A's beliefs about $\Theta_1 = 1$ as $P_A(\Theta_1|d_1, a_1) \sim Be(\sigma_1, \sigma_2)$, with parameters (σ_1, σ_2) for each d_1 and a_1 displayed in Table 4a. Similarly, D assesses A's beliefs about $\Theta_2 = 1$ as $P_A(\Theta_2|d_2, a_2, \theta_1) \sim Be(\sigma_3, \sigma_4)$, with parameters (σ_3, σ_4) in Tables 4b and 4c. When there is only one number (0 or 1) for the corresponding combination of arguments in a table, it means that a degenerate distribution is used for that number. Additionally, the Defender models the Attacker's assumptions regarding her decisions to install anti-air missiles through $P_A(d_1) \sim Dir(1, 7500, 2499)$. Correspondingly, the Defender assesses the Attacker's beliefs about D_2 using $P_A(d_2|d_1, a_1, \theta_1) \sim Dir(\beta_1, \beta_2, \beta_3)$, with parameters in Tables 5a and 5b.

Table 3. Successful attack probabilities to anti-aerial missiles and infrastructure.

(a) $p_D(\Theta_1=1|d_1,a_1)$

	a_1^1	a_1^2	a_1^3
d_1^1	0.00	0.70	0.80
d_1^2	0.00	0.35	0.45
d_1^3	0.00	0.30	0.40

(b) $p_D(\Theta_2=1|d_1,a_1,\Theta_1=1)$

	a_2^1	a_2^2	a_2^3
d_2^1	0.00	0.85	0.95
d_2^2	0.00	0.40	0.60
d_2^3	0.00	0.05	0.15

(c) $p_D(\Theta_2=1|d_1,a_1,\Theta_1=0)$

	a_2^1	a_2^2	a_2^3
d_2^1	0.00	0.25	0.35
d_2^2	0.00	0.05	0.07
d_2^3	0.00	0.01	0.02

Table 4. Beta parameters for D's modeling of A's beliefs on $\Theta_1 = 1$, $\Theta_2 = 1$.

(a) $P_A(\Theta_1=1|d_1,a_1)$

	a_1^1	a_1^2	a_1^3
d_1^1	0	(50,50)	(85,15)
d_1^2	0	(20,80)	(30,70)
d_1^3	0	(10,90)	(25,75)

(b) $P_A(\Theta_2=1|d_2,a_2,\Theta_1=1)$

	a_2^1	a_2^2	a_2^3
d_2^1	0	1	1
d_2^2	0	(70,30)	(90,10)
d_2^3	0	(50,50)	(85,15)

(c) $P_A(\Theta_2=1|d_1,a_1,\Theta_1=0)$

	a_2^1	a_2^2	a_2^3
d_2^1	0	(5,95)	(10,90)
d_2^2	0	(2,98)	(8,92)
d_2^3	0	(1,99)	(5,95)

Table 5. Dirichlet parameters for D's modelling of A's beliefs on D_2.

(a) $P_A(d_2|d_1,a_1,\Theta_1=1)$

	a_1^2	a_1^3
d_1^1	(1,499,9500)	(1,999,9000)
d_1^2	(1,299,9700)	(1,399,9600)
d_1^3	(1,99,9900)	(1,199,9800)

(b) $P_A(d_2|d_1,a_1,\Theta_1=0)$

	a_1^1	a_1^2	a_1^3
d_1^1	(1000,1000,1000)	(1,4499,4500)	(1,1999,8000)
d_1^2	(5000,2500, 2500)	(3500,3500,2000)	(1,1399,8600)
d_1^3	(7500,1500,1000)	(2000,3000,8000)	(1,1199,8800)

Utilities.

– Defender. We use a constant risk-averse utility function $u_D(d_1,a_1,\theta_1,d_2,a_2,\theta_2) = -\exp(\rho_D \times c_D(d_1,a_1,\theta_1,d_2,a_2,\theta_2))$. c_D denotes the monetary costs related to d_1, a_1, θ_1, d_2, a_2 and θ_2 and ρ_D is the risk-aversion coefficient. We consider that $\rho_D = 0.2$ and scale the utility to ensure it remains positive.
– Attacker. The Defender posits that the pirates exhibit risk-prone behavior, characterized by the utility function $u_A(d_1,a_1,\theta_1,d_2,a_2,\theta_2) = \exp(\rho_A \times c_A(d_1,a_1,\theta_1,d_2,a_2,\theta_2))$. c_A indicates the costs linked to d_1, a_1, θ_1, d_2, a_2 and θ_2. To address uncertainties regarding the Attacker's preferences, ρ_A is modeled as $\rho_A \sim \mathcal{U}(1,2)$.

Assume that $c_D(a_1,d_1,a_2,d_2,\theta_1,\theta_2) = f_D(\theta_2) + f_D(\theta_1) + (1+f_D(a_1))f_D(d_1) + (1+f_D(a_2))f_D(d_2)$, where f_D is a monetary value function for all decisions and uncertainties for the Defender. Similarly, $c_A(a_1,d_1,a_2,d_2,\theta_1,\theta_2) = f_A(\theta_2) + f_A(\theta_1) - (1+f_A(d_1))f_A(a_1) - (1+f_A(d_2))f_A(a_2)$ with f_A being the equivalent of f_D for the Attacker. The expressions for c_D and c_A are in M €. The values of f_D and f_A are displayed in Table 6.

Table 6. Values of f_D and f_A for decisions and uncertainties.

	d_1^1	d_1^2	d_1^3	a_1^1	a_1^2	a_1^3	$\Theta_1=0$	$\Theta_1=1$	d_2^1	d_2^2	d_2^3	a_2^1	a_2^2	a_2^3	$\Theta_2=0$	$\Theta_2=1$
f_D	0.0	0.3	0.6	0.0	0.4	0.5	0.0	2.0	0.0	0.3	1.5	0.0	0.5	0.8	0.0	7.5
f_A	0.0	0.4	0.9	0.0	0.2	0.5	0.0	1.5	0.0	0.45	0.7	0.0	0.85	1.0	0.0	3.5

Results. Optimal defender decisions are found utilizing the approach in Sect. 3. For the first stage, the optimal decision is to build a moderate anti-air defense ($d_1^* = d_1^2$). Then, for the second stage, we observe that the best decision for the Defender just depends on whether the attack targeting the anti-air defense is successful or not. Specifically, if $\Theta_1 = 1$, the best decision for the Defender would be to protect the infrastructure with maximum defense ($d_2^* = d_2^3$) to prevent its destruction regardless her first decision and the first stage attack. Otherwise, if the attack is not successful, the Defender will always opt to defend with a moderate defense ($d_2^* = d_2^2$) in anticipation of a future attack.

We also obtain probabilistic forecasts for the Attacker's decisions a_1 and a_2. The distribution $p_D(a_1)$ of the Attacker's first-stage action is $(0.000, 0.715, 0.285)$. The probability distribution over the Attacker's second-stage action, conditioned on the first-stage decisions and uncertainty, $p_D(a_2|d_1, a_1, \Theta_1)$, is presented in Table 7. We observe that the Attacker will always try to destroy the anti-air missiles to facilitate an easier subsequent attack, with the highest probability being that this attack is moderate (a_1^2). Regarding the Attacker's second stage, we observe different behavior depending on whether the attack on the anti-air missiles is successful or not. If the attack is successful, the Attacker is more likely to decide to attack the infrastructure with maximum capacity to complete its destruction. Conversely, if the attack is not successful, the behavior depends on the intensity of the first attack. If this one involved an intense assault (a_1^3), the attacker might decide to either cease the offensive or continue with the attack with similar probability. Alternatively, if the attacker did not target the infrastructure initially, which is unlikely given the above forecast, they will more likely attack with maximum capacity. If the first attack was moderate (a_1^2), the attacker is also slightly more likely to follow up with a powerful attack.

Table 7. Attacker's optimal decision distribution for A_2.

(a) $A_2^*(d_1^*, a_1, \Theta_1 = 1)$			(b) $A_2^*(d_1^*, a_1, \Theta_1 = 0)$		
	a_1^2	a_1^3	a_1^1	a_1^2	a_1^3
d_1^1	(.000,.175,.825)	(.000,.225,.775)	(.078,.100,.822)	(.320,.025,.655)	(.485,.047,.468)
d_1^2	(.000,.173,.827)	(.000,.217,.783)	(.004,.145,.815)	(.068,.102,.830)	(.513,.045,.442)
d_1^3	(.000,.145,.855)	(.000,.190,.810)	(.012,.188,.800)	(.200,.085,.715)	(.500,.045,.455)

5.3 Assessment of Decision Outcomes in the Examples

Similar patterns seem to emerge in both examples. In D_1, the optimal defense strategy for both models involves using a moderate defense, rather than opting for none/minimal defense or a more protective approach that would incur in additional costs. This moderate defense not only provides a good likelihood of protection against attacks but also serves as a deterrent in the first case concerning international piracy. Furthermore, D_2 is contingent on both scenarios to the outcome θ_1. In the international piracy example, if $\Theta_1 = 0$, the defender does not take further action as the game is concluded; however, if $\Theta_1 = 1$, then D will always pay the ransom. In the air combat model, if $\Theta_1 = 1$, the defender implements maximum protection, while if $\Theta_1 = 0$, the defense in D_2 remains moderate.

6 Discussion

We have provided APS algorithms to solve multi-stage sequential and simultaneous games with incomplete information from an ARA perspective, illustrating them with two examples related to international piracy and air-combat modeling. The algorithms provide not only optimal decisions for the Defender at various stages but also forecasts of the Attacker's responses.

The use of APS algorithms for expected utility maximization tend to be advantageous in relevant situations, as their performance usually scales better with the cardinality of decision sets compared to directly optimizing Monte Carlo approximations of expected utilities, as shown in [10]. Hence, the methodology proposed could be relevant in security games that model scenarios in Adversarial Machine Learning [9], where decision sets are usually continuous and assumptions of complete information are questionable. In such settings, ARA provides a natural methodology that effectively handles uncertainty about the attackers' actions. Similarly, an APS algorithmic approach based on ARA might enhance the modeling of uncertainties related to competitive agents in multi-agent reinforcement learning [18]. In future research, we plan to extend this work to general security games modeled as proper bi-agent influence diagrams, beyond the stylized sequential and simultaneous templates analyzed here. We hypothesize that these can be solved by implementing an APS for each Defender decision node in the diagram and a random APS for each Attacker decision node. Additionally, we shall develop new approaches to handle continuous domains, overcoming the limitations of the current approach, laying the foundation for a general algorithmic framework for decision analysis and games using APS. Finally, other avenues for future research include extending the framework to scenarios with multiple attackers and replacing Metropolis-Hastings samplers with Hamiltonian Monte Carlo methods within APS.

Acknowledgments. EU's Horizon 2020 project No. 101021797(STARLIGHT), the AMALFI FBBVA project, AFOSR award FA-9550-21-1-0239, AFOSR-EOARD award FA8655-21-1-7042, and the Spanish Ministry of Science program PID2021-124662OB-I00. DRI supported by the AXA-ICMAT Chair. JMC supported by a fellowship from "la Caixa" Foundation (ID100010434), whose code is LCF/BQ/DI21/11860063.

Disclosure of Interests. No conflict of interests to be declared.

References

1. Kadane, J.B., Larkey, P.D.: Subjective probability and the theory of games. Manag. Sci. **28**(2), 113–120 (1982)
2. Raiffa, H.: The Art and Science of Negotiation. Harvard University Press, Cambridge, MA (1982)
3. Raiffa, H., Richardson, J., Metcalfe, D.: Negotiation Analysis: The Science and Art of Collaborative Decision Making. Harvard University Press, Cambridge, MA (2002)
4. Hargreaves-Heap, S., Varoufakis, Y.: Game Theory: A Critical Introduction. Routledge, New York (2004)
5. Angeletos, G.M., Lian, C.: Forward guidance without common knowledge. Am. Econ. Rev. **108**(9), 2477–2512 (2018)
6. Banks, D., Gallego, V., Naveiro, R., Insua, D.R.: Adversarial risk analysis: an overview. Wiley Interdiscip. Rev. Comput. Stat. **14**(1), e1530 (2022)
7. Gil, C., Parra-Arnau, J.: An adversarial-risk-analysis approach to counterterrorist online surveillance. Sens. **19**(3) (2019)
8. Roponen, J., Salo, A.: Adversarial risk analysis for enhancing combat simulation models. J. Mil. Stud. **6**(2), 82–103 (2015)
9. Insua, D.R., Naveiro, R., Gallego, V., Poulos, J.: Adversarial machine learning: Bayesian perspectives. J. Am. Stat. Assoc. **118**(543), 2195–2206 (2023)
10. Ekin, T., Naveiro, R., Insua, D.R., Torres-Barrán, A.: Augmented probability simulation methods for sequential games. Eur. J. Oper. Res. **306**(1), 418–430 (2023)
11. Bielza, C., Müller, P., Insua, D.R.: Decision analysis by augmented probability simulation. Manag. Sci. **45**(7), 995–1007 (1999)
12. Stahl, D.O., Wilson, P.W.: On players' models of other players: theory and experimental evidence. Games Econ. Behav. **10**(1), 218–254 (1995)
13. Stahl, D.O., Wilson, P.W.: Experimental evidence on players' models of other players. J. Econ. Behav. Organ. **25**(3), 309–327 (1994)
14. Chacon, J.: The modal age of statistics. Int. Stat. Rev. **88**(1), 122–141 (2020)
15. French, S., Insua, D.R.: Statistical Decision Theory. Wiley, Hoboken (2000)
16. Roberts, G.O., Smith, A.F.: Simple conditions for the convergence of the Gibbs sampler and Metropolis-Hastings algorithms. Stoch. process. appl. **49**(2), 207–216 (1994)
17. Virtanen, K., Karelahti, J., Raivio, T.: Modeling air combat by a moving horizon influence diagram game. J. Guid. Control Dyn. **29**(5), 1080–1091 (2006)
18. Gallego, V., Naveiro, R., Insua, D.R.: Reinforcement learning under threats. In: Proc. AAAI Conf. Artif. Intell. vol. 33, pp. 9939–9940 (2019)
19. Müller, P., Sansó, B., De Iorio, M.: Optimal Bayesian design by inhomogeneous Markov chain simulation. J. Am. Stat. Assoc. **99**(467), 788–798 (2004)
20. Sevillano, J.C., Insua, D.R., Rios, J.: Adversarial risk analysis: the Somali pirates case. Decis. Anal. **9**(2), 86–95 (2012)

Non-maximizing Policies that Fulfill Multi-criterion Aspirations in Expectation

Simon Dima[1], Simon Fischer[2], Jobst Heitzig[3], and Joss Oliver[4(✉)]

[1] École Normale Supérieure, Paris, France
simon.dima@ens.psl.eu
[2] Cologne, Germany
[3] Potsdam Institute for Climate Impact Research, Potsdam, Germany
heitzig@pik-potsdam.de
[4] London, UK
joss.oliver62@gmail.com

Abstract. In dynamic programming and reinforcement learning, the policy for the sequential decision making of an agent in a stochastic environment is usually determined by expressing the goal as a scalar reward function and seeking a policy that maximizes the expected total reward. However, many goals that humans care about naturally concern multiple aspects of the world, and it may not be obvious how to condense those into a single reward function. Furthermore, maximization suffers from specification gaming, where the obtained policy achieves a high expected total reward in an unintended way, often taking extreme or nonsensical actions.

Here we consider finite acyclic Markov Decision Processes with multiple distinct evaluation metrics, which do not necessarily represent quantities that the user wants to be maximized. We assume the task of the agent is to ensure that the vector of expected totals of the evaluation metrics falls into some given convex set, called the aspiration set. Our algorithm guarantees that this task is fulfilled by using simplices to approximate feasibility sets and propagate aspirations forward while ensuring they remain feasible. It has complexity linear in the number of possible state–action–successor triples and polynomial in the number of evaluation metrics. Moreover, the explicitly non-maximizing nature of the chosen policy and goals yields additional degrees of freedom, which can be used to apply heuristic safety criteria to the choice of actions. We discuss several such safety criteria that aim to steer the agent towards more conservative behavior.

Keywords: multi-objective decision-making · planning · Markov decision processes · AI safety · satisficing · convex geometry

---

S. Fischer and J. Oliver—Independent Researcher

Supplementary Information The online version contains supplementary material available at https://doi.org/10.1007/978-3-031-73903-3_8.

© The Author(s), under exclusive license to Springer Nature Switzerland AG 2025
R. Freeman and N. Mattei (Eds.): ADT 2024, LNAI 15248, pp. 113–127, 2025.
https://doi.org/10.1007/978-3-031-73903-3_8

1 Introduction

In typical reinforcement learning (RL) and dynamic programming problems an agent is trained or programmed to solve tasks encoded by a single real-valued reward function that it shall maximize. However, many tasks are not easily expressed by such a function [12], human preferences are hard to learn and may not be easy to aggregate across stakeholders [5], and maximizing a misspecified objective may fall prey to reward hacking [1] and Goodhart's law [11], leading to unintended side-effects and potentially harmful consequences.

In this work, we study a particular *aspiration-based* approach to agent design. We assume an existing task-specific world model in the form of a fully observed Markov Decision Process (MDP), where the task is not encoded by a reward function but instead a *multi-criterion evaluation function* and a bounded, convex subset of its range, called an *aspiration set*, that can be thought of as an "instruction" [4] to the agent. Aspiration-type goals can also naturally arise from subtasks in complex environments even if the overall goal is to maximize some objective, when the complexity requires a hierarchical decision-making approach whose highest level selects subtasks that turn into aspiration sets for lower hierarchal levels.

In our version of aspiration-based agents, the goal is to make the *expected value* of the total with respect to this evaluation function fall within the aspiration set, and select from this set according to certain performance and safety criteria. They do so step-wise, exploiting recursion equations similar to the Bellman equation. Thus our approach is like multi-objective reinforcement learning (MORL), with a primary aspiration-based objective and at least one secondary objective incorporated via action-selection criteria [16]. Unlike MORL, the components of the evaluation function (called *evaluation metrics*) are not objectives in the sense of targets for maximization. Rather, an aspiration formulated w.r.t. several evaluation metrics might correspond to a single objective (e.g., "make a cup of tea"). Also, at no point does an aspiration-based agent aggregate the evaluation metrics into a single value. Instead, any trade-offs are built into the aspiration set itself, similar to what [6] call a "safety specification". For example, aspiring to buy a total of 10 oranges and/or apples for at most EUR 1 per item could be encoded with the aspiration set $\{(o, a, c) : o, a \geq 0; c \leq o + a = 10\}$.

A similar set-up to ours has been used in [9] which has used reinforcement learning to find a policy whose expected discounted reward vector lies inside a convex set. Instead of a reinforcement *learning* perspective, we use a model-based *planning* perspective and design an algorithm that explicitly calculates a policy for solving the task, based on a model of the environment.[1] Also, the approach in [9] is concerned with guaranteed bounds for the distance between received rewards and the convex constraints in terms of the number of iterations, whereas we focus on guaranteeing aspiration satisfaction in a fixed number of computational steps, providing a verifiable guarantee in the sense of [6].

---

[1] Nevertheless, our algorithm can also be straightforwardly adapted to learning.

Other agent designs that follow a non-maximization-goal-based approach include quantilizers, decision transformers and active inference. Quantilizers are agents that use random actions in the top $n\%$ of a "base distribution" over actions, sorted by expected return [13]. The goal for decision transformers is to make the expected return equal a particular value R_{target} [3]. The goal for active inference agents is to produce a particular probability distribution of observations [14]. While the goal space for quantilizers and decision transformers, being based on a single real-valued function, is often too restricted for many applications, that of active inference agents (all probability distributions) appears too wide for the formal study of many aspects of aspiration-based decision-making. Our approach is of intermediary complexity.

An important consideration in this work, to ensure tractability in large environments, is also the *computational* complexity in the number of actions, states, and evaluation metrics. We will see that for our algorithm, the preparation of an episode has linear complexity in the number of possible state–action–successor transitions and (conjectured and numerically confirmed) linear average complexity in the number of evaluation metrics, and then the additional per-time-step complexity of the policy is linear in the number of actions, constant in the number of states, and polynomial in the number of evaluation metrics.

Our work also affects the emerging AI safety/alignment field, which views unintended consequences from maximization, e.g., reward hacking and Goodhart's law, as a major source of risk once agentic AI systems become very capable [1].

2 Preliminaries

Environment. An *environment* $E = (\mathcal{S}, s_0, \mathcal{S}_\top, \mathcal{A}, \mathcal{T})$ is a finite Markov Decision Process without a reward function, consisting of a finite *state space* $\mathcal{S}$, an *initial state* $s_0 \in \mathcal{S}$, a nonempty subset $\mathcal{S}_\top \subseteq \mathcal{S}$ of *terminal states*, a nonempty finite *action space* $\mathcal{A}$, and a function $\mathcal{T} : (\mathcal{S} \setminus \mathcal{S}_\top) \times \mathcal{A} \to \Delta(\mathcal{S} \setminus \{s_0\})$ specifying *transition probabilities*: $\mathcal{T}(s,a)(s')$ is the probability that taking action a from state s leads to state s'. We assume that the environment is acyclic, i.e., that it is impossible to reach a given state again after leaving it. We fix some environment E and write $s' \sim s, a$ to denote that s' is distributed according to $\mathcal{T}(s, a)$.

Policy. A *(memory-based) policy* is given by some nonempty finite set $\mathcal{M}$ of memory states internal to the agent, an initial memory state $m_0 \in \mathcal{M}$ and a function $\pi : \mathcal{M} \times (\mathcal{S} \setminus \mathcal{S}_\top) \to \Delta(\mathcal{A} \times \mathcal{M})$ that maps each possible combination of memory state $m \in \mathcal{M}$ and (environment) state $s \in \mathcal{S} \setminus \mathcal{S}_\top$ to a probability distribution over combinations of actions $a \in \mathcal{A}$ and successor memory states $m' \in \mathcal{M}$. Let $\Pi_{\mathcal{M}}$ be the set of all policies with memory space $\mathcal{M}$. The special class of *Markovian* or memoryless policies is obtained when $\mathcal{M}$ is a singleton. Policies which are both Markovian and deterministic are called *pure Markov policies*, and amount to a function $(\mathcal{S} \setminus \mathcal{S}_\top) \to \mathcal{A}$. We denote by Π^0 the set of Markovian policies and by Π^p the set of all pure Markov policies.

Evaluation, Delta, Total. A *(multi-criterion) evaluation function* for the environment is a function $f : (\mathcal{S}\setminus\mathcal{S}_\mathsf{T})\times\mathcal{A}\times(\mathcal{S}\setminus\{s_0\}) \to \mathbb{R}^d$ where $d \geq 1$. The quantity $f(s, a, s')$ is called the *Delta* received under transition $(s, a) \to s'$. It represents by how much certain evaluation metrics change when the agent takes action a in state s and the successor state is s'. Let us fix f for the rest of the paper. The *(received) Total* of a trajectory $h = (m_0, s_0, a_1, m_1, s_1, \ldots, a_T, m_T, s_T)$ is then the cumulative Delta received along the trajectory,

$$\tau(h) = \sum_{t=1}^{T} f(s_{t-1}, a_t, s_t). \tag{1}$$

Value Functions. Given a policy π and an evaluation function f, the state value function $V^\pi : \mathcal{M} \times \mathcal{S} \to \mathbb{R}^d$ is defined as the expected Total accumulated in future steps while following policy π. In particular, $V^\pi(m_0, s_0)$ is the expected Total of the whole trajectory: $V^\pi(m_0, s_0) = \mathbb{E}(\tau)$. Likewise, we define the action value function $Q^\pi : \mathcal{S} \times \mathcal{A} \times \mathcal{M} \to \mathbb{R}^d$. These satisfy the Bellman equations

$$V^\pi(m, s) = \mathbb{E}_{a, m' \sim \pi(m, s)} \left(Q^\pi(s, a, m') \right), \tag{2}$$

$$Q^\pi(s, a, m) = \mathbb{E}_{s' \sim s, a} \left(f(s, a, s') + V^\pi(m, s') \right). \tag{3}$$

They can be calculated by backwards induction since the environment is finite and acyclic, with the base case $V^\pi(m, s) = 0$ for terminal states $s \in \mathcal{S}_\mathsf{T}$. For memoryless policies, we will elide the argument m since $\mathcal{M}$ is a singleton.

The Delta and Total are analogous to reward and return in an MDP with a reward function, and the value functions V and Q are defined as usual, albeit with vector instead of scalar arithmetic. However, while we do assume that the evaluation metrics represent some aspects relevant to the task at hand, we do *not* assume that they represent a form of utility which it is always desirable to increase. Accordingly, the agent's goal will be specified by the user not as a maximization task, but rather as a set of linear constraints on the expected sums of the evaluation metrics, which we call an *aspiration*.

3 Fulfilling Aspirations

Aspirations, Feasibility. An *(initial) aspiration* is a convex polytope $\mathcal{E}_0 \subset \mathbb{R}^d$, representing the values of the expected total $\mathbb{E}\tau$ which are considered acceptable. We say that a policy π *fulfills* the aspiration when it satisfies $V^\pi(m_0, s_0) \in \mathcal{E}_0$. To answer the question of whether it is possible to fulfill a given aspiration, we introduce *feasibility sets*. The *state-feasibility set* of $s \in \mathcal{S}$ is the set of possible values for the expected future Total from s, under any memory-based policy: $\mathcal{V}(s) = \{V^\pi(m, s) \mid \mathcal{M} \text{ finite set}; m_0, m \in \mathcal{M}; \pi \in \Pi_\mathcal{M}\}$; likewise, we define the *action-feasibility set* $\mathcal{Q}(s, a)$. It is straightforward to verify that $\mathcal{V}(s \in \mathcal{S}_\mathsf{T}) = \{0\}$, and that the following recursive equations hold for $s \notin \mathcal{S}_\mathsf{T}$:

$$\mathcal{Q}(s, a) = \mathbb{E}_{s' \sim s, a}\left(f(s, a, s') + \mathcal{V}(s')\right) = \sum_{s'} \mathcal{T}(s, a)(s') \cdot (f(s, a, s') + \mathcal{V}(s')), \tag{4}$$

$$\mathcal{V}(s) = \bigcup_{p \in \Delta(\mathcal{A})} \mathbb{E}_{a \sim p} \mathcal{Q}(s, a) = \bigcup_{p \in \Delta(\mathcal{A})} \sum_{a \in \mathcal{A}} p(a)\mathcal{Q}(s, a). \tag{5}$$

In this, we use set arithmetic: $r\mathcal{X} + r'\mathcal{X}' = \{rx + r'x' \mid x \in \mathcal{X}, x' \in \mathcal{X}'\}$ for $r, r' \in \mathbb{R}$ and $\mathcal{X}, \mathcal{X}' \subset \mathbb{R}^d$. It is clear that feasibility sets are convex polytopes.

Algorithm 1 General scheme for fulfilling feasible aspiration sets

1: **procedure** FULFILLSTATEASPIRATION($s \in \mathcal{S} \setminus \mathcal{S}_T$, nonempty $\mathcal{E} \subseteq \mathcal{V}(s)$)
2:   Find suitable $\mathcal{E}_a \subseteq \mathcal{Q}(s,a)$ for all $a \in \mathcal{A}$, and $p \in \Delta(\mathcal{A})$, s.t. $\mathbb{E}_{a \sim p}(\mathcal{E}_a) \subseteq \mathcal{E}$.
3:   Draw action a from distribution p and do FULFILLACTIONASPIRATION($s, a, \mathcal{E}_a$).
4: **procedure** FULFILLACTIONASPIRATION($s \in \mathcal{S}, a \in \mathcal{A}$, nonempty $\mathcal{E}_a \subseteq \mathcal{Q}(s,a)$)
5:   Find suitable $\mathcal{E}_{s'} \subseteq \mathcal{V}(s')$ for all $s' \in \mathcal{S}$ s.t. $\mathbb{E}_{s' \sim s, a}(f(s, a, s') + \mathcal{E}_{s'}) \subseteq \mathcal{E}_a$.
6:   Execute action a and observe successor state s'.
7:   If s' is terminal, stop; else do FULFILLSTATEASPIRATION($s', \mathcal{E}_{s'}$).

Aspiration Propagation. Algorithm scheme 1 shows a general manner of fulfilling a feasible initial aspiration, starting from a given state or state-action pair. It memorizes and updates aspirations $\mathcal{E}$ and $\mathcal{E}_a$, initially equalling $\mathcal{E}_0$. The agent alternates between being in a certain state s with a *state-aspiration* $\mathcal{E} \subseteq \mathcal{V}(s)$, and being in a state s and having chosen, but not yet performed, an action a, with an *action-aspiration* $\mathcal{E}_a \subseteq \mathcal{Q}(s,a)$. Although this algorithm is written as two mutually recursive functions, it can be formally implemented by a memory-based policy that memorizes the current aspiration $\mathcal{E}$ or $\mathcal{E}_a$.

The way the aspiration set $\mathcal{E}$ is *propagated* between steps is the key part. The two directions of aspiration propagation are slightly different: in state-aspiration to action-aspiration propagation, shown on line 2, the agent may choose the probability distribution (p) over actions, whereas in action-aspiration to state-aspiration propagation, shown on line 5, the next state is determined by the environment with fixed probabilities ($\mathcal{T}$).

The correctness of Algorithm 1 follows from the requirements of lines 2 and 5; that these are possible to fulfill is a consequence of equations (5) and (4). Feasibility of aspirations is maintained as an invariant.

To implement this scheme, we have to specify how to perform aspiration propagation. The procedure used to select action-aspirations $\mathcal{E}_a$ and state-aspirations $\mathcal{E}_{s'}$ should preferably allow some control over how the size of these sets changes over time. On one hand, preventing aspiration sets from shrinking too fast preserves a wider range of acceptable behaviors in later steps[2], but on the other hand, keeping the aspiration sets somewhat smaller than the feasibility sets also provides immediate freedom in the choice of the next action, as detailed in Sect. 4.2. An additional challenge is posed by the complex shape of the feasibility sets, which we must handle in a tractable way.

Approximating Feasibility Sets by Simplices. Let $\mathrm{conv}(X)$ denote the convex hull of any set $X \subseteq \mathbb{R}^d$. Given any tuple $\mathcal{R}$ of $d+1$ memoryless *reference policies* $\pi_1, \ldots, \pi_{d+1} \in \Pi^0$, we define *reference simplices* in evaluation-space as $\mathcal{V}^{\mathcal{R}}(s) = \mathrm{conv}\{V^\pi(s) \mid \pi \in \mathcal{R}\}$ and $\mathcal{Q}^{\mathcal{R}}(s,a) = \mathrm{conv}\{Q^\pi(s,a) \mid \pi \in \mathcal{R}\}$. It is immediate that these are subsets of the convex feasibility sets $\mathcal{V}(s)$ resp. $\mathcal{Q}(s,a)$, and that

---

[2] However, algorithm 1 remains correct even if the aspiration sets shrink to singletons.

$$\mathcal{V}^{\mathcal{R}}(s) \subseteq \bigcup_{p \in \Delta(\mathcal{A})} \mathbb{E}_{a \sim p} \left(\mathcal{Q}^{\mathcal{R}}(s, a) \right), \tag{6}$$

$$\mathcal{Q}^{\mathcal{R}}(s, a) \subseteq \mathbb{E}_{s' \sim s, a} \left(f(s, a, s') + \mathcal{V}^{\mathcal{R}}(s') \right). \tag{7}$$

These imply that we can replace every occurrence of $\mathcal{V}$ and $\mathcal{Q}$ in algorithm 1 with $\mathcal{V}^{\mathcal{R}}$ resp. $\mathcal{Q}^{\mathcal{R}}$, obtaining a correct algorithm to guarantee fulfillment of any initial aspiration $\mathcal{E}_0$ provided it intersects the reference simplex $\mathcal{V}^{\mathcal{R}}(s_0)$.

It turns out that the latter can be guaranteed by a proper choice of reference policies, and that we can always use pure Markov policies for this:

Lemma 1. *For any state s, we have $\mathcal{V}(s) = \text{conv}\{V^\pi(s) \mid \pi \in \Pi^p\}$.*

Proof. Any memory-based policy admits a Markovian policy with the same occupancy measure and hence the same expected Total [7]. Hence $\mathcal{V}(s) = \{V^\pi(s) \mid \pi \in \Pi^0\}$. A convex set is the convex hull of its vertices, so $\mathcal{V}(s) = \text{conv}\{V^\pi(s) \mid \exists y \in \mathbb{R}^d, \pi = \arg\max_{\pi' \in \Pi^0}(y \cdot V^{\pi'}(s))\}$. Finally, maximal policies may be taken to be deterministic, which concludes the argument. □

As a consequence, for any aspiration set $\mathcal{E}$ intersecting the feasibility set $\mathcal{V}(s_0)$, there exists a tuple $\mathcal{R}$ of pure Markov reference policies such that $\mathcal{V}^{\mathcal{R}}(s_0) \cap \mathcal{E} \neq \emptyset$. Section 5 describes a heuristic algorithm for finding such reference policies.

We now turn to explaining a way to enact the aspiration-propagation steps needed in lines 2 and 5 of algorithm 1, based on shifting and shrinking.

4 Propagating Aspirations

4.1 Propagating Action-Aspirations to State-Aspirations

To implement algorithm schema 1, we first focus on line 5 in procedure FULFILLACTIONASPIRATION, which is the easier part. Given a state-action pair s, a and an action-aspiration set $\mathcal{E}_a \subseteq \mathcal{Q}^{\mathcal{R}}(s, a)$, we must construct nonempty state-aspiration sets $\mathcal{E}_{s'} \subseteq \mathcal{V}^{\mathcal{R}}(s')$ for all possible successor states s', such that $\mathbb{E}_{s' \sim s, a}(f(s, a, s') + \mathcal{E}_{s'}) \subseteq \mathcal{E}_a$. We assume that all reference simplices are non-degenerate, i.e. have full dimension d. This is almost surely the case if there are sufficiently many actions and these have enough possible consequences.

Tracing Maps. Under this assumption, we define tracing maps $\rho_{s,a,s'}$ from the reference simplex $\mathcal{Q}^{\mathcal{R}}(s, a)$ to $\mathcal{V}^{\mathcal{R}}(s')$. Since domain and codomain are simplices, we can choose $\rho_{s,a,s'}$ to be the unique affine linear map that maps vertices to vertices, $\rho_{s,a,s'}(Q^{\pi_i}(s, a)) = V^{\pi_i}(s')$. For any point $e \in \mathcal{Q}^{\mathcal{R}}(s, a)$, it follows from equation (3) that $\mathbb{E}_{s' \sim s, a}(f(s, a, s') + \rho_{s,a,s'}(e)) = e$. Accordingly, to propagate aspirations of the form $\mathcal{E}_a = \{e\}$, it is sufficient to just set $\mathcal{E}_{s'} = \{\rho_{s,a,s'}(e)\}$. However, for general subsets $\mathcal{E}_a \subseteq \mathcal{Q}^{\mathcal{R}}(s, a)$, the set $\mathbb{E}_{s' \sim s, a}(f(s, a, s') + \rho_{s,a,s'}(\mathcal{E}_a))$ is in general strictly larger than $\mathcal{E}_a$, and hence we map $\mathcal{E}_a$ in a different way.

For this, choose an arbitrary "anchor" point $e \in \mathcal{E}_a$; here we let e be the average of the vertices, $C(\mathcal{E}_a)$, but any other standard type of center would also work (e.g. analytic center, center of mass/centroid, Chebyshev center). Now, let

$\mathcal{X}_{s'} = \mathcal{E}_a - e + \rho_{s,a,s'}(e)$ be shifted copies of the action-aspiration. We would like to use the $\mathcal{X}_{s'}$ as state-aspirations, and indeed they have the property that

$$\mathbb{E}_{s'\sim s,a}(f(s,a,s') + \mathcal{X}_{s'}) = \mathbb{E}_{s'\sim s,a}(f(s,a,s') + \rho_{s,as'}(e) - e + \mathcal{E}_a) \quad (8)$$
$$= e - e + \mathbb{E}_{s'\sim s,a}(\mathcal{E}_a) = \mathcal{E}_a \quad \text{as } \mathcal{E}_a \text{ is convex.} \quad (9)$$

This is almost what we want, but it might be that $\mathcal{X}_{s'}$ is not a subset of $\mathcal{V}^{\mathcal{R}}(s')$. To rectify this, we opt for a shrinking approach, setting $\mathcal{E}_{s'} = r_{s'} \cdot (\mathcal{E}_a - e) + \rho_{s,a,s'}(e)$ for the largest $r_{s'} \in [0,1]$ such that the result is a subset of $\mathcal{V}^{\mathcal{R}}(s')$. As $\mathcal{E}_{s'}$ does not depend on any other successor state $s'' \neq s'$, we can wait until the true s' is known and only compute $\mathcal{E}_{s'}$ for that s', which saves computation time.

Proposition 1. *Given all values $V^{\pi_i}(s)$ and $Q^{\pi_i}(s,a)$ and both the $k_c \geq d+1$ constraints and $k_v \geq d+1$ vertices defining $\mathcal{E}_a$, the shrinking version of action-to-state aspiration propagation has time complexity $O([k_c^{1.5}d + (dk_v)^{1.5}]L)$, where L is the precision parameter defined in [15]. If $\mathcal{E}_a$ is a simplex, this is $O(d^3 L)$.*

Proof. A linear program (LP) with m constraints and n variables has time complexity $O(f(m,n)L)$, where $f(m,n) = (m+n)^{1.5}n$ and L is a precision parameter [15]. $C(\mathcal{E}_a)$ can be calculated with an LP with $m = k_c + 1$ and $n = d+1$. Finding the convex coefficients for $\rho_{s,a,s'}$ requires solving a system of $d+1$ linear equations, needing time $O(d^\omega)$, where $2 \leq \omega < 2.5$ is the exponent of matrix multiplication. Finding the shrinking factor r is an LP with $m = (d+1)k_v$ and $n = 1$, and then computing the constraints and vertices of $\mathcal{E}_{s'}$ is $O(d(k_c + k_v))$. In all, this gives a time complexity of $O(Lf(k_c, d) + d^\omega + Lf(dk_v, 1) + d(k_c + k_v)) \leq O(L(k_c + d)^{1.5}d + d^\omega + L(dk_v)^{1.5}) \leq O(L(k_c^{1.5}d + (dk_v)^{1.5}))$. □

4.2 Choosing Actions and Action-Aspirations

This is the core of our construction. In state s with state-aspiration $\mathcal{E}$, the policy probabilistically selects an action a and action-aspiration $\mathcal{E}_a$ as follows:

(i) For each of the $d+1$ vertices $V^{\pi_i}(s)$ of the state's reference simplex $\mathcal{V}^{\mathcal{R}}(s)$, find a *directional action set* $\mathcal{A}_i \subseteq \mathcal{A}$ containing those actions whose reference simplices $\mathcal{Q}^{\mathcal{R}}(s,a)$ "lie between" $\mathcal{E}$ and the vertex $V^{\pi_i}(s)$.
(ii) From the full action set $\mathcal{A}_0 = \mathcal{A}$ and from each directional action set $\mathcal{A}_i$ ($i = 1 \ldots d+1$) independently, use some arbitrary, potentially probabilistic procedure to select one element, giving *candidate actions* $a_0, \ldots, a_{d+1}$.
(iii) For each candidate a_i, compute an action-aspiration $\mathcal{E}_{a_i}$ by shifting and shrinking the state-aspiration $\mathcal{E}$ into the reference simplex $\mathcal{Q}^{\mathcal{R}}(s, a_i)$.
(iv) Compute a probability distribution $p \in \Delta(\{0, \ldots, d+1\})$ that makes $\mathbb{E}_{i\sim p}\mathcal{E}_{a_i} \subseteq \mathcal{E}$ and has as large a p_0 as possible.
(v) Finally execute candidate action a_i with probability p_i ($i = 0 \ldots d+1$) and memorize its action-aspiration $\mathcal{E}_{a_i}$.

We now describe steps (i)–(iv) in detail. Algorithm 2 has the corresponding pseudocode, and Fig. 1 illustrates the involved geometric operations.

Algorithm 2 Action selection ("shrinking" variant)
───
Require: reference simplex vertices $V^{\pi_i}(s)$, $Q^{\pi_i}(s,a)$; state s; state-aspiration $\mathcal{E}$;
shrinking schedule $r_{\max}(s)$.
1: $x \leftarrow C(\mathcal{E}_s);\ \mathcal{E}' \leftarrow \mathcal{E} - x$ ▷ *center, shape*
2: **for** $i = 0, \ldots, d+1$ **do**
3:     $\mathcal{A}_i \leftarrow$ DIRECTIONALACTIONSET(i)
4:     $a_i \leftarrow$ SAMPLECANDIDATEACTION$(\mathcal{A}_i)$ ▷ *any (probabilistic) choice from $\mathcal{A}_i$*
5:     $\mathcal{E}_{a_i} \leftarrow$ SHRINKASPIRATION(a_i, i) ▷ *action-aspiration*
6: solve linear program $p_0 = \max!,\ \sum_{i=0}^{d+1} p_i \mathcal{E}_{a_i} \subseteq \mathcal{E}$ for $p \in \Delta(\{0,\ldots,d+1\})$
7: sample $i \sim p$ and execute a_i
8: memorize $m \leftarrow (s, a_i, \mathcal{E}_{a_i})$

9: **procedure** DIRECTIONALACTIONSET(i)
10:     **if** $i = 0$, **return** $\mathcal{A}$
11:     $\mathcal{X}_i \leftarrow \mathrm{conv}\{x, V^{\pi_i}(s)\}$ ▷ *segment from m to ith vertex*
12:     **return** $\{a \in \mathcal{A} \mid Q^{\mathcal{R}}(s,a) \cap \mathcal{X}_i \neq \emptyset\}$ ▷ *actions with feasible points on $\mathcal{X}_i$*

13: **procedure** SHRINKASPIRATION(a_i, i)
14:     $v \leftarrow V^{\pi_i}(s)$ if $i > 0$, else $C(Q^{\mathcal{R}}(s,a_i))$ ▷ *vertex or center of target*
15:     $y \leftarrow v - x$ ▷ *shifting direction*
16:     $r \leftarrow \max\{r \in [0, r_{\max}(s)] \mid \exists \ell \geq 0, \ell y + x + r\mathcal{E}' \subseteq Q^{\mathcal{R}}(s,a)\}$ ▷ *size of largest shifted and shrunk copy of $\mathcal{E}$ that fits into $Q^{\mathcal{R}}(s,a)$*
17:     $\ell \leftarrow \min\{\ell \geq 0 \mid \ell y + x + r\mathcal{E}' \subseteq Q^{\mathcal{R}}(s,a)\}$ ▷ *shortest dist. that makes it fit in*
18:     **return** $\ell y + x + r\mathcal{E}'$ ▷ *shift and shrink*
───

(i) Directional Action Sets. Compute the average of the vertices of $\mathcal{E}$, $x = C(\mathcal{E})$, and the "shape" $\mathcal{E}' = \mathcal{E} - x$. For $i = 0$, put $\mathcal{A}_i = \mathcal{A}$. For $i > 0$, let $\mathcal{X}_i$ be the segment from x to the vertex $V^{\pi_i}(s)$ and check for each action $a \in \mathcal{A}$ whether its reference simplex $Q^{\mathcal{R}}(s,a)$ intersects $\mathcal{X}_i$, using linear programming. Let $\mathcal{A}_i$ be the set of those a, which is nonempty since $\pi_i(s) \in \mathcal{A}_i$ by definition of $V^{\pi_i}(s)$.

(ii) Candidate Actions. For each $i = 0 \ldots d+1$, use an arbitrary, possibly probabilistic procedure to select a candidate action $a_i \in \mathcal{A}_i$. Since $\mathcal{A}_0 = \mathcal{A}$, any possible action might be among the candidate actions. This freedom can be used to improve the policy in terms of the evaluation metrics, e.g., by choosing actions that are expected to lead to a rather low variance of the received Total. It can also be used to incorporate additional action-selection criteria that are unrelated to these evaluation metrics, to increase overall safety, e.g., by preferring actions that avoid unnecessary side-effects. We'll discuss this in Sect. 6.

(iii) Action-aspirations. Given direction i and action a_i, we now aim to select a large subset $\mathcal{E}_{a_i} \subseteq Q^{\mathcal{R}}(s,a_i)$ that fits into a shifted version $z_i + \mathcal{E}'$. We determine a direction y towards which to shift $\mathcal{E}_s$: if $i = 0$, towards the average of the vertices, $C(Q^{\mathcal{R}}(s,a_i))$, otherwise towards the reference vertex $V^{\pi_i}(s)$. This is lines 14 to 15 of Alg. 2. As before, we use shrinking: find the largest shrinking factor $r \in [0, r_{\max}(s)]$ for which there is a shifting distance $\ell \geq 0$ so that $\ell y + r\mathcal{E}' \subseteq Q^{\mathcal{R}}(s,a_i)$, using a linear program with two variables r, ℓ, and then find the smallest such ℓ for that r using another linear program. The "shrinking schedule"

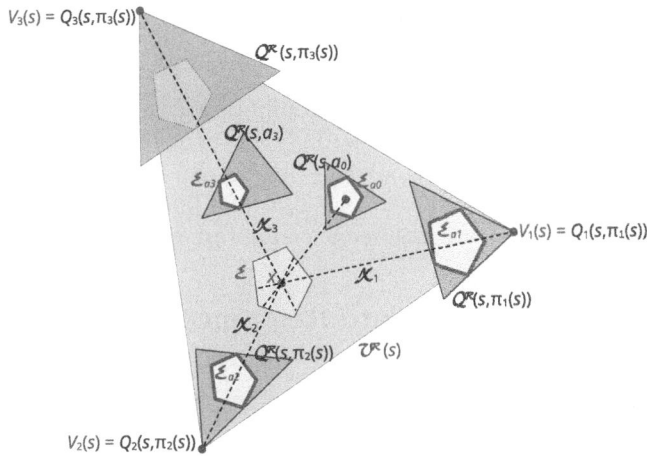

Fig. 1. Construction of action-aspirations $\mathcal{E}_a$ from state-aspiration $\mathcal{E}$ and reference simplices $\mathcal{V}^{\mathcal{R}}(s)$ and $\mathcal{Q}^{\mathcal{R}}(s,a)$ by shifting and shrinking. See main text for details.

$r_{\max}(s) \in [0,1]$ might be used to enforce some amount of shrinking to increase the freedom for choosing the action mixture, which is the next step.

(iv) Suitable Mixture of Candidate Actions. We next find probabilities $p_0, \ldots, p_{d+1}$ for the candidate actions so that the corresponding mixture of aspirations $\mathcal{E}_{a_i}$ is a subset of $\mathcal{E}_s$, $\sum_{i=0}^{d+1} p_i \mathcal{E}_{a_i} \subseteq \mathcal{E}_s$. We show below that this equation has a solution. Because we want the action a_0 that was chosen freely from the whole action set $\mathcal{A}$ to be as likely as possible, we maximize its probability p_0. This is done in line 6 of Algorithm 2 using linear programming. Note that the smaller the sets $\mathcal{E}_a$, the looser the set inclusion constraint and thus the larger p_0. We can influence the latter via the shrinking schedule $r_{\max}(s)$, for example by putting $r_{\max}(s) = (1 - 1/T(s))^{1/d}$ where $T(s)$ is the remaining number of time steps at state s, which would reduce the amount of freedom (in terms of the volume of the aspiration set) linearly and end in a point aspiration for terminal states.

Lemma 2. *The linear program in line 6 of Algorithm 2 has a solution.*

Proof. Because $x \in \mathcal{V}^{\mathcal{R}}(s)$, there are convex coefficients p' with $\sum_{i=1}^{d+1} p'_i(V^{\pi_i}(s) - x) = 0$. As each shifting vector z_i is a positive multiple of $V^{\pi_i}(s) - x$, there are also convex coefficients p with $\sum_{i=1}^{d+1} p_i z_{a_i} = 0$. Since $\mathcal{E}_{a_i} \subseteq \mathcal{E}_s + z_{a_i}$ and $\mathcal{E}_s$ is convex, we then have

$$\sum_i p_i \mathcal{E}_{a_i} \subseteq \sum_i p_i (\mathcal{E}_s + z_{a_i}) \subseteq \mathcal{E}_s + \sum_i p_i z_{a_i} = \mathcal{E}_s. \tag{10}$$

□

Proposition 2. *Given all values $V^{\pi_i}(s)$ and $Q^{\pi_i}(s,a)$ and both the $k_c \geq d+1$ many constraints and $k_v \geq d+1$ many vertices defining $\mathcal{E}$, this part of the* construction (Algorithm 2) has time complexity $O([k_v^{1.5} d^{2.5}|\mathcal{A}| + (k_v k_c)^{1.5} d] L)$. If $\mathcal{E}$ is a simplex, this is $O(d^4 |\mathcal{A}| L)$.

Proof. Using the notation from Prop. 1, computing x is $O(f(k_v, d) L)$. For each i, a, verifying whether $a \in \mathcal{A}_i$ and computing r, ℓ is done by LPs with $m \leq O(d k_v)$ constraints and $n \leq 2$ variables, giving $O(d|\mathcal{A}| f(d k_v, 2) L) = O(d^{2.5} k_v^{1.5} |\mathcal{A}| L)$. The LP for calculating p has $m = d + 4 + k_v k_c$ and $n = d + 2$, hence complexity $O((k_v k_c)^{1.5} d L)$. The other arithmetic operations are of lower complexity. □

5 Determining Appropriate Reference Policies

First, find some feasible aspiration point $x \in \mathcal{E}_0 \cap \mathcal{V}(s_0)$ (e.g., using binary search). We now aim to find policies whose values are likely to contain x in their convex hull, by using backwards induction and greedily minimizing the angle between the vector $\mathbb{E}\tau - x$ and a suitable direction in evaluation space.

More precisely: Pick an arbitrary direction (unit vector) $y_1 \in \mathcal{R}^d$, e.g., uniformly at random. Then, for $k = 1, 2, \ldots$ until stopping:

1. Let π_k be the pure Markovian policy π defined by backwards induction as

$$\pi(s) = \underset{a \in \mathcal{A}}{\operatorname{argmax}} \frac{y_k^\top (Q^\pi(s,a) - x)}{\|Q^\pi(s,a) - x\|_2}. \tag{11}$$

Let $v_k = V^{\pi_k}(s_0)$ be the resulting candidate vertex for our reference simplex. If $k \geq d + 1$, run a primal-sparse linear program solver [18] to determine if $x \in \operatorname{conv}\{v_1, \ldots, v_k\}$. If so, the solver will return a basic feasible solution, i.e. $d + 1$ many vertices that contain x in their convex hull. Let the policies corresponding to the vertices that are part of the basic feasible solution be our reference policies and stop. Otherwise, continue to step 2 below:

2. Let $e_k = (x - v_k)/\|x - v_k\|_2$ be the unit vector in the direction from v_k to x.
3. Let $y_{k+1} = \sum_{i=1}^{k} e_i / k$ be the average of all those directions. Assuming that $x \in \mathcal{V}(s_0)$ and because of the hyperplane separation theorem, choosing directions like this ensures that the algorithm doesn't loop with $v_{l+1} = v_l$ for some l. Also, note that to check $x \in \operatorname{conv}\{v_1, \ldots, v_k, v_{k+1}\}$ it is sufficient to check $v_{k+1} - x \in \operatorname{cone}\{x - v_1, \cdots, x - v_k\}$, the cone generated by the negative of the vertices centred at x. So hunting for a policy whose value lies approximately in the direction y_{k+1} from x gives us a good chance of finding a vertex in the aforementioned cone.

We were not able to prove any complexity bounds for this part, but performed numerical experiments with random binary tree-shaped environments of various time horizons, with only two actions (making the directions $v_k - x$ deviate rather much from y and thus presenting a kind of worst-case scenario) and two possible successor states per step, and uniformly random transition probabilities and Deltas $\in [0,1]^d$. These suggest that the expected number of iterations of the above is $O(d)$, which we thus conjecture to be also the case for other sufficiently well-behaved random environments. Indeed, even if the policies π_k were chosen uniformly *at random* (rather than targeted like here) and the corresponding points $V^{\pi_k}(s_0)$ were distributed uniformly in all directions around x (which is a plausible uninformative prior assumption), then one can show easily (using [17]) that the expected number of iterations would be exactly $2d + 1$.[3]

6 Selection of Candidate Actions

As we have seen in Sect. 4.2 (ii), when choosing actions, we still have many remaining degrees of freedom. Thus, we can use additional criteria to choose actions while still fulfilling the aspirations. We discuss a few candidate criteria here which are related either to gaining information, improving performance, or reducing potential safety-related impacts of implementing the policy.

For many of the criteria, there are *myopic* versions, which only rely on quantities that are already available at each step in the algorithms presented so far, or *farsighted* versions which depend on the continuation policy and thus have to be specifically computed recursively via Bellman-style equations.

Information-Related Criteria. If the used world model is imperfect, one might want the agent to aim to gain knowledge by exploration, e.g. by considering some measure of expected information gain such as the evidence lower bound.

6.1 Performance-Related Criteria

For now, the task of the agent in this paper has been given by specifying aspiration sets for the expected total of the evaluation function. It is natural to consider extensions of this approach to further properties of the trajectory distribution, e.g. by specifying that the variance of the total should be small.

A simple, myopic approach to reducing variance is preferring actions and action-aspirations that are somehow close to the state aspiration $\mathcal{E}$, e.g. by choosing action-aspirations where the Hausdorff distance $d_H(\mathcal{E}_a, \mathcal{E})$ is small. A more principled, farsighted approach would be choosing actions and action-aspirations such that the variance of the resulting total is small. Based on equation (2), the variance can be computed from the total raw second moment M^π as

$$M^\pi(s, a, \mathcal{E}_a) = \mathbb{E}_{s', \mathcal{E}_{s'} \sim s, a, \mathcal{E}_a}\left[\|f(s, a, s')\|_2^2 + 2f(s, a, s') \cdot V^\pi(s', \mathcal{E}_{s'})\right.$$
$$\left. + \mathbb{E}_{a', \mathcal{E}_{a'} \sim \pi(s', \mathcal{E}_{s'})} M^\pi(s', a', \mathcal{E}_{a'})\right], \quad (12)$$
$$\mathrm{Var}(s, a, \mathcal{E}_a) = M^\pi(s, a, \mathcal{E}_a) - \|Q^\pi(s, a, \mathcal{E}_a)\|_2^2. \quad (13)$$

Note that computing this farsighted metric requires knowing the continuation policy π, for which algorithm 2 does not suffice in its current form as it only samples actions. It is however easy to convert it to an algorithm for computing the whole local policy $\pi(s, \mathcal{E})$, which is described in the Supplement.

---

[3] According to [17], the probability that we need exactly $k \geq d+1$ iterations is $f(k-1) - f(k)$ with $f(k) = 2^{1-k} \sum_{\ell=0}^{d-1} \binom{k-1}{\ell}$. Hence $\mathbb{E}k = \sum_{k=d+1}^{\infty} k(f(k-1) - f(k)) = (d+1)f(d) + \sum_{k=d+1}^{\infty} f(k)$. It is then reasonably simple to prove by induction that $\sum_{k=d+1}^{\infty} f(k) = d$, and hence $\mathbb{E}k = 2d+1$.

6.2 Safety-Related Criteria

As mentioned in the introduction, unintended consequences of optimization can be a source of safety problems, thus we suggest to not use any of the criteria introduced in this section as maximization/minimization goals to completely determine the chosen actions; instead, they can be combined into a loss for a softmin action selection policy $\pi_i(a \in \mathcal{A}_i) \propto \exp\big(-\beta \sum_j \alpha_j g_j(a)\big)$, where $g_j(a)$ are the individual criteria. Indeed, in analogy to quantilizers, choosing among adequate actions at random can by itself be considered a useful safety measure, as a random action is very unlikely to be special in a harmful way.

Disordering Potential. Our first safety criterion is related to the idea of "fail safety", and (somewhat more loosely) to "power seeking". More precisely, it aims to prevent the agent from moving the system into a state from which it could make the environment's state trajectory become very unpredictable (bring the environment into "disorder") because of an internal failure, or if it wanted to. We define the *disordering potential* at a state to be the Shannon entropy $H^\pi(s_t)$ of the stochastic state trajectory $S_{>t} = (S_{t+1}, S_{t+2}, \dots)$ that would arise from the policy π which maximizes that entropy:

$$H^\pi(s_t) := \mathbb{E}_{s_{>t}|s_t,\pi}(-\log \Pr(s_{>t}|s_t,\pi)). \tag{14}$$

It is straightforward to compute this quantity using the Bellman-type equations

$$H^\pi(s) = 1_{s \notin S_\top} \mathbb{E}_{a \sim \pi(s)}(-\log \pi(s)(a) + H^\pi(s,a)), \tag{15}$$

$$H^\pi(s,a) = \mathbb{E}_{s' \sim s,a}(-\log \mathcal{T}(s,a)(s') + H^\pi(s')). \tag{16}$$

To find the maximally disordering policy π, we assume $\pi(s')$ and thus $H^\pi(s')$ is already known for all potential successors s' of s. Then $H^\pi(s,a)$ is also known for all a and to find $p_a = \pi(s)(a)$ we need to maximize $f(p) = \sum_a p_a(\mathcal{H}^\pi(s,a) - \log p_a)$ such that $\sum_a p_a = 1$. Using Lagrange multipliers, we find that for all a, $\partial_{p_a} f(p) = \mathcal{H}^\pi(s,a) - \log p_a - 1 = \lambda$ for some constant λ, hence $p_a \propto \exp(H^\pi(s,a))$ is a softmax policy w.r.t. future expected Shannon entropy. Therefore

$$\pi(s)(a) = p_a = \exp(H^\pi(s,a))/Z, \qquad Z = \sum_a \exp(H^\pi(s,a)), \tag{17}$$

$$H^\pi(s) = \log Z = \log \sum_a \exp(H^\pi(s,a)). \tag{18}$$

Deviation from Default Policy. If we have access to a default policy π^0 (e.g. a policy that was learned by observing humans or other agents performing similar tasks), we might want to choose actions in a way that is similar to this default policy. An easy way to measure this is by using the Kullback–Leibler divergence from the default policy π^0 to the agent's policy π. Given that we do not know the local policy $\pi(s)$ yet when we decide how to choose the action in the state s, we use an estimate $\hat{p}_a$ (e.g. $\hat{p}_a = 1/(2+d)$) instead to compute the expected total Kullback–Leibler divergence like

$$\text{KLdiv}(s, \hat{p}_a, a, \mathcal{E}_a) = \log(\hat{p}_a/\pi^0(s)(a)) + \mathbb{E}_{s' \sim s,a} \text{KLdiv}(s, g(s, a, \mathcal{E}_a, s')), \tag{19}$$

$$\text{KLdiv}(s, \mathcal{E}) = 1_{s \notin S_\top} \mathbb{E}_{(a, \mathcal{E}_a) \sim \pi(s, \mathcal{E})} \text{KLdiv}(s, \pi(s, \mathcal{E})(a), a, \mathcal{E}_a), \tag{20}$$

where $g : (s, a, \mathcal{E}_a, s') \mapsto \mathcal{E}_{s'}$ implements action-to-state aspiration propagation.

7 Discussion and Conclusion

7.1 Special Cases

A Single Evaluation Metric. It is natural to ask what our algorithm reduces to in the single-criterion case $d = 1$. The reference simplices can then simply be taken to be the intervals $\mathcal{V}^{\mathcal{R}}(s) = [V^{\pi_{\min}}(s), V^{\pi_{\max}}(s)]$ and $\mathcal{Q}^{\mathcal{R}}(s, a) = [Q^{\pi_{\min}}(s, a), Q^{\pi_{\max}}(s, a)]$, where $\pi_{\min}, \pi_{\max}$ are the minimizing and maximizing policies for the single evaluation metric. Aspiration sets are also intervals, and action-aspirations $\mathcal{E}_a$ are constructed by shifting the state-aspiration $\mathcal{E}$ upwards or downwards into $\mathcal{Q}^{\mathcal{R}}(s, a)$ and shrinking it to that interval if necessary. To maximize p_{a_0}, the linear program for p will assign zero probability to that "directional" action a_1 or a_2 whose $\mathcal{E}_a$ lies in the same direction from $\mathcal{E}$ as $\mathcal{E}_{a_0}$ does. In other words, the agent will mix between the "freely" chosen action a_0 and a suitable amount of a "counteracting" action a_1 or a_2 in the other direction.

Relationship to Satisficing. A subcase of the $d = 1$ case is when the upper bound of the initial state-aspiration interval coincides with the maximal possible value, $\mathcal{E}_0 = [e, V^{\pi_{\max}}(s_0)]$, i.e., when the goal is to achieve an expected Total of at least e. The agent then starts out as a form of "satisficer" [10]. However, due to the shrinking of aspirations over time, aspiration sets of later states s' might no longer be of the same form but might end at values strictly lower than $V^{\pi_{\max}}(s')$ if the interval $[V^{\pi_{\min}}(s'), V^{\pi_{\max}}(s')]$ is wider than the interval $[Q^{\pi_{\min}}(s, a), Q^{\pi_{\max}}(s, a)]$. In other words, even an initial satisficer can turn into a "proper" aspiration-based agent in our algorithm that avoids maximization in more situations than a satisficer would. In particular, also the form of satisficing known as "quantilization" [13], where all feasible expected Totals above some threshold get positive probability, is not a special case of our algorithm. One can however change the algorithm to quantilization behaviour by constructing successor state aspirations differently, by simply applying the tracing map to the interval, $\mathcal{E}_{s'} = \rho_{s,a,s'}[\mathcal{E}_a]$ (which is not feasible for $d > 1$).

Probabilities of Desired or Undesired Events. Another special case is when $d > 1$ but the d evaluation metrics are simply indicator functions for certain events. E.g., assume all Deltas are zero except when reaching a terminal state $s' \in \mathcal{S}_\top$, in which case $f_i(s, a, s') = \mathbb{1}(s' \in E_i)$ for some subset of desirable or undesirable states $E_i \subseteq \mathcal{S}_\top$. If the first $k \le d$ many events are desirable in the sense that we want each probability $\Pr(E_i)$ to be $\ge \alpha$ for some $\alpha < 1$, and the other $d - k$ many events are undesirable in the sense that we want each probability $\Pr(E_j)$ to be $\le \beta$ for some $\beta > 0$, then we can encode this goal as the initial aspiration set $\mathcal{E}_0 = [\alpha, 1]^k \times [0, \beta]^{d-k}$. Note that the different events need not be independent or mutually exclusive, as long as the aspiration is feasible. Aspirations of this type might be especially natural in combination with methods of inductive reasoning and belief revision that are also based on this type of encoding [8]. This could eventually be useful for a "provably safe" approach to AI [6].

7.2 Relationship to Reinforcement Learning

Even though we formulated our approach in a planning framework where the environment's transition probabilities are known and simple enough to admit dynamic programming, it is clear from Eq. (11) that the required reference policies π and corresponding reference vertices $V^\pi(s)$, $Q^\pi(s,a)$ can in principle also be approximated by reinforcement learning techniques such as (deep) expected SARSA in more complex environments or environments that are given only as samplers without access to transition probabilities. For the single-criterion case, preliminary results from numerical experiments suggest that this is indeed a viable approach.[4] Future work should explore this further and also consider using approximate dynamic programming methods (e.g., [2]).

If the expected number of learning passes needed to find the necessary reference policies is indeed $O(d)$ as conjectured (see end of Sect. 5)[5], our approach might turn out to have much lower average complexity than the alternative reinforcement learning approach to convex aspirations from [9], which appears to require up to $O(\epsilon^{-2})$ many learning passes to achieve an error of less than ϵ.

7.3 Invariance Under Reparameterization

For many applications there will be several possible parameterizations of the d-dimensional evaluation space into d different evaluation metrics, so the question arises which parts of our approach are invariant under which types of reparameterizations of evaluation space. It is easy to see that all parts are invariant under affine transformations, except for the algorithm for finding reference policies which is only invariant under orthogonal transformations since it makes use of angles, and except for certain safety criteria such as total variance.

Supplementary Materials. This article is accompanied by a supplementary text, containing alternative versions of the main algorithm, and a supplementary video illustrating the evolution of action-aspirations over a sample episode with $d = 2$. The text, which contains a link to the video, is available at https://doi.org/10.5281/zenodo.13318993, and there is Python code available at https://doi.org/10.5281/zenodo.13221511.

Acknowledgments. We thank the members of the SatisfIA project, AI Safety Camp, the Supervised Program for Alignment Research, and the organizers of the Virtual AI Safety Unconference.

Disclosure of Interests. The authors have no competing interests to declare.

---

[4] Note however that some safety-related action selection criteria, especially those based on information-theoretic concepts, require access to transition probabilities which would then have to be learned in addition to the reference simplices.

[5] Farsighted action selection criteria would require an additional learning pass to also learn the actual policy and the resulting action evaluations.

CRediT Author Statement. Authors are listed in alphabetical ordering and have contributed equally. Simon Dima: Formal analysis, Writing - Original Draft, Writing - Review & Editing. Simon Fischer: Formal analysis, Writing - Original Draft, Writing - Review & Editing. Jobst Heitzig: Conceptualization, Methodology, Software, Writing - Original Draft, Writing - Review & Editing, Supervision. Joss Oliver: Formal analysis, Writing - Original Draft, Writing - Review & Editing.

References

1. Amodei, D., Olah, C., Steinhardt, J., Christiano, P., Schulman, J., Mané, D.: Concrete problems in AI safety. arXiv preprint arXiv:1606.06565 (2016)
2. Bonet, B., Geffner, H.: Solving POMDPs: RTDP-Bel vs. point-based algorithms. In: IJCAI, pp. 1641–1646. Pasadena CA (2009)
3. Chen, L., et al.: Decision transformer: reinforcement learning via sequence modeling (2021)
4. Clymer, J., et al.: Generalization analogies (GENIES): a testbed for generalizing AI oversight to hard-to-measure domains. arXiv preprint arXiv:2311.07723 (2023)
5. Conitzer, V., et al.: Social choice for AI alignment: dealing with diverse human feedback. arXiv preprint arXiv:2404.10271 (2024)
6. Dalrymple, D., et al.: Towards guaranteed safe AI: a framework for ensuring robust and reliable AI systems. arXiv preprint arXiv:2405.06624 (2024)
7. Feinberg, E.A., Sonin, I.: Notes on equivalent stationary policies in Markov decision processes with total rewards. Math. Meth. Oper. Res. **44**(2), 205–221 (1996). https://doi.org/10.1007/BF01194331
8. Kern-Isberner, G., Spohn, W.: Inductive reasoning, conditionals, and belief dynamics. J. Appl. Log. **2631**(1), 89 (2024)
9. Miryoosefi, S., Brantley, K., Daumé, H., Dudík, M., Schapire, R.E.: Reinforcement learning with convex constraints. In: Proceedings of the 33rd International Conference on Neural Information Processing Systems (2019)
10. Simon, H.A.: Rational choice and the structure of the environment. Psychol. Rev. **63**(2), 129 (1956)
11. Skalse, J.M.V., Farrugia-Roberts, M., Russell, S., Abate, A., Gleave, A.: Invariance in policy optimisation and partial identifiability in reward learning. In: International Conference on Machine Learning, pp. 32033–32058. PMLR (2023)
12. Subramani, R., et al.: On the expressivity of objective-specification formalisms in reinforcement learning. arXiv preprint arXiv:2310.11840 (2023)
13. Taylor, J.: Quantilizers: a safer alternative to maximizers for limited optimization (2015). https://intelligence.org/files/QuantilizersSaferAlternative.pdf
14. Tschantz, A., et al.: Reinforcement learning through active inference (2020)
15. Vaidya, P.: Speeding-up linear programming using fast matrix multiplication. In: 30th Annual Symposium on Foundations of Computer Science, pp. 332–337 (1989)
16. Vamplew, P., Foale, C., Dazeley, R., Bignold, A.: Potential-based multiobjective reinforcement learning approaches to low-impact agents for AI safety. Eng. Appl. Artif. Intell. **100**, 104186 (2021)
17. Wendel, J.G.: A problem in geometric probability. Math. Scand. **11**(1), 109–111 (1962)
18. Yen, I.E.H., Zhong, K., Hsieh, C.J., Ravikumar, P.K., Dhillon, I.S.: Sparse linear programming via primal and dual augmented coordinate descent. In: Advances in Neural Information Processing Systems, vol. 28 (2015)

Adversarial Risk Analysis for Automated Lane-Changing in Heterogeneous Traffic

Roi Naveiro[1], David Ríos Insua[2], and William N. Caballero[3] ([✉])

[1] CUNEF Universidad, Madrid, Spain
[2] ICMAT-CSIC, Madrid, Spain
[3] Air Force Institute of Technology, Dayton, OH, USA
william.caballero@afit.edu

Abstract. The global transition from manned to automated vehicles is anticipated to occur incrementally. As such, interactions between automated driving systems (ADS) and manned vehicles motivate related decision-support research. This manuscript develops a novel modeling framework based on adversarial risk analysis focusing on lane-changing maneuvers. An empirical evaluation is provided within a simulated environment serving to validate the modeling approach and solution methodology under a specified traffic scene. Additional model extensions to alternative traffic scenes and different driver-rationality assumptions are provided. In so doing, we showcase the potential for decision theory to manage ADS behavior in heterogeneous traffic. This research also highlights the need for an ADS to prudently balance computational resources between perception and decision tasks.

Keywords: Automated Driving Systems · Adversarial Risk Analysis · Bayesian Decision Analysis

1 Introduction

Due to advancements in computational hardware and artificial intelligence, *automated driving systems* (ADSs) are primed to revolutionize human transportation. However, transitioning to fully automated roadway systems will be a gradual process based on existent technological and societal challenges (Caballero et al., 2023). Herein, we systematically explore one such major issue: the interactions of ADSs and manned vehicles (MVs) in heterogeneous traffic. We set forth an adversarial risk analysis (ARA) framework (Banks et al., 2015) for a subset of interactions occurring under these conditions.

The composition of vehicles on a roadway significantly impacts ADS operations. A homogeneous fleet of ADSs can communicate to optimize and coordinate maneuvers. For example, Hult et al. (2020) demonstrate how level-5 ADSs (see Society of Automobile Engineers, 2018) can collaborate at intersections based on mixed-integer quadratic and non-linear programs. Mariani et al. (2020) extensively discuss coordination research in high-level ADSs. In contrast, the gradual penetration of ADSs onto global roadways introduces complex challenges. ADSs

© The Author(s), under exclusive license to Springer Nature Switzerland AG 2025
R. Freeman and N. Mattei (Eds.): ADT 2024, LNAI 15248, pp. 128–143, 2025.
https://doi.org/10.1007/978-3-031-73903-3_9

and MVs will interact for a significant time posing substantial barriers to traffic system efficiency and numerous unresolved issues. Di and Shi (2021) offer a thorough survey focusing on different scenarios based on the number of ADSs and MVs in a traffic scene. Of their scenarios, we focus on the 1ADS+1MV case, a complex yet manageable setting that highlights key aspects of the heterogeneous-traffic problem priming future decision-theoretic studies in more complex scenarios.

Czarnecki (2018) presents a catalog of potential ADS maneuvers one may consider, e.g., car following, and lane changing. Since lane changing is the least well-studied and most complex (Yu et al., 2018), it is the focus of this paper. In the existing literature, a distinction is typically made between discretionary and mandatory lane changes. The former occurs based upon preference, whereas the latter is compelled by environmental imperatives (e.g., lane mergers). This paper primarily explores discretionary lane changes, thereby considering a wider spectrum of behaviors.

Ahmed (1999) provides a detailed description of lane-changing problems and their initial rule-based solutions. However, our motivation stems from game-theoretic alternatives. Notably, Talebpour et al. (2015) model lane-change decisions based on simultaneous two-agent, non-zero-sum, non-cooperative games involving a vehicle that performs the lane change (the target vehicle) and another behind them on the roadway (the lag vehicle); under strong common-knowledge assumptions, the authors consider complete and incomplete-information games depending upon whether or not vehicle-to-vehicle (V2V) communication is available. Conversely, other authors have modeled the lane-changing problem as a Stackelberg game whereby the target vehicle (leader) makes a lane-change decision and, after observing this decision, the lag vehicle (follower) selects an optimal response. Yu et al. (2018) propose one such Stackelberg game in which the target vehicle makes a binary-lane-change decision on a two-lane road, and the lag vehicle subsequently decides upon their acceleration. The follower's utilities are considered to be known by the leader, except for one parameter that assesses the follower's aggressiveness. A heuristic approach to estimate this parameter is proposed by the authors. Zhang et al. (2019) also consider lane-changing in a two-lane road, but focus on mandatory maneuvers, analyzing the interaction between a single target vehicle and multiple lag vehicles in the target lane. They propose a solution technique that involves solving independent Stackelberg games for each follower. The final solution is based upon the follower that provides the leader with the greatest expected utility. Similar to the work of Yu et al. (2018), lag-vehicle aggressiveness is the only unknown parameter, and is estimated heuristically.

Such game-theoretic models provision important, incremental advancements to the management of heterogeneous traffic. However, they are underpinned by optimistic assumptions: the authors alternatively assume that (1) lag vehicles are expected utility maximizers, and (2) lag vehicle utilities are either known or minimally uncertain. Whereas these approaches may be directly applicable to situations involving multiple ADSs in which V2V communications are active, we contend that, in heterogeneous traffic, the common-knowledge assumption is too strong and the modeling of lag vehicles as perfectly rational decision-makers may be unrealistic. Thus, we propose an alternative decision theoretic strategy based upon ARA that weakens such assumptions. This approach also differs from rigid rule-based standards for lane-changing decision-aiding systems, e.g., ISO 17387. Our goal is to provision prescriptive support to an ADS encountering a lane-changing situation, accounting for conflicting objectives and realistic uncertainty about the lag vehicle's knowledge, beliefs, and preferences. We also extend earlier work by ensuring the lane-change decision depends on the traffic scene.

2 ADS Lane-Changing in Heterogeneous Traffic

This section sets forth our lane-changing problem, reducing the interaction to individual decision problems in accordance with ARA. We alternately expound upon the ADS and MV decision problems and explore their resolutions.

2.1 Traffic Scene

Consider a two-lane road wherein an ADS (A) is the target vehicle facing a lane-change decision. A Stackelberg-game structure is adopted with the ADS as the leader. The ADS's behavior is observed by (the driver of) a lag MV in the adjacent lane. This MV (M) is the follower who, after observing A's action, chooses its preferred response. Our goal is to prescribe decision-making for A.

Figure 1a represents this problem as a bi-agent influence diagram (BAID). The diagram features circular nodes representing uncertainties, hexagonal nodes modeling preferences over consequences, and square nodes illustrating decisions. Arrows pointing to decision nodes indicate that decisions are made with knowledge of predecessor values; those pointing to chance and utility nodes indicate that events or consequences are influenced by predecessors. Different colors highlight issues specific to each agent: white for the ADS and gray for the MV. Notationally, random variables are represented by uppercase letters, while specific instances of these variables are denoted by lowercase letters.

Figure 1a depicts that A makes a lane-change decision, $a \in \mathcal{A}$, which is observed by the lag vehicle who afterwards makes its decision, $m \in \mathcal{M}$. Each vehicle bases its decision on its understanding of the roadway's state. This state is denoted by $\theta \in \Theta$ and summarizes myriad structural, physical, and perceptual conditions. The ADS infers θ from sensor output, Y_A. Similarly, the lag vehicle captures Y_M. These inputs are non-deterministic and assume values $y_A \in \mathcal{Y}_A$ and $y_M \in \mathcal{Y}_M$. The decision spaces, sensor output domains, and state domains vary based on the complexity of the traffic scene model. A specific example is provided in Sect. 3.1. This generic state-observation model allows for the tractable consideration of sensor (perception) error within ADS (MV) decision-making. Interactions between a and m, in conjunction with θ, lead to a probabilistic outcome S assuming some value $s \in \mathcal{S}$. A and M perceive utilities u_A and u_M that depend upon their actions and the realized outcome s.

Based on Fig. 1a, we deduce the problems faced by the ADS (Fig. 1b) and the MV (Fig. 1c). Subsequent sections formalize these ARA reductions and present accompanying algorithmic approaches. The relative advantages and disadvantages of higher-fidelity models are discussed later in this paper.

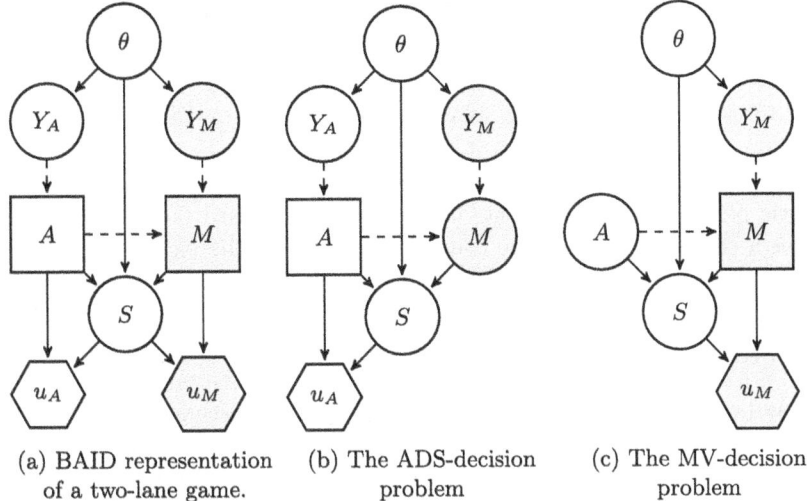

(a) BAID representation of a two-lane game.

(b) The ADS-decision problem

(c) The MV-decision problem

Fig. 1. Influence diagrams of the two-lane game

For simplicity, the remainder of this section considers all random variables to be continuous and uses the term *probability density function* (pdf), but our formulation applies to both continuous and discrete cases via its use of the Riemann-Stieltjes integral for expectations, see, e.g., Berger (2013).

2.2 Formulating the ADS's Decision Problem

Given the observed y_A, to identify its optimal decision-theoretic lane-change, designated $a^*(y_A) \in \mathcal{A}$, the ADS should discern the following elements.

- $u_A(a, s)$: The ADS's realized utility from making decision $a \in \mathcal{A}$, and inducing the outcome $s \in \mathcal{S}$.
- $p_A(\theta|y_A)$: The ADS's pdf of the roadway's state, given the sensor data $y_A \in \mathcal{Y}_A$.
- $p_A(m|a, y_M)$: The ADS's pdf associated with MV making decision $m \in \mathcal{M}$ after having observed $a \in \mathcal{A}$ and given y_M.
- $p_A(y_M|\theta)$: The ADS's pdf of the lag vehicle's sensors output y_M given state θ.
- $p_A(s|m, a, \theta)$: The ADS's pdf for outcome $s \in \mathcal{S}$ given the decisions m and a as well as the state θ.

The cumulative distribution functions (cdfs) corresponding to each probability density function will be denoted as $P_A(\cdot)$. Using the previously defined elements, the ADS goal is to select an $a \in \mathcal{A}$ that maximizes their conditional expected utility, $\mathbb{E}_A[u_A(a, s)|y_A]$[1], 67275

---

[1] $\mathbb{E}_A$ indicates the expectation is taken over quantities unknown to the ADS.

$$a^*(y_A) = \arg\max_{a \in \mathcal{A}} \left[\mathbb{E}_A[u_A(a,s)|y_A]\right] = \arg\max_{a \in \mathcal{A}} \left[\int_\mathcal{S} u_A(a,s)\,\mathrm{d}P_A(s|a,y_A)\right]. \quad (1)$$

However, since $P_A(s|a, y_A)$ is not available, A solves an equivalent representation, based on standard influence diagrams computations (Shachter, 1986)

$$\mathbb{E}_A[u_A(a,s)|y_A] = \iiiint_\Omega u_A(a,s)\,\mathrm{d}P_A(s|m,a,\theta)\,\mathrm{d}P_A(m|a,y_M)\,\mathrm{d}P_A(y_M|\theta)\,\mathrm{d}P_A(\theta|y_A). \quad (2)$$

where Ω is the product of the domains $\Theta, \mathcal{Y}_M, \mathcal{M}$, and $\mathcal{S}$. From a decision-theoretic perspective, all elements in Eq. (2) are standard but $P_A(m|a, y_M)$. It involves a complicated strategic component based on the lag vehicle's decision problem. Additional analysis will facilitate an estimate $\hat{P}_A(m|a, y_M)$.

2.3 Forecasting the MV's Behavior

Per standard ARA procedures, $\hat{P}_A(m|a, y_M)$ can be estimated via the MV's problem. Its exact parameterization is unknown to the ADS, but suppose for the moment that the following elements are known with certainty:

- $u_M(m, s)$: The lag vehicle's utility for taking action $m \in \mathcal{M}$ resulting in the outcome $s \in \mathcal{S}$.
- $p_M(\theta|y_M)$: The MV's posterior probability density function for the roadway's state, given the data y_M perceived by its driver.
- $p_M(s|m, a, \theta)$: MV's probability density function of the outcome $s \in \mathcal{S}$ given the decisions m and a as well as the roadway's state θ.

Based on these and, assuming the lag vehicle's driver is an expected utility maximizer, upon observing a and y_M, the driver will select an action $m^*(a, y_M)$ that maximizes the conditional expected utility,

$$m^*(a, y_M) = \arg\max_{m \in \mathcal{M}} \left[\mathbb{E}_M[u_M(m,s)|a,y_M]\right] = \arg\max_{m \in \mathcal{M}} \left[\int_\mathcal{S} u_M(m,s)\,\mathrm{d}P_M(s|m,a,y_M)\right],$$

or, equivalently, using standard influence diagram reductions (Shachter, 1986),

$$m^*(a, y_M) = \arg\max_{m \in \mathcal{M}} \left[\int_\Theta \int_\mathcal{S} u_M(m,s)\,\mathrm{d}P_M(s|m,a,\theta)\,\mathrm{d}P_M(\theta|y_M)\right].$$

This reduction excludes the MV's beliefs about y_A given a and y_M. Such a simplification is a modeling choice rooted in the assumption that an MV is unable to conceptualize the ADS sensor outputs (e.g., from a convolutional neural network) and is likely to ignore them.

Unfortunately, the ADS does not know the MV's parameterization with certainty, and this equation cannot be resolved exactly. However, if uncertainty about this parameterization is modeled in a Bayesian manner, the ADS may codify their perception of the MV's preferences via a random utility function $U_M(m, s)$ and beliefs via random probability distributions $\Pi_M(\cdot)$, leading to

$$M^*(a, y_M) = \arg\max_{m \in \mathcal{M}} \left[\int_\Theta \int_\mathcal{S} U_M(m, s) \, d\Pi_M(s|m, a, \theta) \, d\Pi_M(\theta|y_M) \right]. \quad (3)$$

One may generate samples of $M^*(a, y_M)$ by simulating random variates from the random functions and solving M's expected utility maximization accordingly. These samples can be used to form an estimate $\hat{P}_A(m|a, y_M)$ of $P_A(m|a, y_M)$.

2.4 Solving the ADS's Decision Problem

Algorithm 1 presents a generic solution approach for Eq. (1); it uses Algorithm 2, considering discrete or discretized action sets, to estimate the probabilities of the MV's actions via Eq. (3). The respective action spaces $(\mathcal{A}, \mathcal{M})$ and supports $(\Theta, \mathcal{S}, \mathcal{Y}_A, \mathcal{Y}_M)$ of the decision and random variables are presented generally. In real-world operations, these sets may be of high-dimensionality and infinite cardinality but, for practical use, it is unlikely that all relevant uncertainties can be considered. An ADS has, at most, a few seconds to perform computations before a decision must be made.

Algorithm 1. Solve ADS problem
---
input y_A, $u_A(a, s)$, $P_A(s|m, a, \theta)$, $P_A(y_M|\theta)$ and $P_A(\theta|y_A)$
for $a \in \mathcal{A}$ and $y_M \in \mathcal{Y}_M$ **do**
   Estimate $P_A(m|a, y_M)$ using, e.g., Algorithm 2
end for
Compute

$$a^*(y_A) = \arg\max_{a \in \mathcal{A}} \iiiint_\Omega u_A(a, s) \, dP_A(s|m, a, \theta) \, d\hat{P}_A(m|a, y_M) \, dP_A(y_M|\theta) \, dP_A(\theta|y_A).$$

return $a^*(y_A)$

Algorithm 2. Predict MV actions
---
input a, y_M, $U_M(m, s)$, $\Pi_M(\theta|y_M)$ and $\Pi_M(s|m, a, \theta)$
for $i = 1$ to N **do**
   Sample $u_M^i(m, s) \sim U_M(m, s)$, $P_M^i(\theta|y_M) \sim \Pi_M(\theta|y_M)$, $P_M^i(s|m, a, \theta) \sim \Pi_M(s|m, a, \theta)$
   Compute

$$m^{*,i}(a, y_M) = \arg\max_{m \in \mathcal{M}} \left[\int_\Theta \int_\mathcal{S} u_M^i(m, s) \, dP_M^i(s|m, a, \theta) \, dP_M^i(\theta|y_M) \right].$$

end for
for $m \in \mathcal{M}$ **do**

$$\hat{P}_A(m|a, y_M) = \frac{1}{m} \sum_{i=1}^N \mathbb{I}(m^{*,i} \leq m)$$

end for
return $\hat{P}_A(m|a, y_M)$

Should an ADS manufacturer allocate substantial computing resources to the decision making problem, higher fidelity representations of the action spaces and supports may be utilized. This affects how $\hat{P}_A(m|a, y_M)$ is estimated and the optimization method in Algorithm 1. For example, if $\mathcal{M}$ and $\mathcal{A}$ are n-dimensional connected sets, for some sampled action $a \in \mathcal{A}$, Algorithm 2 may be adapted for standard optimization techniques. If multiple samples of $M^*(a, y_M)$ have been collected, a kernel density estimate may be leveraged to identify $\hat{P}_A(m|a, y_M)$. In turn, standard techniques may be used to search $\mathcal{A}$ for a high-quality solution, e.g., the coordinate descent or Nelder-Mead approaches (Bazaraa et al., 2013).

Unfortunately, sensor systems require significant computational power and only the most-expensive ADSs will be capable of real-time, high-fidelity modeling. Thus, the effective utilization of lower-fidelity models is a necessity and, akin to standard stochastic optimization techniques, discretizing the action spaces and supports is an efficient means to do so. Assuming the action spaces and supports are of moderate cardinality, resolving Eqs. (1) and (3) can be accomplished expeditiously through parallel computation; Algorithm 1 clearly exhibits this feature. Under a discretized setting, the evaluation of such equations reduces to standard decision-theoretic computations. This technique allows most of the ADS's computational power to be reserved for perception tasks, thereby ensuring that the inputs into Eq. (1) are of the highest possible quality.

In either setting, after the ADS and MV make their decisions and some s is observed, additional evidence is available to the ADS to inform its beliefs. The probability distributions characterizing them can be updated in a Bayesian manner, and such information may underpin subsequent decisions. In this manner, the ADS can constantly learn, adapt, and improve its decisions.

3 A Numerical Evaluation

This section considers a simplified, yet realistic, traffic scene to illustrate the performance of the proposed methodology and demonstrate that the results align with intuitive expectations. Assuming the bulk of computational resources are required for perception tasks, we leverage the discretization technique discussed in Sect. 2 to resolve Eqs. (1) and (3). In our scenario, we posit that an obstacle in the ADS's lane prompts the decision to change lanes.

3.1 Model Parameterization

We begin by specifying foundational elements of the game. We first set forth the action spaces, supports, and outcome spaces before provisioning the parameters required for each of the ADS and MV decision problems reflected in Fig. 1.

We consider $\mathcal{A} = \{a_1, a_2, a_3\}$ such that a_1, a_2 and a_3 correspond to *change lane*, *remain in lane*, and *perform an emergency maneuver*, respectively, whereas $\mathcal{M} = \{m_1, m_2, m_3\}$ such that m_1, m_2 and m_3 correspond to *accelerate*, *decelerate*, and *change lane*, respectively. A four-dimensional state characterizes the roadway such that $\theta = (\theta_1, \theta_2, \theta_3, \theta_4)$ where θ_1 refers to the pavement's wetness, and θ_2, θ_3 and θ_4 represent the number of people in the ADS, the number of people in the MV, and the number of pedestrians at risk in an emergency. Sensor measurements are taken on each θ_i, thereby implying similar supports; however, as discussed subsequently, only measurements on θ_1 are stochastic.

We set $\mathcal{S} = \{s_1, s_2, s_3, s_4, s_5\}$ whereby s_1, s_2, s_3, s_4 and s_5 are associated with a *major accident, minor accident, safely executed interaction, pedestrian casualty* and *crash with obstacle*. A major accident assumes the MV and ADS are essentially destroyed with fatalities occurring in both vehicles; a minor accident assumes minor physical damage to the vehicles and their passengers; a safely executed interaction results in no physical damage or injuries; pedestrian casualties assume the death of all pedestrians in the driving scene; a crash with obstacle injures all ADS passengers and incurs physical damage to the vehicle.

Parameterizing the ADS Decision Problem. Herein, we define the ADS's multi-objective preference model, the outcome consequences upon which its model is based, the ADS's component utility functions, and the ADS's beliefs.

ADS Preference Model. Caballero et al. (2022) present a generic preference model for ADSs. Herein, for the purposes of illustrating the validity of the ARA model, we consider a simplified version taking into account *internal safety* (damage to ADS and injuries/fatalities to its passengers), *external safety* (damage to MV and injuries/fatalities outside ADS) and *trip duration* as the relevant objectives. The consequences for any $s \in \mathcal{S}$ are described via a vector $c_A(s)$ having elements $c_{A,i}(s), i = 1, ..., 8$, that respectively refer to the number of internal injuries and fatalities, the proportion of ADS damage, the number of external injuries and fatalities, the proportion of MV damage, the number of pedestrian fatalities, and the ADS's speed.

Outcome Consequences. Table 1 provides consequences associated with $s \in \mathcal{S}$. For example, when there is a major crash (s_1 occurs), θ_2 people will die in the ADS and θ_3 will die in the MV, both vehicles will be totally damaged but no pedestrian casualties occur and the ADS stops. We note that, for vehicle damage (e.g., $c_{A,3}(s)$), 0.5 designates partial damage. For the ADS's speed (i.e., $c_{A,8}(a, s)$), the three columns designate the corresponding speed depending on the action; 1 is the maximum speed, 0.5 an intermediate speed and 0 designates stopped. These values appear in negative as consequences are expressed as costs.

Table 1. ADS consequences by lane-change outcome.

Outcome	$c_{A,1}(s)$	$c_{A,2}(s)$	$c_{A,3}(s)$	$c_{A,4}(s)$	$c_{A,5}(s)$	$c_{A,6}(s)$	$c_{A,7}(s)$	$c_{A,8}(a_1, s)$	$c_{A,8}(a_2, s)$	$c_{A,8}(a_3, s)$
s_1	0	θ_2	1	0	θ_3	1	0	0	0	0
s_2	θ_2	0	0.5	θ_3	0	0.5	0	0	0	0
s_3	0	0	0	0	0	0	0	-1	-0.5	0
s_4	0	0	0	0	0	0	θ_4	0	0	0
s_5	θ_2	0	0.5	0	0	0	θ_4	0	0	0

ADS Utility. Following Couce-Vieira et al. (2020), we aggregate the consequences additively and use a constant absolute risk averse (CARA) utility leading to an ADS preference model $u_A(a, s) = 1 - \exp(\rho_A \sum_{i=1}^{8} w_{A,i} c_{A,i}(s))$ where ρ_A is the ADS's risk aversion coefficient and $w_{A,i}$ homogenize the criteria.

ADS Beliefs. The ADS must discern two non-strategic probabilistic elements, $p_A(\theta|y_A)$ and $p_A(s|m, a, \theta)$, specified below, along with other undefined elements.

The ADS's sensor measurement y_A is four-dimensional having elements $y_{A,1}$ through $y_{A,4}$, corresponding to each θ_i. We consider two cases depending on whether the pavement is dry, i.e., $\theta_1 = 0$, or wet, i.e., $\theta_1 = 1$. It is assumed that $y_{A,1}$ is a noisy observation of the pavement state, such that the classifier's positive and negative recall, $p(y_{A,1} = 1|\theta_1 = 1)$ and $p(y_{A,1} = 0|\theta_1 = 0)$, both equal 0.95. Combining this likelihood with a uniform prior over θ_1 we arrive at the following posterior probabilities given the sensor measurements

$$p(\theta_1 = 1|y_{A,1} = 1) = 0.95, \qquad p(\theta_1 = 1|y_{A,1} = 0) = 0.05. \tag{4}$$

Moreover, we assume that the ADS is able to precisely capture θ_2, θ_3, and θ_4 with negligible error that need not be expressed stochastically.

We present $p_A(s|m, a, \theta)$ in Table 2. For example, if the road is dry when the ADS changes lanes and the MV accelerates, there is a probability of 2/3 of a major accident (s_1), 1/6 of a minor accident (s_2), 1/6 of a safely executed maneuver (s_3), and 0 of a pedestrian casualty or a crash with the obstacle (s_4 or s_5). Alternatively, if the road is wet when the ADS changes lane and the MV accelerates there is a probability of 5/6 of a major crash, 1/8 of a minor crash, 1/24 of no crash, 0 of a pedestrian casualty, and 0 of a crash with an obstacle.

Table 2. ADS conditional beliefs over outcomes.

		m_1	m_2	m_3
$\theta_1 = 0$	a_1	(2/3, 1/6, 1/6, 0, 0)	(1/6, 1/6, 2/3, 0, 0)	(1/24, 1/24, 11/12, 0, 0)
	a_2	(0, 0, 0.95, 0, 0.05)	(0, 0, 0.95, 0, 0.05)	(2/3, 1/3, 0, 0, 0)
	a_3	(0, 0, 0, 1, 0)	(0, 0, 0, 1, 0)	(0, 0, 0, 1, 0)
$\theta_1 = 1$	a_1	(5/6, 1/8, 1/24, 0, 0)	(1/3, 1/3, 1/3, 0, 0)	(1/3, 1/3, 1/3, 0, 0)
	a_2	(0, 0, 0.7, 0, 0.3)	(0, 0, 0.7, 0, 0.3)	(2/3, 1/3, 0, 0, 0)
	a_3	(0, 0, 0, 1, 0)	(0, 0, 0, 1, 0)	(0, 0, 0, 1, 0)

*Dry pavement, first three rows. Wet pavement, second three rows.

Parameterizing the MV Decision Problem. The previous section developed all ADS-decision-problem parameters but $\hat{p}_A(m|a, y_M)$. Such is the focus herein.

MV Preference Model. The ADS assumes that the MV bases its decision on internal safety and trip duration factors. It is assumed that M acts more selfishly and less comprehensively than an ADS.

Table 3. MV consequences by lane-change outcome.

Outcome	$c_{M,1}(s)$	$c_{M,2}(s)$	$c_{M,3}(s)$	$c_{M,4}(m_1,s)$	$c_{M,4}(m_2,s)$	$c_{M,4}(m_3,s)$
s_1	0	θ_3	1	0	0	0
s_2	θ_3	0	0.5	0	0	0
s_3	0	0	0	-1	-0.5	0
s_4	0	0	0	-1	-0.5	0
s_5	0	0	0	-1	-0.5	0

Outcome Consequences. Table 3 provides the MV consequences. Similar notation to the ADS decision problem is used, with appropriate subscripts for the MV.

MV Utility. A CARA preference model is also used for the MV. Namely, we have $u_M(m,s) = 1 - \exp\left(\rho_M \sum_{i=1}^{4} w_{M,i} c_{M,i}(s)\right)$, wherein ρ_M and $w_{M,i}$ represent a risk-aversion coefficient and homogenizing weights. However, the ADS is uncertain about these parameters. A typical model for $(w_{M,1}, w_{M,2}, w_{M,3}, w_{M,4})$ is a Dirichlet distribution. Under limited information, a defensible model for ρ_M is a uniform distribution. Collectively, these form an uncertain utility model for the MV from the ADS's perspective.

MV Beliefs. The ADS must also determine the random probability models from Sect. 2.3. The quantity $\Pi_M(\theta|y_M)$ is complex to determine directly because it requires the ADS to model the cognitive processing of the MV's driver, but a relevant heuristic bases this quantity on $p_A(\theta|y_A)$ with some additional uncertainty. Thus, we root $\Pi_M(\theta_1|y_M)$ on Eq. (4) such that its values are reflected as means of beta distributions; for the other θ_j-values, previous assumptions enable simplification and the exclusion of their attendant beliefs. Finally concerning $\Pi_M(s|m,a,\theta)$, these beliefs are similarly based on the probabilities $p_A(s|m,a,\theta)$ in Table 2 but with uncertainty described via Dirichlet distributions. For example, when we condition on m_1, a_1, and dry conditions, we use a Dirichlet distribution with parameters $100(2/3, 1/6, 1/6, 0, 0)$.

3.2 Experimental Results

Several experiments[2] were performed utilizing the aforementioned parameterization. A reference evaluation serves to benchmark later experimentation on the impacts of the roadway conditions and the ADS's preference model.

The experiments are constructed similarly to those of Ríos Insua et al. (2022) via a simulation model. It comprises a routine to estimate the probabilities of different MV reactions given the ADS decision; based on these probabilities, a routine maximizes the ADS expected utility. Expected utilities for both the ADS and the MV are approximated through Monte Carlo (MC) simulations. Based on preliminary tests, the number of MC samples required for approximating the MV and ADS expected utility to the desired precision was 1000. Moreover, $p(m|a)$ is also approximated through MC simulation with 1000 samples.

---

[2] Code is available at https://github.com/roinaveiro/ads_lane_changing.

The $w_{A,i}$-values are initially set to $(0.03, 0.21, 0, 0.03, 0.21, 0, 0.21, 0.31)$, reflecting prioritization of fatalities over injuries and egalitarian ADS ethics. By assigning higher weights to fatalities, we underscore the gravity of life-threatening outcomes compared to injuries. Moreover, by treating all individuals with equal importance, we equally value preserving the lives of all individuals in the traffic scene. The ADS risk-aversion coefficient ρ_A is set to 0.5. The number of passengers in the ADS ranges from 0 to 5, whereas the number of passengers in the MV ranges from 1 to 5. The MV utility weights follow a Dirichlet distribution with parameters $\alpha(0.1, 0.5, 0.05, 0.35)$, and ρ_M follows a uniform distribution over $(0.5, 1.5)$. The number of people in the street ranges from 0 to 5. This entails a high-risk setting because the ADS and the MV value speed over safety.

Reference Experiment. The first experiment checks the soundness of our approach with one person in the ADS, one person in the MV and one person in the driving scene. Table 4 provides the ADS's forecasts of the MV actions given the ADS decisions, i.e., with Y_M marginalized.

Table 4. Forecasts $\hat{p}(m|a)$ of MV actions given ADS decisions.

	m_1	m_2	m_3
a_1	0.00	0.86	0.14
a_2	1.00	0.00	0.00
a_3	1.00	0.00	0.00

When the ADS changed lanes, A thinks that the MV will most likely decelerate or (with less probability) change lanes. In turn, when the ADS remains in the same lane or makes an emergency stop, it is certain that the MV will accelerate.

Table 5 presents how often each ADS action is optimal under wet and dry conditions. When the pavement is dry, the ADS changes lanes more often than not. If the pavement is wet, it is more conservative and often decides not to change lanes. Table 6 provides the proportion of attained outcomes, suggesting reasonable performance of the model. When the pavement is dry, the ADS changes lanes more often and risky scenarios s_1 and s_2 occur more frequently. The outcome s_4 never attains, as the ADS never makes an emergency maneuver.

Table 5. ADS-action-optimality proportion in reference experiment.

Action	a_1	a_2	a_3
Dry	0.95	0.05	0.00
Wet	0.05	0.95	0.00

Table 6. Proportion of observed outcomes in reference experiment.

Outcome	s_1	s_2	s_3	s_4	s_5
Dry	0.16	0.16	0.68	0.0	0.01
Wet	0.02	0.02	0.68	0.0	0.29

Impact of Passengers and Pedestrians. We now consider the effect of passengers and pedestrians on ADS behavior, examining the following configurations:

1. 0 ADS passengers, 5 MV passengers, and 1 pedestrian.
2. 4 ADS passengers, 1 MV passenger, and 1 pedestrian.
3. 5 ADS passengers, 5 MV passengers, and 0 pedestrians.

Table 7 provides the percentage of time each action was taken and the proportion of time that each scenario occurred for each configuration.

Table 7. Impact of passengers/pedestrians on ADS behavior and outcomes

Configuration	Pavement	Optimal ADS Action			Observed Outcome of Interaction				
		a_1	a_2	a_3	s_1	s_2	s_3	s_4	s_5
1	Dry	0.95	0.05	0.00	0.04	0.04	0.92	0.00	0.01
	Wet	0.05	0.95	0.00	0.02	0.02	0.68	0.00	0.29
2	Dry	0.00	1.00	0.00	0.00	0.00	0.95	0.00	0.05
	Wet	0.00	1.00	0.00	0.00	0.00	0.70	0.00	0.30
3	Dry	0.00	1.00	0.00	0.00	0.00	0.95	0.00	0.05
	Wet	0.00	1.00	0.00	0.00	0.00	0.70	0.00	0.30

In configuration 1, the ADS decides to change lanes 95% of the time. This is similar to the reference experiment, but the probability of a safely executed maneuver is higher. This occurs because, as the number of MV passengers increases, it tends to be more self-protective and, instead of accelerating, it changes lanes. Alternatively, in configuration 2, the MV has fewer passengers and is less self-protective. Since the ADS cares equally about all passengers and pedestrians, it acts more conservatively, inducing it to remain in the same lane. A similar phenomenon occurs in configuration 3.

Impact of ADS Decision Weights and Number of Pedestrians. We now explore the effect of decision weights on ADS behavior. Namely, we consider an ADS whose weight given to speed is much smaller, and study how the number of pedestrians affects its behavior. Both the ADS and MV have 5 passengers. The ADS weights are $(0.05, 0.31, 0, 0.05, 0.31, 0, 0.2, 0.07)$ and the MV weights are $(0.05, 0.1, 0.05, 0.8)$. Two configurations are considered such that there are zero (configuration 1) and five (configuration 2) pedestrians in the driving scene. Results are compiled in Table 8.

Table 8. Impact of varied decision weights and pedestrians.

Configuration	Pavement	Optimal ADS Action			Observed Outcome of Interaction				
		a_1	a_2	a_3	s_1	s_2	s_3	s_4	s_5
1	Dry	0.00	0.95	0.05	0.00	0.00	0.90	0.05	0.05
	Wet	0.00	0.05	0.95	0.00	0.00	0.04	0.95	0.02
2	Dry	0.00	1.00	0.00	0.00	0.00	0.95	0.00	0.05
	Wet	0.00	1.0	0.00	0.00	0.00	0.70	0.00	0.30

In configuration 1, the ADS often chooses to stay in its lane when the pavement is dry. This is because the probability of a collision with an obstacle is low. However, if the pavement is wet, the ADS opts for an emergency stop. This decision can be made safely since no pedestrians are present. Conversely, configuration 2 has five pedestrians in the road, and the ADS does not consider the emergency stop to be viable.

Impact of the ADS Risk Aversion Coefficient. Utilizing the initial decision weights, we study the effect of the ADS's risk aversion coefficient. We compare $\rho_A = 0.5$ and $\rho_A = 2.5$. Results are presented in Table 9. As expected, when $\rho_A = 2.5$, the ADS acts more conservatively and remains in the lane.

4 Problem and Model Variants

The traffic scene and MV model presented in Sect. 2.1 are but one of multitudinous varieties. We present a few such model variants herein.

Table 9. Impact of ρ_A on ADS behavior and outcomes.

ρ_A	Pavement	Optimal ADS Action			Observed Outcome of Interaction				
		a_1	a_2	a_3	s_1	s_2	s_3	s_4	s_5
0.5	Dry	0.95	0.05	0.00	0.16	0.19	0.68	0.00	0.01
	Wet	0.05	0.95	0.00	0.02	0.02	0.68	0.00	0.29
2.5	Dry	0.0	1.00	0.00	0.00	0.00	0.95	0.00	0.05
	Wet	0.00	1.00	0.00	0.00	0.00	0.70	0.00	0.30

ADS as Lag Vehicle. Consider the case where vehicle roles are reversed: the MV is the target vehicle and the ADS is the lag vehicle. This is akin to a car-following problem, another major ADS structured maneuver (Czarnecki, 2018). The ADS decision problem is much simpler in this setting. As the ADS observes m, forecasting is no longer necessary. Using the notation of Sect. 2.2, the ADS can identify an optimal action by solving

$$a^*(m, y_A) = \arg\max_{a \in \mathcal{A}} \iint u_A(a, s) \, \mathrm{d}P_A(s|m, a, \theta) \, \mathrm{d}P_A(\theta|y_A).$$

This decision problem is analogous to Eq. (3) because the MV and ADS have switched roles. However, because we know the ADS's beliefs, this optimization can be performed without the introduction of additional uncertainties.

Descriptive Model for the Lag Vehicle. Modeling the MV's choices via maximum expected utility may be disputed due to its human control. The same cannot be said of the ADS because its design is user-controlled. Whereas Wang et al. (2022) account for an agent's bounded rationality via the quantal response equilibrium, we propose the MV's actions be discretized and modeled via cumulative prospect theory (Tversky and Kahneman, 1992).

To calculate an action's worth the following elements must be available: $p_M(\theta|y_M)$ and $p_M(s|m,a,\theta)$, defined as in Sect. 2.3, the lag vehicle's value function $v_M(m,s)$; and a probability weighting function w_M that overweighs low probabilities and underweighs high probabilities. Based on these functions, upon observing a and y_M, the MV will select an action $m^*(a, y_M)$ solving

$$\arg\max_{m \in \mathcal{M}} \left[\sum_{\theta \in \Theta} \sum_{s \in \mathcal{S}} v_M(m,s) w_M \left(p_M(s|m,a,\theta) p_M(\theta|y_M) \right) \right]. \quad (5)$$

However, the ADS does not know the MV's parameterization, and this problem cannot be resolved exactly. If uncertainty about this parameterization is modeled in a Bayesian manner, the ADS may codify their perception of the MV's preferences via a random value function $V_M(m,s)$, beliefs via random probability mass functions $\Pi_M(\cdot)$, and a random weighting function W_M leading to the random optimal action $M^*(a, y_M)$ defined by

$$\arg\max_{m \in \mathcal{M}} \left[\sum_{\theta \in \Theta} \sum_{s \in \mathcal{S}} V_M(m,s) W_M \left(\Pi_M(s|m,a,\theta) \Pi_M(\theta|y_M) \right) \right]. \quad (6)$$

Algorithm 2 may solve Eq. (6) using $\hat{P}_A^{PT}(m|a, y_M)$ to estimate $P_A(m|a, y_M)$.

5 Discussion

The gradual introduction of ADSs on roadways ensures the importance of managing ADS and MV interactions. With a focus on lane-changing, we provided a framework for such management based on ARA and multicriteria decision analysis. In so doing, we proposed a tractable means to more faithfully model the underlying uncertainty and address the multitudinous objectives associated with ADS and MV interactions. Numerical experiments were conducted and extensions were provided based on roadway composition and driver rationality.

Our framework performed well empirically; however, to more thoroughly test our methodology, a higher-fidelity simulation must be examined with more in-depth sensitivity analysis. Such testing will enable the modeling elements to be appropriately tuned for real-world implementation. We also mentioned various bottlenecks that need to be resolved by ADS manufacturers to allocate available computational resources. The scale of this issue, as well as the means of its rectification, cannot be identified without conducting testing that closely mirrors real-world conditions.

The empirical testing conducted herein also discretized the environment and action spaces. Extensions to continuous or hybrid cases, possibly based on the work of Cobb (2007), may be fruitful. If the computational burden is manageable, such approaches engender higher fidelity to the associated models and solutions.

Moreover, we used a powerful, but heuristic, method to learn about the lag vehicle's unknown parameters; if the target vehicle can interact with the lag vehicle, e.g. by activating the turn signal and observing the reaction of the lag vehicle, learning about these unknown parameters can be accomplished via Bayesian updating. It is also worthwhile to explore whether ARA approaches with partial information about preferences and beliefs, see e.g. Roponen et al. (2020), are computationally feasible in this domain. Finally, while we only considered the 1ADS+1MV case herein, they serve as the basis for more complex scenarios (e.g., the 1ADS+nMV, mADS+nMV, and three-lane changing problems) to be considered in future research.

Acknowledgments/Disclaimer. Research supported by EU's Horizon 2020 project 815003 Trustonomy, the AMALFI FBBVA project, the Air Force Scientific Office of Research (AFOSR) award FA-9550-21-1-0239, and AFOSR European Office of Aerospace Research and Development award FA8655-21-1-7042. DRI is supported by the AXA-ICMAT Chair and the Spanish Ministry of Science program PID2021-124662OB-I00. Views expressed do not reflect the official position of the US Government; cleared for public release by the USAFA PA office.

References

Ahmed, K.: Modeling drivers' acceleration and lane changing behavior. Ph.D. thesis, MIT (1999). Intelligent Transport

Banks, D., Rios, J., Rios Insua, D.: Adversarial Risk Analysis. Francis Taylor (2015)

Bazaraa, M.S., Sherali, H.D., Shetty, C.M.: Nonlinear Programming: Theory and Algorithms. Wiley, Hoboken (2013)

Berger, J.O.: Statistical Decision Theory and Bayesian Analysis. Springer, New York (2013). https://doi.org/10.1007/978-1-4757-4286-2

Caballero, W., Rios Insua, D., Banks, D.: Decision support issues in automated driving systems. Int. Trans. Oper. Res. **30**, 1216–1244 (2023)

Caballero, W.N., Naveiro, R., Rios Insua, D.: Modeling ethical and operational preferences in automated driving systems. Decis. Anal. **14**, 21–43 (2022)

Cobb, B.R.: Influence diagrams with continuous decision variables and non-gaussian uncertainties. Decis. Anal. **4**(3), 136–155 (2007)

Couce-Vieira, A., Insua, D.R., Kosgodagan, A.: Assessing and forecasting cybersecurity impacts. Decis. Anal. **17**(4), 356–374 (2020)

Czarnecki, K.: Automated driving system task analysis: structured road maneuvers. Technical report, U. Waterloo (2018)

Di, X., Shi, R.: A survey on autonomous vehicle control in the era of mixed-autonomy: from physics-based to AI-guided driving policy learning. Transp. Res. Part C: Emerg. Technol. **125**, 103008 (2021)

Hult, R., Zanon, M., Gros, S., Wymeersch, H., Falcone, P.: Optimisation-based coordination of connected, automated vehicles at intersections. Veh. Syst. Dyn. **58**(5), 726–747 (2020)

Mariani, S., Cabri, G., Zambonelli, F.: Coordination of autonomous vehicles: taxonomy and survey. arXiv preprint arXiv:2001.02443 (2020)

Ríos Insua, D., Caballero, W.N., Naveiro, R.: Managing driving modes in automated driving systems. Transp. Sci. **56**, 1259–1278 (2022)

Roponen, J., Ríos Insua, D., Salo, A.: Adversarial risk analysis under partial information. Eur. J. Oper. Res. **287**(1), 306–316 (2020)

Shachter, R.D.: Evaluating influence diagrams. Oper. Res. **34**(6), 871–882 (1986)

Society of Automobile Engineers: Taxonomy and definitions for terms related to driving automation systems for on-road motor vehicles. Technical report, SAE (2018). https://doi.org/10.4271/j3016_201806

Talebpour, A., Mahmassani, H.S., Hamdar, S.H.: Modeling lane-changing behavior in a connected environment: a game theory approach. Transp. Res. Procedia **7**, 420–440 (2015)

Tversky, A., Kahneman, D.: Advances in prospect theory: cumulative representation of uncertainty. J. Risk Uncertain. **5**(4), 297–323 (1992)

Wang, B., Li, Z., Wang, S., Li, M., Ji, A.: Modeling bounded rationality in discretionary lane change with the quantal response equilibrium of game theory. Transp. Res. Part B: Methodol. **164**, 145–161 (2022)

Yu, H., Tseng, H.E., Langari, R.: A human-like game theory-based controller for automatic lane changing. Transp. Res. Part C: Emerg. Technol. **88**, 140–158 (2018)

Zhang, Q., Langari, R., Tseng, H.E., Filev, D., Szwabowski, S., Coskun, S.: A game theoretic model predictive controller with aggressiveness estimation for mandatory lane change. IEEE Trans. Intell. Veh. **5**(1), 75–89 (2019)

A Fully Bayesian Approach to Bilevel Problems

Vedat Dogan[✉], Steven Prestwich, and Barry O'Sullivan

Insight SFI Research Centre for Data Analytics, School of Computer Science and IT,
University College Cork, Cork, Ireland
{vedat.dogan,steven.prestwich,barry.osullivan}@insight-centre.org

Abstract. The mathematical models of many real-world decision-making problems contain two levels of optimization. In these models, one of the optimization problems appears as a constraint of the other one, called follower and leader, respectively. These problems are known as bilevel optimization problems (BOPs) in mathematical programming and are widely studied by both classical and evolutionary optimization communities. The nested nature of these problems causes many difficulties such as non-convexity and disconnectedness for traditional methods, and requires a huge number of function evaluations for evolutionary algorithms. This paper proposes a fully Bayesian optimization approach, called FB-BLO. We aim to reduce the necessary function evaluations for both upper and lower level problems by iteratively approximating promising solutions with Gaussian process surrogate models at both levels. The proposed FB-BLO algorithm uses the other decision-makers' observations in its Gaussian process model to leverage the correlation between decisions and objective values. This allows us to extract knowledge from previous decisions for each level. The algorithm has been evaluated on numerous benchmark problems and compared with existing state-of-the-art algorithms. Our evaluation demonstrates the success of our proposed FB-BLO algorithm in terms of both effectiveness and efficiency.

Keywords: Bilevel Decision-Making · Bayesian Optimization · Gaussian Process · Stackelberg Games

1 Introduction

The hierarchical and nested nature of real-world decision-making problems has increased the interest of researchers in bilevel optimization. Various applications in different domains are implemented as bilevel optimization problems (BOPs). One of the recent applications is in the domain of operational research for solving the retailer-consumer problem considering multiple strategic retailer involvement and customers' switching behaviours [19]. Another application of BOPs is for modelling the multiple decision-makers behaviour. For instance, in the context of homeland security the objective is to ensure that the country is secure, safe and resilient against terrorism and other possible attacks. Therefore, a government needs to model its own decision-making model to prevent terrorist attacks, considering cost efficiency without causing any harm to the public during operations. The government also needs to take into account the potential outcome of its decisions, such as terrorists' possible reactions after the operation. So, the nested decision-making problem must be modelled considering both levels' objectives and constraints for ease of government. In these kinds of applications, one decision-maker acts as a leader and the other one acts as a follower where follower objectives and constraints appear as a constraint of leaders' objective problems. The government acts as the leader in our example but it could be the contrary in some situations. Mathematically BOPs can be formulated as follows:

$$\begin{aligned}
&\underset{x_u, x_l}{\text{Minimize}} \ F(x_u, x_l') \\
&\text{s.t.} \ \ x_l' \in \underset{x_l}{\text{argmin}} \{ f(x_u, x_l) : g_j(x_u, x_l) \leq 0, \ j = 1, 2, \ldots, J \} \\
&\quad \quad G_k(x_u, x_l') \leq 0, \ k = 1, 2, \ldots, K
\end{aligned} \quad (1)$$

where x_u and x_l' represent the upper level (leader) decisions and optimal lower level (follower) reactions. F and f represent the leader's and follower's objective values, respectively. G and g are leader's and follower's constraints. As we can see from Eq. 1, the follower's optimization problem is nested within the leader's. This structure causes computational challenges compared with single-level optimization problems. For instance, it is proven that BOPs are NP-hard even with linear objective functions [22]. Also, it is difficult to ensure lower level optimality while solving the bilevel problem, which is generally vital as non-optimal lower level solutions may mislead the algorithm. Further, the necessary function evaluation is expected to be extremely high as each upper level solution requires solving the lower level optimization problem. Such problems are relevant to different domains such as environmental economics literature [40] which commonly appears as Stackelberg games in Game Theory [33]. Over the years, many applications have been modelled as BOPs such as machine learning [43], supply chain management [42], security [15], defence industry [17], etc.

BOPs have been widely investigated over the years by mathematical programming literature. For instance, BOPs are simplified using Karush-Kuhn-Tucker (KKT) conditions [3] to reduce the two levels into a single level and solve the resulting problem using standard methods. The disadvantage of these kind of classical approaches is that they require strong assumptions such as smoothness, nonlinearity or convexity of one or both levels. However, these assumptions do not always correspond to real-world problems. For this reason the literature has focused on evolutionary algorithms (EA) to solve BOPs [28] due to their ability to handle the discreteness and nonlinearity of the optimization problems at both levels [31]. Nevertheless, EAs require a large number of function evaluations because of their nature of population-based search strategies to obtain a competitive result. To tackle this problem, EAs have been improved with new technologies that have developed over the years [18,26]. Surrogate-assisted EAs are developed to reduce the number of function evaluations required by EAs and they achieved competitive performance for solving BOPs [6,37]. The majority of them focused on modelling the lower level problem to decrease the necessary evaluation, such as [8]. The authors proposed a method to model with surrogate-assisted EAs for both levels in [20] and they achieved competitive results compared with others.

Even though both proposed EAs and surrogate-assisted EAs at both levels decrease the computational cost, the nature of the population-based search strategy still requires a high number of function evaluations. Regarding that, a few Bayesian approach-based black-box optimization algorithms have been applied to solve BOPs. In [21], standard Bayesian optimization (BO) is used to solve upper level problems and EA is used to solve lower level problems. However, they only used the upper level decisions and objective values to train the upper level model. [12] improved the algorithm by a conditional BO approach to decrease the computational cost by conditioning lower level decisions while approximating upper level objectives. Even though they both showed that the necessary function evaluation decreased while approximating the bilevel global optima, the computational cost at the lower level was still high because of the population-based search strategy at the lower level. To this end, in this paper we propose an algorithm that uses a fully Bayesian optimization approach to solve bilevel optimization problems, called FB-BLO. In contrast to the proposed black-box approaches methods above, we developed a Gaussian process ($\mathcal{GP}$) based surrogate model for both upper and lower level optimization problems. The combination of using $\mathcal{GP}$ models at both upper and lower levels and applying BO significantly reduces the necessary function evaluations and improves the optimization performance. The FB-BLO algorithm is analysed and compared to the well-established algorithm in the literature. Our contributions are as follows:

- The proposed FB-BLO method for bilevel optimization models both upper and lower level problems as black-box. It eliminates the necessity of assumptions generally required by classical approaches and reduces the necessary function evaluations which are high in general by population-based search strategies.
- Assuming both upper and lower level optimization problems as black-box and modelling by a surrogate-assisted model makes the proposed approach promising when upper level objective function F or lower level objective function f or both are unavailable, unreliable or impractical to obtain.
- We also provide the theoretical development of the proposed FB-BLO algorithm with competitive results in terms of accuracy and function evaluations after the experiment by two popular benchmarks for BOPs in the literature.

The rest of this paper is organized as follows. In Sect. 2, we provide a brief literature review by analysing existing methods for solving BOPs and the motivation of this study. Then in Sect. 3, we propose the FB-BLO algorithm and provide the theoretical details. Empirical studies are presented and discussed with results in Sect. 4. Finally, Sect. 5 concludes this article by discussing limitations and future works.

2 Related Work and Motivation

Over the years, many approaches have been developed to tackle BOPs. H. von Stackelberg introduced Stackelberg games [33] connected to Game Theory in 1952 in mathematical programming literature, and the first formulation was proposed in [4] by J. Bracken and J. McGill in 1973. They also proposed the defence application formulation in the following year [5] and used "mathematical programs with optimization problems in the constraints" instead of bilevel problems. BOPs were modelled as mathematical programs at this time and are difficult to handle mathematically because of the hierarchical optimization structure. It may introduce difficulties such as non-convexity and disconnectedness between the upper level and lower level problems, even for simple instances. It has been shown that bilevel programming is strongly NP-hard [16], and it has been proven that just evaluating a solution is also an NP-hard task [35].

Even though exact approaches are not efficient and require strong assumptions, they are the main focus in the bilevel optimization literature. One of the oldest is reformulation from bilevel to single level by using KKT conditions [3]. This has been widely used to solve linear [38] and quadratic [13] bilevel problems. The other classical approaches to BOPs are gradient descent [35], trust region [9] and penalty methods [39]. Because of the necessity of strong assumptions for applying classical approaches, the literature focused on EAs to solve BOPs, and the first EA proposed to solve BOPs was in [41]. They used EAs to solve the upper level and classical approaches for solving the lower level problem. Later, several works applied KKT conditions to reduce the bilevel problem into a single level and apply EAs to solve the reduced version of the problem [32]. Several nested algorithms were also proposed which use EAs at both levels with different approaches. For instance [1] applied a differential evolution algorithm at both levels and [23] implemented a particle swarm optimization approach to solve BOPs. EAs reached competitive results compared with classical approaches, but the required number of function evaluations is still huge because of the nature of the population-based search strategy. Meta-modelling approaches have been developed over the years to handle this problem [29]. This is a commonly used method for expensive-to-evaluate optimization problems [36]. It uses a surrogate model to approximate the actual model to reduce the function evaluations required to solve the actual problem. Even though using a surrogate model for the lower level problem decreases the function evaluation significantly, still EAs at the upper level require a high number of function evaluations to achieve competitive results. For further details about theoretical developments of classical and evolutionary algorithms for BOPs and more applications, the readers may refer to [7,31].

Recently, [21] proposed a nested approach to decrease function evaluations by approximating the optimal values. They applied a general Bayesian approach at the upper level by using upper level decisions and objective values, and an evolutionary algorithm to solve the lower level problem. [11] presented a multi-objective acquisition optimization approach to improve BO at the upper level. Following that, [12] presented a conditional BO-based approach to solve the bilevel problems by conditioning lower level decisions at the upper level, and they presented competitive results. They decreased the upper level function evaluations significantly, but it was still high at the lower level.

Based on the above observations, we develop a fully BO algorithm for BOPs, called FB-BLO. Unlike existing approaches, the FB-BLO algorithm uses $\mathcal{GP}$ surrogate model for both upper and lower levels to decrease at both upper and lower level function evaluations while approximating the optimal objective values. It selects a small number of candidates where the uncertainty is high during the optimization at both levels by the BO process and uses both decisions and its own objective values to model. Therefore, the number of necessary function evaluations can be reduced significantly while ensuring approximation to the global optima. Because both levels' surrogate models use both levels' decisions, the upper and lower level models cooperate to improve the optimization process. Moreover, the proposed FB-BLO algorithm can be implemented in several fields where the objective functions are expensive to evaluate, unavailable and impractical to model because of complexity or other reasons. It makes it easier to model practical problems as the proposed algorithm works only on decisions and the outcomes of the decisions, without requiring function characteristics such as nonconvexity, differentiability, etc.

3 Proposed Method

In this section, we first briefly describe BO and $\mathcal{GP}$, and then we propose the main steps of the FB-BLO by explaining how upper and lower level optimizations are implemented with the Bayesian method.

BO is a global optimization method to optimize expensive-to-evaluate blackbox functions. It is guided by a statistical model of the objective function, which is called a surrogate model, generally the $\mathcal{GP}$ [14] is selected. The $\mathcal{GP}$ model takes a prior distribution to be a multivariate normal with a mean function and covariance function, represented by $\mu_0(\cdot)$ and $\sigma_0^2(\cdot)$ respectively. Let us assume that we have n decisions $x_{1:n} = \{x_i\}_{i=0}^n$ and observations $f(x_{1:n}) = \{f(x_i)\}_{i=0}^n$, then the $\mathcal{GP}$ posterior distribution is calculated as in Eq. 2;

$$f(x')|f(x_{1:n}) \sim \mathcal{N}(\mu_n(x'), \sigma_n^2(x'))$$
$$\mu_n(x') = \Sigma_0(x', x_{1:n})\Sigma_0(x_{1:n}, x_{1:n})^{-1}(f(x_{1:n}) - \mu_0(x_{1:n}))) + \mu_0(x') \quad (2)$$
$$\sigma_n^2(x') = \Sigma_0(x', x') - \Sigma_0(x', x_{1:n})\Sigma_0(x_{1:n}, x_{1:n})^{-1}\Sigma_0(x_{1:n}, x')$$

where x' is a new candidate and $\Sigma_0(\cdot, \cdot)$ is a covariance matrix. Because it is expensive to evaluate the true objective function, the $\mathcal{GP}$ uses acquisition functions which are much cheaper and easier to optimize. Acquisition functions are designed to evaluate the trade-off between exploration and exploitation of search space and the current promising area. If we represent the acquisition function by $\alpha(\cdot)$ then the $\mathcal{GP}$ selects the next candidate by optimizing the predefined acquisition function, $x_{n+1} = \arg\max \alpha_n(x)$. For further details, the readers may refer to [25].

Algorithm 1: Procedure of the FB-BLO Algorithm

Input: $\{x_{u_i}\}_{i=0}^n$ (the upper level random decision vectors with the size of n)
Output: $x_u^{best}, x_l^{best}, F^{best}, f^{best}$ (the best objective values and decision set)
/* Initialization of the Algorithm   */
1 Set the bilevel data set $\mathcal{D} = \emptyset$;
2 Conduct lower level optimization for each upper level decision;
3 Evaluate the objective function results for each decision pair;
4 Add the decisions and related solutions to the bilevel dataset $\mathcal{D}$;
5 Initialize the bilevel dataset $\mathcal{D} = \{x_{u_i}, x_{l_i}, F_i(x_{u_i}, x_{l_i}), f_i(x_{u_i}, x_{l_i})\}_{i=0}^n$ with random decision vectors and related objective values;
6 **while** *termination criteria not fulfilled* **do**
   /* upper level Optimization   */
7 $\quad \mathcal{GP}_u \leftarrow$ Built upper level Gaussian process model based on all different *upper level* decisions and related upper level and lower level objective values in $\mathcal{D}$;
8 $\quad \mathbf{X}_u^* \leftarrow$ Select the best upper level decision candidate x_u^* by evaluating upper level acquisition optimization;
9 $\quad$ **for** $x_u^* \in \mathbf{X}_u^*$ **do**
      /* lower level Optimization   */
10 $\quad\quad \mathcal{GP}_l \leftarrow$ Built lower level Gaussian process model based on *lower level* solutions and upper level and lower level objective values in $\mathcal{D}$;
11 $\quad\quad \mathbf{X}_l^* \leftarrow$ Select the best lower level decision candidate x_l^* by evaluating lower level acquisition optimization;
12 $\quad$ Evaluate upper level $F(\mathbf{X}_u^*, \mathbf{X}_l^*)$ and lower level $f(\mathbf{X}_u^*, \mathbf{X}_l^*)$ objective results;
13 $\quad$ Insert solutions in into bilevel dataset $\mathcal{D}$;
14 $\quad$ Update upper level and lower level $\mathcal{GP}$ models with new observations;
15 **return** $x_u^{best}, x_l^{best}, F^{best}, f^{best}$ (the best objective values and decision set) in the bilevel dataset $\mathcal{D}$;

We present the main steps of the proposed FB-BLO in Algorithm 1 and share the flowchart in Fig. 1. The FB-BLO first generates n upper level decision samples $\{x_{u_i}\}_{i=0}^n$ using the Latin hypercube sampling [34] to begin. After that, the FB-BLO performs the lower level optimization by BO and uses the $\mathcal{GP}$ as a surrogate model for each initialized upper level decision. Consecutively, the FB-BLO evaluates the upper and lower level objective values and updates the data set with objective values along with optimal lower level decision x_l^* considering upper level value x_u. During upper level optimization, the proposed algorithm initializes the upper level Gaussian process model $\mathcal{GP}_u$ with upper and lower level decision vectors, and upper level objective values x_u, x_l^* and $F(x_u, x_l^*)$. Therefore, the $\mathcal{GP}_u$ is built based on the previous observations to search the following decision candidate for the upper level optimization problem. For each upper level decision candidate, the FB-BLO performs the lower level optimization to approximate the optimal value. During lower level optimization, a $\mathcal{GP}$ model is trained to approximate lower level problem, called $\mathcal{GP}_l$, with both decision vectors and lower level objective values x_u, x_l^* and $f(x_u, x_l^*)$. By

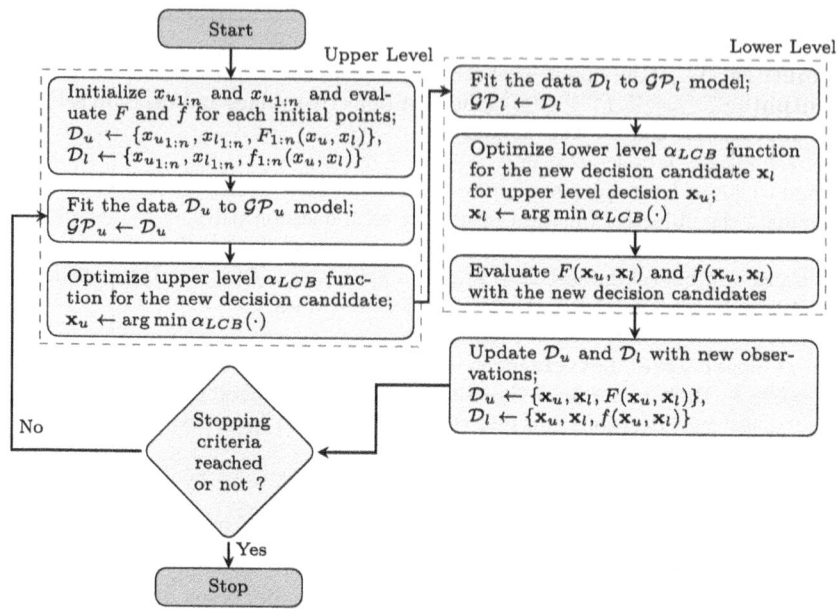

Fig. 1. Flowchart of the FB-BLO

using both decision vectors of the upper and lower level optimization problems to train $\mathcal{GP}$ models, the algorithm takes leverage to use the correlation between decisions at both levels and objective values to converge faster and to perform a smarter search. After lower level optimization, the FB-BLO adds the obtained results to the data set to update upper and lower level $\mathcal{GP}$ models.

We can observe that during upper level optimization, the FB-BLO algorithm selects the promising candidate by approximating the problem considering the previous observations. Then, it approximates the lower level problem considering the upper level decision variable. Using the BO approach and approximating both levels decreases the number of function evaluations significantly while converging the optimal value at both levels.

As we described above, the FB-BLO algorithm creates a Gaussian process surrogate model and it performs BO to select the most promising candidate during the upper level optimization, called $\mathcal{GP}_u$. Assume that we have the bilevel data set $\mathcal{D} = \{x_{u_i}, x_{l_i}^*, F_i(x_{u_i}, x_{l_i}^*)\}_{i=0}^n$ contains n observations and upper level objective value where x_{u_i} is upper level decision vector and $x_{l_i}^*$ best lower level responses. Then, the FB-BLO algorithm initializes the upper level surrogate model $\mathcal{GP}_u$ with a prior mean function and variance defined by μ_u and σ_u^2 respectively. The standard deviation can also be used to measure the uncertainty of predicted results. Let us assume that the decision vector represented by $\mathbf{x_l} = \{(x_{u_i}, x_{l_i}^*)\}_{i=0}^n$. After that, we construct the mean vector by calculating $\mu_{0_u}(\cdot)$ at each $\mathbf{x}_i$ and covariance matrix by calculating Σ_{0_u} at each pair of decision points

$\mathbf{x}_i$ and $\mathbf{x}_j$. Then, the posterior mean and variance are formulated as follows:

$$F(\mathbf{x}')|F(\mathbf{x}_{1:n}) \sim \mathcal{GP}_u(\mu_{n_u}(\mathbf{x}'), \sigma^2_{n_u}(\mathbf{x}'))$$
$$\mu_{n_u}(\mathbf{x}') = \Sigma_0(\mathbf{x}', \mathbf{x}_{1:n})\Sigma_0(\mathbf{x}_{1:n}, \mathbf{x}_{1:n})^{-1}(F(\mathbf{x}_{1:n}) - \mu_0(\mathbf{x}_{1:n}))) + \mu_0(\mathbf{x}')$$
$$\sigma^2_{n_u}(\mathbf{x}') = \Sigma_0(\mathbf{x}', \mathbf{x}') - \Sigma_0(\mathbf{x}', \mathbf{x}_{1:n})\Sigma_0(\mathbf{x}_{1:n}, \mathbf{x}_{1:n})^{-1}\Sigma_0(\mathbf{x}_{1:n}, \mathbf{x}')$$
(3)

where $F(\cdot)$ is the upper level objective function and $\mathbf{x}'$ is the next candidate to be estimated at the upper level. Then the algorithm obtains the next promising point by minimizing the lower confidence bound (LCB) acquisition function [10], $\mathbf{x}' = \arg\min \alpha_{LCB}(\mathbf{x})$, as defined in Eq. 4.

$$\alpha_{LCB}(\mathbf{x}') = \mu_{n_u}(\mathbf{x}') - \beta^{\frac{1}{2}}\sigma_{n_u}(\mathbf{x}')$$
(4)

The FB-BLO trains the $\mathcal{GP}_u$ model based on the entire dataset $\mathcal{D}$. So the upper level model in FB-BLO considers upper and lower level decision variables as inputs to predict the upper level objective values. Therefore, it initializes the $\mathcal{GP}_u$ to approximate the relationship between x_u, x_l^* and $F(x_u, x_l^*)$. In this way, the proposed algorithm can leverage the correlation of both upper and lower level decision sets with the upper level objective values. Then it optimizes the predefined acquisition function which is much cheaper in terms of computation than the true objective value, then selects the next promising candidate. It dramatically reduces the necessary number of function evaluations at the upper level. The optimal previous lower level objective value could also be used to train $\mathcal{GP}_u$ but it could affect the efficiency.

Similar upper level solutions likely lead to similar lower level optima as indicated in [30]. Considering this fact, for lower level optimization problems, FB-BLO initializes another Gaussian process surrogate model called $\mathcal{GP}_l$. The main reason behind this idea is selecting the decision candidates from an area in the search space that has the highest uncertainty. During the lower level optimization, considering a given upper level decision variable, the FB-BLO first creates the model with previous observations x_u, x_l^* and lower level objective $f(x_u, x_l^*)$. Then, the $\mathcal{GP}_l$ model uses x_u, x_l^* and $f(x_u, x_l^*)$ to approximate the lower level objective value. Subsequently, for each upper level decision, the proposed algorithm uses the $\mathcal{GP}_l$ model to select lower level optimum x_l' then updates the bilevel dataset with the set of decisions along with objective values $x_u, x_l', F(x_u, x_l')$ and $f(x_u, x_l')$ respectively. It should be noted that both $\mathcal{GP}_u$ and $\mathcal{GP}_l$ share the same bilevel data set for building the model. The only difference is that each model uses its own objective values to predict. Also, note that the proposed FB-BLO algorithm is developed for optimistic BOPs and tested only with continuous variables.

4 Empirical Studies

This section presents empirical results and compares the FB-BLO algorithm with six other algorithms. First, we briefly describe the compared algorithms along with the test problems. Then, we share the parameter settings for the experiments and performance metrics respectively. We also assess the performance of the proposed FB-BLO algorithm to verify the effectiveness and efficiency of both $\mathcal{GP}$ models and investigate the performance on selected benchmark problems.

4.1 Experimental Settings

The performance of the proposed algorithm is compared with six state-of-the-art algorithms selected from the bilevel optimization literature. These are ConBaBo [12], BOBP [21], MOTEA [8], BL-CMA-ES [18] and BLEAQ2 [26]. The results for the comparison are extracted from the related references in the literature. The ConBaBo uses a conditional Bayesian approach for solving upper level problems during the optimization process. An evolutionary algorithm is implemented to solve the lower level problems and they focused on decreasing function evaluations by surrogating *only* upper level optimization problems by conditioning lower level decisions. The BOBP algorithm applies a standard Bayesian approach to solve bilevel problems without conditioning and they used only upper level variables and objective values to solve the upper level problems. The MOTEA uses a multi-objective transformation-based evolutionary algorithm to solve multiple lower level optimization in a parallel and collaborative way. In the BL-CMA-ES algorithm, a bilevel covariance matrix adaptation is used to solve BOPs and a specific sharing mechanism is developed to use prior knowledge for lower level problems. The NBLEA algorithm is designed as a nested algorithm that solves both level problems with genetic algorithms. The BLEAQ2 algorithm uses a quadratic approximation to obtain both levels' optimum values to reduce the number of necessary function evaluations.

We evaluate the performance of the proposed FB-BLO algorithm on sixteen popular benchmark problems including linear, nonlinear, constrained and unconstrained minimization problems, called TP [28] and SMD [27] problems. Each problem has different convexity considering the interaction between the upper level and lower level decision-makers. The TP benchmark contains one three-dimensional, six four-dimensional, one five-dimensional and two ten-dimensional bilevel problems, and the SMD benchmark has six unconstrained bilevel problems. For more details about the benchmarks and the mathematical formulations, please see Sect. 1 in the supplementary material.

All experiments were conducted on a single core of 1.4 GHz Quad Core i5, 8 Gb 2133 Mhz LPDDR3 RAM. The algorithm is executed 31 times for each test function and the results shown are medians. The termination criteria are selected as the accuracy difference $\epsilon = 10^{-6}$ for each level and for not converging for a certain iteration, the closest accuracy and step are chosen as a result. Later, the best decision is selected after convergence. All experiments were conducted on a single core of 1.4 GHz Quad BO and $\mathcal{GP}$ are implemented in Python via the GPyOpt library [2] and *optimize_restarts* selected as 10 with the parameters of *verbose=False*, and *exact_feval=True*. The method initialized n Sobol points to construct the initial GP model. The LCB acquisition function has been selected for the $\mathcal{GP}$ and the Radial Basis Function (RBF) kernel is used.

Table 1. Median accuracy table for both TP and SMD benchmarks.

		FB-BLO	CONBABO	BOBP	MOTEA	BL-CMA-ES	NBLEA	BLEAQ2
TP1	ULFAcc	0.465	1.02	23.6	0.185	0.000001	0.426	0.000002
	LLFAcc	0.277	1.99	29.6	0.298	0.00002	0.702	0.000001
TP2	ULFAcc	0.000032	0.000001	0.0007	0.000227	0.000001	0.000001	0.000001
	LLFAcc	10.0	10.0	83.8	0.00643	0.00113	100.0	100.0
TP3	ULFAcc	0.000124	0.663	0.120	1.011	0.364	0.00652	0.000011
	LLFAcc	0.000417	0.0001	0.663	0.00216	0.219	0.00132	0.000025
TP4	ULFAcc	0.000016	1.76	1.57	0.000001	0.000001	0.293	0.000008
	LLFAcc	0.00002	0.528	0.101	0.000001	0.000001	0.00568	0.000001
TP5	ULFAcc	0.944	0.710	0.251	0.854	1.12	0.000349	0.00571
	LLFAcc	0.214	0.445	0.231	0.877	5.45	0.00140	0.265
TP6	ULFAcc	0.000016	0.132	0.0004	0.000001	0.000001	0.000006	0.00544
	LLFAcc	0.000045	0.606	0.00230	0.000026	0.000028	0.0000532	0.193
TP7	ULFAcc	0.000262	0.6	0.285	0.000679	0.00066	0.00157	0.00136
	LLFAcc	0.000262	0.6	0.285	0.000679	0.00066	0.00157	0.00136
TP8	ULFAcc	0.000001	0.000001	0.0008	0.000177	0.000002	0.000001	0.00254
	LLFAcc	10.0	10.0	80.6	0.00245	0.000406	100.0	100.0
TP9	ULFAcc	0.000001	0.000001	0.000001	0.000001	0.000001	0.000032	0.000007
	LLFAcc	0.000001	0.000001	0.000001	0.000001	0.000001	0.000001	0.000001
TP10	ULFAcc	0.000001	0.000001	0.000001	0.000001	0.000001	3.02	0.00186
	LLFAcc	0.000001	0.000001	0.000001	0.000001	0.000001	0.000001	0.000001
SMD1	ULFAcc	0.000059	0.0000018	–	0.000001	0.000001	0.000033	0.00018
	LLFAcc	0.000001	0.000001	–	0.000001	0.000001	0.000015	0.00018
SMD2	ULFAcc	0.000038	0.000009	–	0.000001	0.000001	0.000021	0.000119
	LLFAcc	0.000001	0.000001	–	0.000001	0.000001	0.000013	0.000117
SMD3	ULFAcc	0.000063	0.000002	–	0.000001	0.000001	0.000043	0.00647
	LLFAcc	0.000001	0.000001	–	0.000001	0.000001	0.000014	0.00297
SMD4	ULFAcc	0.000027	0.000006	–	0.000001	0.000001	0.000016	0.000024
	LLFAcc	0.000001	0.000001	–	0.000004	0.000005	0.000011	0.000333
SMD5	ULFAcc	0.000003	0.000003	–	0.000001	0.000001	0.000039	0.000724
	LLFAcc	0.000001	0.000001	–	0.000002	0.000002	0.000014	0.000657
SMD6	ULFAcc	0.000007	0.000002	–	0.000001	0.000001	0.000653	0.000001
	LLFAcc	0.000001	0.000001	–	0.000001	0.000001	0.000126	0.000001

For optimizing the acquisition function, the L-BFGS method [24] has been used. The exploration-exploitation parameter for LCB, w is set to 2.

4.2 Performance Comparison

Table 1 shows the accuracy of the different algorithms on both TP and SMD benchmarks. For BOPs, the absolute difference between the true optimum value

Table 2. Median function evaluation table for both TP and SMD benchmarks.

		FB-BLO	CONBABO	BOBP	MOTEA	BL-CMA-ES	NBLEA	BLEAQ2
TP1	ULFEval	33	18	211	320	214	523	136
	LLFEval	64	190	1558	2265	2084	2568	242
TP2	ULFEval	28	17	35	293	224	763	255
	LLFEval	52	148	383	2186	1986	4237	440
TP3	ULFEval	61	10	89	307	333	468	158
	LLFEval	88	111	1128	1838	2124	5472	224
TP4	ULFEval	54	14	16	294	367	746	198
	LLFEval	92	1063	334	1733	2165	8649	788
TP5	ULFEval	44	23	57	339	430	1021	272
	LLFEval	80	222	319	2135	2406	9842	967
TP6	ULFEval	43	8	12	414	330	624	161
	LLFEval	71	323	182	19300	2178	6736	323
TP7	ULFEval	57	13	72	356	134	748	112
	LLFEval	99	274	320	4608	6445	5532	287
TP8	ULFEval	20	14	37	134	314	364	241
	LLFEval	42	114	413	2389	2018	9638	467
TP9	ULFEval	36	8	16	313	2508	2262	1512
	LLFEval	128	1020	396	13800	12800	32460	14130
TP10	ULFEval	42	15	21	187	9122	2458	1847
	LLFEval	264	3689	974	24400	46410	4868	24510
SMD1	ULFEval	65	11	–	3200	321	839	266
	LLFEval	162	380	–	22600	20800	546000	27500
SMD2	ULFEval	73	14	–	293	297	1010	290
	LLFEval	174	381	–	21800	19800	610000	25100
SMD3	ULFEval	53	23	–	307	333	938	311
	LLFEval	187	770	–	18300	21200	589000	35700
SMD4	ULFEval	73	11	–	294	334	1350	410
	LLFEval	168	258	–	17300	21800	604000	26400
SMD5	ULFEval	61	24	–	339	367	1040	286
	LLFEval	139	905	–	21300	21600	584000	36200
SMD6	ULFEval	82	25	–	414	430	1310	274
	LLFEval	197	769	–	21000	24000	36700	92900

and the best-obtained value by the proposed algorithm is used for accuracy comparison. The upper and lower level accuracy, called ULAcc and LLAcc respectively, can be calculated by $ULAcc = |F^{best} F^*|$ and $LLAcc = |f^{best} f^*|$ where F^* and f^* represent the true optimum value of upper and lower level optimization

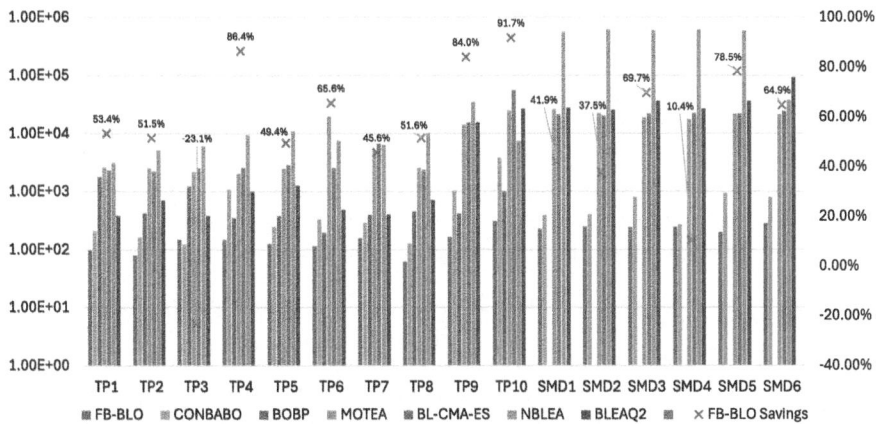

Fig. 2. The number of total function evaluation comparisons in logarithmic scale (left y-axis) and the FB-BLO savings comparing with the CONBABO (right y-axis).

problems, and F^{best} and f^{best} are the best obtained values from the proposed algorithm, respectively. Also, according to the nested structure of the BOPs, it is hard to determine the computational resources of other experiments. Therefore we share the necessary number of function evaluations for upper and lower level problems in Table 2, denoted by $ULFEval$ and $LLFEval$ respectively.

As shown in Table 1, the FB-BLO algorithm achieved competitive accuracy scores compared with the others. Both CONBABO and BOBP are based on the Bayesian approach to bilevel problems and the proposed FB-BLO algorithm reached better accuracy on the majority of the TP problems. The BLEAQ2 and MOTEA algorithms achieved $1E-06$ accuracy in several instances of the TP benchmark which is fairly better than the proposed FB-BLO algorithm. TP9 and TP10 problems are relatively higher dimension problems and the FB-BLO algorithm achieved competitive accuracy on these problems. The SMD benchmark problems are unconstrained problems and we can see from Table 1 our proposed algorithm performed well with the other compared methods.

As shown in Table 2, the FB-BLO algorithm surpasses all other algorithms in terms of function evaluations by requiring the minimum number of function evaluations for all problems in both benchmarks. We can observe that the MOTEA, BL-CMA-ES, NBLEA and BLEAQ2 algorithms consume much more function evaluations compared with the Bayesian-based approaches. The main reason behind this is the integrated population-based search strategy of EAs. The CONBABO and BOBP algorithms require more function evaluations compared with the proposed FB-BLO algorithm because they both used BO at the upper level and EAs at the lower level problem. The FB-BLO algorithm achieved competitive results with much more fewer function evaluations compared with the other Bayesian-based algorithms, CONBABO and BOBP because of applying a fully Bayesian approach at both levels. From the experimental results in Tables 1 and

2, we conclude that the combination of using BO at both upper and lower level significantly reduces the necessary function evaluations when solving BOPs.

Solving BOPs effectively requires good performance at both levels, therefore we also consider the total function evaluations in the whole process. Figure 2 compares the total function evaluations in logarithmic scale on the left y-axis considering the sum of upper and lower level optimization. It also contains the FB-BLO savings compared with the second-best algorithm CONBABO on the right y-axis. The saving of the total function evaluation is calculated as follows:

$$\text{FB-BLO Savings}(\%) = \frac{\text{TotalFEval}_{CONBABO} - \text{TotalFEval}_{FB-BLO}}{\text{TotalFEval}_{CONBABO}} \quad (5)$$

As we can see from Fig. 2, the FB-BLO algorithm reached competitive accuracies in all test problems but TP3. The best improvement is TP10 with 91.7% compared with the CONBABO. We also run the Wilcoxon rank-sum test at a significance level of 0.05 to verify if there is a significant difference between the total function evaluation of the FB-BLO algorithm and the CONBABO algorithm. The obtained p-value from the test is 0.00058 ($p < 0.05$) which is smaller than the significance level, and it shows that the performance of the proposed algorithm is significant. We can observe that the proposed algorithm achieved competitive results with significantly fewer total function evaluations and is successful in solving both constraint and unconstraint bilevel problems.

5 Conclusion and Future Works

BOPs are a specific kind of optimization problem that features two or more decision-makers and models a hierarchical decision-making process. In this study, we developed a fully BO approach to solve BOPs, called FB-BLO. We model the upper and lower level problems with two different $\mathcal{GP}s$ and used both upper and lower level decisions with their objectives to use the cooperation between both decisions and objective values. Also, the black-box approach at both levels made it possible to model expensive-to-evaluate functions that may be unavailable or impractical to real-world applications. Based on the experimental results, it was demonstrated that the proposed algorithm can improve the performance in terms of computational cost while obtaining competitive results. It significantly reduced the number of function evaluations and handled the complexities without any assumptions such as nonconvexity, differentiability, etc. Hence, we believe that the proposed fully Bayesian approach is a promising way to improve the ability to solve BOPs and model real-world problems more easily and effectively.

In this study, we investigated and experimented with the presented approach only with the continuous BOPs. It would be interesting to explore and verify the performance of this approach in combinatorial search space. We used the basic acquisition function during BO and it would also be interesting to see more advanced acquisition function performances at both levels. Many practical applications have been modelled as BOPs as we mentioned above, so there are numerous practical problems to solve and explore the performance without any assumptions about the characteristics of objective functions.

Acknowledgments. This publication has emanated from research conducted with the financial support of Science Foundation Ireland under Grant number 12/RC/2289-P2 at Insight the SFI Research Centre for Data Analytics at UCC, which is co-funded under the European Regional Development Fund. For Open Access, the author has applied a CC BY public copyright licence to any Author Accepted Manuscript version arising from this submission. The authors have no competing interests to declare that are relevant to the content of this article.

References

1. Angelo, J.S., Krempser, E., Barbosa, H.J.: Differential evolution for bilevel programming. In: IEEE Congress on Evolutionary Computation, pp. 470–477 (2013)
2. Authors, T.G.: GPyOpt: a Bayesian optimization framework in python (2016)
3. Bard, J.F., Falk, J.E.: An explicit solution to the multi-level programming problem. Comput. Oper. Res. **9**(1), 77–100 (1982)
4. Bracken, J., McGill, J.T.: Mathematical programs with optimization problems in the constraints. Oper. Res. **21**(1), 37–44 (1973)
5. Bracken, J., McGill, J.T.: Defense applications of mathematical programs with optimization problems in the constraints. Oper. Res. **22**(5), 1086–1096 (1974)
6. Cai, X., Gao, L., Li, X.: Efficient generalized surrogate-assisted evolutionary algorithm for high-dimensional expensive problems. IEEE Trans. Evol. Comput. **24**(2), 365–379 (2020)
7. Camacho, J.F., Corpus, C., Villegas, J.G.: Metaheuristics for bilevel optimization: a comprehensive review. Comput. Oper. Res. **161**, 106410 (2024)
8. Chen, L., Liu, H.L., Li, K., Tan, K.C.: Evolutionary bi-level optimization via multi-objective transformation-based lower level search. IEEE Trans. Evol. Comput. 1 (2023)
9. Colson, B., Marcotte, P., Savard, G.: A trust-region method for nonlinear bilevel programming: algorithm and computational experience. Comput. Optim. Appl. **30**, 211–227 (2005)
10. Cox, D.D., John, S.: SDO: a statistical method for global optimization. In: Multidisciplinary Design Optimization: State-of-the-Art, pp. 315–329 (1997)
11. Dogan, V., Prestwich, S.: Bayesian optimization with multi-objective acquisition function for bilevel problems. In: Longo, L., O'Reilly, R. (eds.) AICS 2022. CCIS, vol. 1662, pp. 409–422. Springer, Cham (2023). https://doi.org/10.1007/978-3-031-26438-2_32
12. Dogan, V., Prestwich, S.: Bilevel optimization by conditional Bayesian optimization. In: Nicosia, G., Ojha, V., La Malfa, E., La Malfa, G., Pardalos, P.M., Umeton, R. (eds.) LOD 2023, Part I. LNCS, vol. 14505, pp. 243–258. Springer, Heidelberg (2024). https://doi.org/10.1007/978-3-031-53969-5_19
13. Edmunds, T., Bard, J.: Algorithms for nonlinear bilevel mathematical programs. IEEE Trans. Syst. Man Cybern. **21**(1), 83–89 (1991)
14. Frazier, P.: A tutorial on Bayesian optimization. ArXiv abs/1807.02811 (2018)
15. Girigoudar, K., Roald, L.A.: Identifying secure operating ranges for der control using bilevel optimization. IEEE Trans. Smart Grid **15**(3), 2921–2933 (2024)

16. Hansen, P., Jaumard, B., Savard, G.: New branch-and-bound rules for linear bilevel programming. SIAM J. Sci. Stat. Comput. **13**(5), 1194–1217 (1992)
17. Haywood, A.B., Lunday, B.J., Robbins, M.J.: Intruder detection and interdiction modeling: a bilevel programming approach for ballistic missile defense asset location. Omega **110**, 102640 (2022)
18. He, X., Zhou, Y., Chen, Z.: Evolutionary bilevel optimization based on covariance matrix adaptation. IEEE Tran. Evol. Comput. **23**(2), 258–272 (2019)
19. Hong, Q., Meng, F., Liu, J., Bo, R.: A bilevel game-theoretic decision-making framework for strategic retailers in both local and wholesale electricity markets. Appl. Energy **330**, 120311 (2023)
20. Jiang, H., Chou, K., Tian, Y., Zhang, X., Jin, Y.: Efficient surrogate modeling method for evolutionary algorithm to solve bilevel optimization problems. IEEE Trans. Cybern. 1–13 (2023)
21. Kieffer, E., Danoy, G., Bouvry, P., Nagih, A.: Bayesian optimization approach of general bi-level problems. In: Proceedings of the Genetic and Evolutionary Computation Conference Companion, GECCO 2017, pp. 1614–1621. Association for Computing Machinery, New York (2017)
22. Kleinert, T., Labbé, M., Plein, F., Schmidt, M.: Technical note—there's no free lunch: on the hardness of choosing a correct big-M in bilevel optimization. Oper. Res. **68** (2020)
23. Li, X., Tian, P., Min, X.: A hierarchical particle swarm optimization for solving bilevel programming problems. In: Rutkowski, L., Tadeusiewicz, R., Zadeh, L.A., Żurada, J.M. (eds.) ICAISC 2006. LNCS (LNAI), vol. 4029, pp. 1169–1178. Springer, Heidelberg (2006). https://doi.org/10.1007/11785231_122
24. Liu, D.C., Nocedal, J.: On the limited memory BFGS method for large scale optimization. Math. Program. **45**(1), 503–528 (1989)
25. Rasmussen, C.E.: Gaussian processes in machine learning. In: Bousquet, O., von Luxburg, U., Rätsch, G. (eds.) ML -2003. LNCS (LNAI), vol. 3176, pp. 63–71. Springer, Heidelberg (2004). https://doi.org/10.1007/978-3-540-28650-9_4
26. Sinha, A., Lu, Z., Deb, K., Malo, P.: Bilevel optimization based on iterative approximation of mappings. J. Heuristics **26** (2020)
27. Sinha, A., Malo, P., Deb, K.: Test problem construction for single-objective bilevel optimization. Evol. Comput. **22** (2013)
28. Sinha, A., Malo, P., Deb, K.: Efficient evolutionary algorithm for single-objective bilevel optimization (2013)
29. Sinha, A., Malo, P., Deb, K.: Evolutionary algorithm for bilevel optimization using approximations of the lower level optimal solution mapping. Eur. J. Oper. Res. **257**(2), 395–411 (2017)
30. Sinha, A., Malo, P., Deb, K.: Evolutionary algorithm for bilevel optimization using approximations of the lower level optimal solution mapping. Eur. J. Oper. Res. **257**, 395–411 (2017)
31. Sinha, A., Malo, P., Deb, K.: A review on bilevel optimization: from classical to evolutionary approaches and applications. IEEE Trans. Evol. Comput. **22**(2), 276–295 (2018)
32. Sinha, A., Soun, T., Deb, K.: Evolutionary bilevel optimization using KKT proximity measure. In: IEEE Congress on Evolutionary Computation, pp. 2412–2419 (2017)
33. von Stackelberg, H.: Marktform und Gleichgewicht. Die Handelsblatt-Bibliothek "Klassiker der Nationalökonomie" (1934)
34. Stein, M.: Large sample properties of simulations using Latin hypercube sampling. Technometrics **29**(2), 143–151 (1987)

35. Vicente, L., Savard, G., Júdice, J.: Descent approaches for quadratic bilevel programming. J. Optim. Theory Appl. **81**(2), 379–399 (1994)
36. Wang, G., Shan, S.: Review of metamodeling techniques in support of engineering design optimization. J. Mech. Design **129** (2007)
37. Wang, H., Feng, L., Jin, Y., Doherty, J.: Surrogate-assisted evolutionary multitasking for expensive minimax optimization in multiple scenarios. IEEE Comput. Intell. Mag. **16**(1), 34–48 (2021)
38. Wein, L.: Or forum—homeland security: from mathematical models to policy implementation. Oper. Res. **57**, 801–811 (2009)
39. White, D.J., Anandalingam, G.: A penalty function approach for solving bi-level linear programs. J. Global Optim. **3**, 397–419 (1993)
40. Yan, S., Wang, W., Li, X., Lv, H., Fan, T., Aikepaer, S.: Stochastic optimal scheduling strategy of cross-regional carbon emissions trading and green certificate trading market based on stackelberg game. Renew. Energy **219**, 119268 (2023)
41. Yin, Y.: Genetic-algorithms-based approach for bilevel programming models. J. Transp. Eng.-ASCE **126** (2000)
42. Zhang, Q., Liu, S.Q., D'Ariano, A., Chung, S.H., Masoud, M., Li, X.: A bi-level programming methodology for decentralized mining supply chain network design. Expert Syst. Appl. **250**, 123904 (2024)
43. Zheng, A.Y., He, T., Qiu, Y., Wang, M., Wipf, D.: BloomGML: graph machine learning through the lens of bilevel optimization (2024)

Collaborative Information Dissemination with Graph-Based Multi-Agent Reinforcement Learning

Raffaele Galliera[1,2(✉)], Kristen Brent Venable[1,2], Matteo Bassani[1], and Niranjan Suri[1,2,3]

[1] Institute for Human and Machine Cognition, Pensacola, USA
{rgalliera,bvenable,mbassani,nsuri}@ihmc.org
[2] Department of Intelligent Systems and Robotics, The University of West Florida, Pensacola, FL, USA
[3] US Army Research Laboratory, Adelphi, MD, USA

Abstract. Efficient information dissemination is crucial for supporting critical operations across domains like disaster response, autonomous vehicles, and sensor networks. This paper introduces a Multi-Agent Reinforcement Learning (MARL) approach as a significant step forward in achieving more decentralized, efficient, and collaborative information dissemination. We propose a Partially Observable Stochastic Game (POSG) formulation for information dissemination empowering each agent to decide on message forwarding independently, based on the observation of their one-hop neighborhood. This constitutes a significant paradigm shift from heuristics currently employed in real-world broadcast protocols. Our novel approach harnesses Graph Convolutional Reinforcement Learning and Graph Attention Networks (GATs) with dynamic attention to capture essential network features. We propose two approaches to accomplish cooperative information dissemination, L-DyAN and HL-DyAN, differing in terms of the information exchanged among agents. Our experimental results show that our trained policies outperform existing methods, including the state-of-the-art heuristic, in terms of network coverage and communication overhead on dynamic networks of varying density and behavior.

Keywords: Multi-Agent Reinforcement Learning · Information Dissemination · Communication Networks · Broadcast Networks · Graph Neural Networks · DGN

1 Introduction

Group communication, implemented in a broadcast or multicast fashion, finds a natural application in different networking scenarios, such as Vehicular Ad-hoc Networks (VANETs) [1,2], with the necessity to disseminate information about the nodes participating, e.g. identity, status, or crucial events happening in the network. These systems can be characterized by congestion-prone networks and/or different resource constraints, such that message dissemination becomes considerably expensive if not adequately managed. For this matter, message forwarding calls for scalable and distributed solutions that are able to minimize the total number of forwards and, at the same time, achieve the expected coverage. Moreover, modern protocols often require careful adjustments of their parameters before achieving adequate forwarding policies, which would otherwise result in sub-optimal performance in terms of delivery ratio and latency [3].

Recently, researchers have considered learning communication protocols [4] with Multi-Agent Reinforcement Learning (MARL) [5]. At its core, MARL seeks to design systems where multiple agents learn to optimize their objective by interacting with the environment and the other entities involved. Such tasks can be competitive, cooperative, or a combination of both, depending on the scenario. As agents interact within a shared environment, they often find the need to exchange information to optimize their collective performance. This has led to the development of communication mechanisms that are learned rather than pre-defined, allowing agents to cooperate better utilizing their learned signaling system.

Nevertheless, learning to communicate with MARL comes with several challenges. In multi-agent systems, actions taken by one agent can significantly impact the rewards and state transitions of other agents, rendering the environment more complex and dynamic, and ensuring that agents develop a shared and consistent communication protocol, is an area of active research. Methods such as CommNet [6] and BiCNet [7], focus on the communication of local encodings of agents' observations. These approaches allow agents to share a distilled version of their perspectives, enabling more informed collective decision-making. ATOC [8] and TarMAC [9] have ventured into the realm of attention mechanisms. By leveraging attention, these methods dynamically determine which agents to communicate with and what information to share, leading to more efficient and context-aware exchanges. Yet another approach, as exemplified by Graph Convolutional Reinforcement Learning (DGN) [10], harnesses the power of Graph Neural Networks (GNNs) and attention mechanisms to model the interactions, relations, and communications between agents.

However, to the best of our knowledge, no MARL-based method involving proactive communication and GNNs has been proposed to address the unique challenges of optimizing the process of information dissemination within a broadcast dynamic network. In such a scenario, nodes need to cooperate to spread the information by forwarding it to their immediate neighbors, which might change over time, while relying on their limited observation of the entire graph. Furthermore, their collaboration and ability to accomplish dissemination are bound by the limitations of the underlying communication channels. This means that both the forwarding actions and the amount of information exchange needed for effective cooperation are constrained and should be minimized.

Contributions. In this work, we introduce **a novel Partially Observable Stochastic Game (POSG) for optimized information dissemination** in dynamic broadcast networks, forming the basis for our MARL framework.[1] To this end, we design a MARL algorithm to encourage **cooperation within dynamic neighborhoods** where

---
[1] https://github.com/RaffaeleGalliera/melissa.

node connections are frequently changing. Furthermore, we design and test **two distinct architectures**, namely Local Dynamic Attention Network (L-DyAN) and Hyperlocal Dynamic Attention Network (HL-DyAN), which leverage Graph Attention Network (GAT) with dynamic attention [11] and Dueling Q-Networks [12].

Our **experimental study** demonstrates our methods' efficacy in achieving superior network coverage across dynamic graphs in different scenarios, outperforming DGN and the established Multipoint Relay (MPR) [13] heuristic. Moreover, our approach operates on one-hop observations and empowers nodes to take independent forwarding decisions, unlike MPR. By exploring the potential of learning-based approaches for addressing information dissemination in dynamic networks, our work underscores the versatility of MARL in present and future, real-world applications such as information dissemination in social networks [14], space networks [15], and vehicular communications [16].

2 Background

Reinforcement Learning (RL). provides solutions for sequential decision-making problems formulated as Markov Decision Process (MDPs) [17,18]. The Partially Observable Markov Decision Process (POMDP) extends the MDP framework to scenarios where agents have limited or partial observability of the underlying environment and make decisions based on belief states, which are probability distributions over the true states. To this end, several methods have been proposed such as Deep Recurrent Q-Learning [19].

Multi-Agent Reinforcement Learning. For multi-agent systems the RL paradigm extends to MARL [5], where multiple entities, potentially learners and non-learners, interact with the environment. In this context the generalization of POMDPs leads to Decentralized Partially Observable Markov Decision Process (Dec-POMDP), characterized by the tuple $\langle \mathcal{I}, \mathcal{S}, \mathcal{A}^i_{i \in I}, \mathcal{P}, \mathcal{R}, \mathcal{O}^i_{i \in I}, \gamma \rangle$. Here, $\mathcal{I}$ represents the set of agents, $\mathcal{S}$ denotes the state space, $\mathcal{A}^i{}_{i \in \mathcal{I}}$ stands for the action space for each agent, $\mathcal{P}$ is the joint probability distribution governing the environment dynamics given the current state and joint actions, $\mathcal{R}$ denotes the reward function, and $\mathcal{O}^i_{i \in \mathcal{I}}$ represents the set of observations for each agent. Such game-theoretic settings are used to model fully cooperative tasks where all agents have the same reward function and share a common reward.

A more general model, adopted in this work, is the Partially Observable Stochastic Game (POSG), where each agent receives an individual reward $\mathcal{R}^i{}_{i \in \mathcal{I}}$, allowing the definition of fully competitive and mixed tasks such as zero-sum and general-sum games [20]. Several MARL algorithms have been presented in the literature, addressing different tasks (cooperative, competitive, or mixed) and pursuing different learning goals such as stability or adaptation [5,21].

Graph Convolutional Reinforcement Learning. In DGN [10], the dynamics of multi-agent environments are represented as a graph, where each agent is a node with a set of neighbors determined by specific metrics. In this approach, a key role is played by Relation Kernels, which use multi-headed dot product to merge features within an agent's receptive field. Such field, then, increases with the number of graph convolutional layers allowing to capture detailed interactions and relationships between agents. During training, a batch of experiences $\mathcal{B}$ is sampled and the following loss is minimized:

$$L(\theta) = \frac{1}{|\mathcal{B}|} \sum_{\mathcal{B}} \frac{1}{N} \sum_{i=1}^{N} (y_i - Q(O_{i,\mathcal{C}}, a_i; \theta))^2 \qquad (1)$$

where N is the number of agents, and $O_{i,\mathcal{C}}$ is the observation of agent i with the respective adjacency matrix $\mathcal{C}$. We build on DGN and design novel MARL architectures for optimizing information dissemination in dynamic networks.

Optimized Flooding in Broadcast Networks. A dynamic broadcast network can be represented as a dynamic graph $\mathcal{G}(t) = (\mathcal{V}, \mathcal{E}(t))$, where each node represents a (possibly) mobile node and an edge between two nodes at time t represents the two corresponding nodes being within each other's broadcasting range at that time. Hence, for every node $v \in \mathcal{V}$, the set of its neighbors at time t is defined as $\mathcal{N}_v(t) = \{u \in \mathcal{V} | (v, u) \in \mathcal{E}(t)\}$.

A main objective of broadcast communications over connected networks is called Optimized Flooding [22] and it is achieved when the information emitted from a given node $v \in \mathcal{V}$ reaches every other node $u \neq v$, thanks to forwarding actions of a set of nodes $\mathcal{D} \subseteq \mathcal{V}$. While maximizing coverage it is also desirable to minimize redundant transmissions, which might impact resource utilization, such as bandwidth, power consumption, and latency. From a graph-theoretic point of view this can be achieved by identifying a specific subset of nodes, called a Minimum Connected Dominating Set (MCDS), that will be tasked with forwarding the information. This task requires the introduction of a centralized entity with complete knowledge of the network state and has been shown to be NP-complete [23]. A much more efficient and realistic approach is to approximate the MCDS in a distributed manner, relying only on local observations of the network made from each node's perspective. Indeed, this is the approach taken by the MPR heuristic and our MARL approach.

3 Related Work

The MPR selection algorithm is a technique developed to efficiently disseminate information in Mobile Ad-hoc Networks (MANETs) and wireless mesh networks. It achieves this by having each node designate certain one-hop neighbors to forward messages arriving from them, thereby reducing the overall transmission load and preventing excessive network broadcasting. This process involves nodes exchanging "HELLO messages" to identify and select their MPR sets, ensuring network coverage with minimal redundancy.

In real-world protocols, MPR plays an essential role. For instance, in the Optimized Link State Routing (OLSR) protocol [13], MPR is fundamental in managing the distribution of Topology Control (TC) messages. Similarly, in Simplified Multicast Forwarding (SMF) [24], MPR is employed for the efficient forwarding of multicast packets.

In this work, we compare our approach with the MPR selection algorithm, as outlined in the standard OLSR implementation [13], leveraging this algorithm as a baseline for distributed message dissemination in dynamic graph structures. However, we define a completely different approach that leverages MARL and, unlike MPR, only requires an anonymized knowledge of the one-hop neighborhood and empowers each agent to independently decide their message forwarding policy.

Recent work has considered MARL approaches in the context of communication networks [25–27]. The more closely related to the presented work is DeepMPR [27] which addresses the optimization of specific networking metrics in the context of multicast networks. While related, both the problem considered and the approach taken here are different. In particular, we focus on coverage and forwarded message minimization rather than metrics such as goodput, which apply only to such domains. Moreover, in contrast to [27] where the proposed method utilizes PPO [28] and policies trained with observation and action spaces tailored to specific graph scenarios, we design a more general approach, introducing a novel POSG with a dynamic number of participating agents and a scalable graph-based solution, which employs two novel approaches based on DGN and GAT with dynamic attention for capturing essential local features and relations among agents.

4 Method

In this section, we describe our novel MARL approach for optimizing information flooding in dynamic broadcast networks. We start by presenting our POSG formulation and conclude with our two architectures, L-DyAN and HL-DyAN, designed to achieve efficient dissemination while requiring different degrees of communication.

4.1 MARL Formulation

We envision the dissemination process discretized into timesteps and episodes starting with a source node transmitting the information (a message) to its immediate (one-hop) neighbors. Each node in the graph corresponds to an agent observing its one-hop neighborhood and their features. At every timestep, nodes that have received the message will sense their neighborhood and decide whether to forward it to its current one-hop neighbors or stay silent. However, agents do not have any control or information on who will be part of their neighborhood at the next time-step. Finally, the agents' objective is to disseminate the message emitted from the source node, i.e. maximize the network coverage, while minimizing the amount of forwarding actions, i.e. the messages, required.

An agent becomes a meaningful actor, receiving appropriate reinforcement signals only once it receives the message and for a limited number of steps. We capture this by implementing two different elements of our POSG. On the one hand, we distinguish "Graph Episodes" from "Agent Episodes", allowing the agents to dynamically enter and leave the game independently. For this purpose, upon message reception, we limit the Agent's Episode to a fixed number of steps (*local horizon*) during which it decides whether to forward the message to its immediate neighbors or not. Graph Episodes, model the overall dissemination process, and terminate once every agent that has received the message has exhausted its local horizon. Given the agents' asynchronous presence, reward signals are issued individually to each agent, but capture the necessary degree of cooperation within their neighborhoods.

In our formulation agents are anonymous (i.e. not identified by any ID) and sense only their immediate neighborhood, accessing the degree of connectivity of such neighbors and observing their forwarding behavior. This is far more parsimonious than what is required by MPR that requires agents to obtain a complete, identified, two-hop knowledge. More specifically, given the broadcast network represented by graph $\mathcal{G}_0 = (\mathcal{V}, \mathcal{E}_0)$ at time t_0, and node $n_s \in \mathcal{V}$, we define the POSG associated to the optimized flooding of $\mathcal{G}_0$ with source n_s and network update function $\mathcal{U}$, with the tuple $\langle \mathcal{I}, \mathcal{S}, \mathcal{A}^i_{i \in I}, \mathcal{U}, \mathcal{P}, \mathcal{R}^i_{i \in I}, \mathcal{O}^i_{i \in I}, \gamma \rangle$, where:

Agents Set $\mathcal{I}$. Set $\mathcal{I}$ contains one agent for each node in $\mathcal{V}$. $\mathcal{I}$ is divided into three disjoint sets which are updated at every timestep t: the active set $\mathcal{I}_a(t)$, the done set $\mathcal{I}_d(t)$, and the idle set $\mathcal{I}_i(t)$. Agents in $\mathcal{I}_i(t)$ are inactive because they have not received the message yet. At the beginning of the process, $\mathcal{I}_i(t)$ will contain all agents except the one associated with n_s. Agents in $\mathcal{I}_d(t)$ are also inactive, after participating in the game and terminating their Agent Episode once the local horizon is reached. $\mathcal{I}_a(t)$, instead, includes the set of agents actively participating in the game at timestep t. Agents in $\mathcal{I}_i(t)$ are moved to $\mathcal{I}_a$ at time step $t+1$, if they have been forwarded the information, hence starting their Agent Episode.

Actions $\mathcal{A}^i_{i \in I}$. For any time step t, if agent i is in $\mathcal{I}_a(t)$, then, $\mathcal{A}^i$ contains two possible actions: forward the information to their neighbors or stay silent. The action set for agents in $\mathcal{I}_i(t)$ and $\mathcal{I}_d(t)$ is, instead, empty.

Environment Dynamics $\mathcal{P}$ and Network Update $\mathcal{U}$. The environment dynamics are defined by the transition function $\mathcal{P} : \mathcal{S} \times \mathcal{A}^1 \times \cdots \times \mathcal{A}^{|I|} \to \Delta(\mathcal{S})$, where $\Delta(\mathcal{S})$ represents the set of probability distributions over the state space $\mathcal{S}$. In our POSG model, we incorporate a general stochastic network update function, $\mathcal{U}$, controlling how the edges of the network change over time. This element allows us to capture various dynamics such as agent mobility or other factors that may affect a network's connectivity. More formally, at every timestep t, the graph structure is updated such that $\mathcal{G}_{t+1} = \mathcal{G}(\mathcal{V}, \mathcal{U}(\mathcal{E}_t))$. The message-forwarding mechanism is purposefully modeled as deterministic and, at each timestep t, if an active agent i forwards the message, all nodes in $\mathcal{N}_i(t)$ will receive it.

Observations $\mathcal{O}^i_{i \in I}$ and State Set $\mathcal{S}$. Each node in the graph has a set of three features observable by other neighboring agents at each time step t: neighborhood size, the number of messages transmitted, and its last action. The agents' observations are represented as the graph describing their one-hop neighborhood and the features associated with each node in this local structure. In our setting, the state $\mathcal{S}_t$ corresponds to the current graph structure $\mathcal{G}_t$ and the following information for each node: its features, the set to which the agent belongs ($\mathcal{I}_a$, $\mathcal{I}_d$, or $\mathcal{I}_i$), and the remaining steps of the local horizon.

Rewards $\mathcal{R}^i_{i \in I}$. At the end of each step every agent in $\mathcal{I}_a$ is issued with a reward signal with positive and negative components. The positive term rewards the agent based on its two-hop coverage, i.e. how many one- and two-hop neighbors have received the information. One of two penalties might be issued, based on the agent's behavior. If the agent has forwarded during its last action, it will participate in a shared transmission cost, punishing the agent for the number of messages sent by its neighborhood. Otherwise, it will receive penalties based on the unexploited coverage potential of neighbors

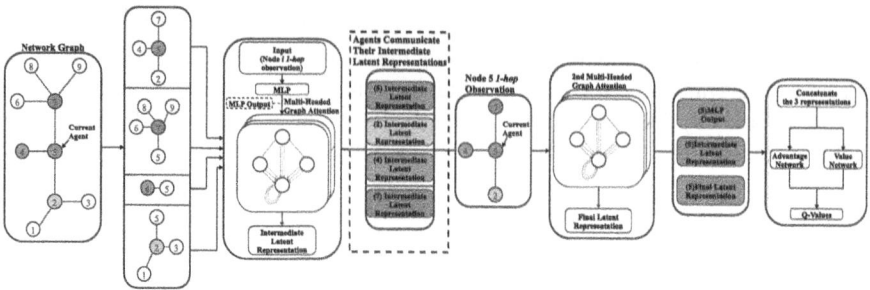

Fig. 1. Information flow from a single agent observation (5) to the produced Q-Values in L-DyAN.

who have not yet received the information. Formally, the reward signal for agent i at time t be defined as follows:

$$r_{i,t} = \frac{\upsilon(\mathcal{M}_i(t), t)}{|\mathcal{M}_i(t)|} - p(i,t), \mathcal{M}_i(t) = \bigcup_{u \in \mathcal{N}_i(t) \cup \{i\}} \mathcal{N}_u(t) \setminus \{i\} \quad (2)$$

$$p(i,t) = \begin{cases} m(\mathcal{N}_i(t), t), & \text{if } i \in \mathcal{T}(t) \\ \mu(\mathcal{N}_i(t), t), & \text{if } i \in \mathcal{I}_a(t) \setminus \mathcal{T}(t) \end{cases} \quad (3)$$

In Eq. 2, $\mathcal{M}_i(t)$ represents the set of two-hop neighbors of agent i at t. $\upsilon(\mathcal{M}_i, t)$ denotes the number of them that by timestep t have already received the message, while $p(i,t)$, defines the penalties assigned to agent i. The latter is further described in Eq. 3, where $\mathcal{T}(t)$ is the set of active agents that have forwarded the message at least once. Here $m(\mathcal{N}_i(t), t)$ denotes the sum of the number of messages transmitted by the current neighborhood of agent i by timestep t. The term $\mu(\mathcal{N}_i(t), t)$ instead defines the Maximum Normalized Coverage Potential of node i:

$$\mu(\mathcal{N}_i(t), t) = \frac{\max(\mathcal{C}_i(t))}{\sum \mathcal{C}_i(t)} \quad \mathcal{C}_i(t) = \{|\mathcal{N}_j(t)| : j \in \mathcal{N}_i(t) \cap \mathcal{I}_i(t)\} \quad (4)$$

On the one hand, we note that by assessing the ability of an agent's neighborhood to reach nodes beyond its immediate neighbors, Eq. 2, encourages agents to collectively cover more nodes through coordination within their vicinity. On the other hand, the neighborhood-shared transmission steers the agents away from redundancy, promoting efficient dissemination. Finally, the Maximum Normalized Coverage Potential counterbalances the shared transmission costs, by hastening transmission to nodes with highly populated neighborhoods that have not yet been reached.

4.2 Learning Approach

The idea behind L-DyAN and HL-DyAN is to encourage cooperation within dynamic neighborhoods, where links between nodes can form and/or disappear over time. We therefore propose a loss function comprising neighborhood experiences, the usage of

GAT layer(s) with dynamic attention [11], and the presence of a dueling network to separately estimate the state-value and the advantages for each action [12]. The choice of a GAT layer with dynamic attention is driven by its capability of capturing expressive attention mechanisms within a graph, a feature shown to be weaker in dot-product attention, as used in DGN [11].

Cooperative Dynamic Neighborhoods. During training, at each timestep t, the tuple $(\mathcal{O}_{\mathcal{I}_a(t)}, \mathcal{A}_{\mathcal{I}_a(t)}, \mathcal{R}_{\mathcal{I}_a(t)}, \mathcal{O}'_{\mathcal{I}_a(t)})$ is stored in a circular replay buffer with a fixed length. $\mathcal{O}_{\mathcal{I}_a(t)}$ indicates the set of observations of all agents in $\mathcal{I}_a(t)$, $\mathcal{A}_{\mathcal{I}_a(t)}$ the set of actions taken by these agents, $\mathcal{R}_{\mathcal{I}_a(t)}$ is the set of rewards, and $\mathcal{O}'_{\mathcal{I}_a(t)}$ the set of observations of agents in $\mathcal{I}_a(t)$ at the next timestep.

At each training step, we sample a random batch $\mathcal{B}$ from the replay buffer, with every sample containing the experience of some agent i and the ones of its current and active neighbors $\mathcal{N}_{i,\mathcal{I}_a(t)} = \mathcal{N}_i(t) \cap \mathcal{I}_a(t)$. The loss for each sample is computed not only based on the agent's own experience but also considering the experiences of its active neighbors. We denote $\mathcal{N}_{i,\mathcal{I}_a(t)}^{+i} = \mathcal{N}_{i,\mathcal{I}_a(t)} \cup \{i\}$ and define the loss function:

$$\mathcal{L}(\theta) = \frac{1}{|\mathcal{B}|} \sum_\mathcal{B} \frac{1}{|\mathcal{N}_{i,\mathcal{I}_a(t)}^{+i}|} \sum_{j \in \mathcal{N}_{i,\mathcal{I}_a(t)}^{+i}} \left(y_t^j - Q(o^j, a^j; \theta) \right)^2, \qquad (5)$$

where, for each agent j, y_t^j is the target return and $Q(o^j, a^j; \theta)$ the predicted Q value, parameterized with θ, given the observation o^j and action a^j. From this point onward, we will drop the superscript j when referring to o, a, r, and y as they will refer to a single experience. Additionally, we take advantage of the agents' short-lived experiences and perform n-step returns, with n equal to the local horizon (k). We note that the replay buffer is temporally sorted and organized such that every individual episode, ongoing or terminated with a length up to k, can be uniquely identified. If the buffer contains the remaining steps until the termination of the agent's episode, the n-step computation serves an unbiased value of the return: $y_t = \sum_{i=0}^{k-t} \gamma^i r_{t+i}$.

If the trajectory stored in the buffer contains only the next j steps before termination, y_t will be estimated as:

$$y_t = \sum_{i=0}^{j-1} \gamma^i r_{t+i} + \gamma^j Q(o_{t+i}, \mathrm{argmax}_{a' \in \mathcal{A}} Q(o_{t+i}, a'; \theta); \bar{\theta}), \qquad (6)$$

where θ is the current network and $\bar{\theta}$ is the target network.

Local-DyAN. The first architecture we propose is depicted in Fig. 1 and consists of an encoder module comprised of three different stages: one Multi Layer Perceptron (MLP) followed by two multi-headed GATs [29] with dynamic attention [11]. The final latent representation will comprise the concatenation of each stage output, which is then fed to a dueling network decoding the final representation into the predicted Q values. After each encoding stage, a ReLU activation function is applied.

We now describe the flow from the agent's observation to the Q values prediction and we show how it can be integrated into broadcast communication protocols. Agent

i's observation at time t is first fed to the MLP encoding stage. This results in a learned representation of the features belonging to agent i and its neighbors, denoted respectively $\mathbf{x}_i$ and $\mathbf{x}_j, \forall j \in \mathcal{N}_i(t)$. Following such encoding stage, the output of each of the M attention heads of the first GAT is:

$$\mathbf{x}_i^m = \alpha_{i,i}^m \mathbf{W} \mathbf{x}_i + \sum_{j \in \mathcal{N}_i(t)} \alpha_{i,j}^m \mathbf{W} \mathbf{x}_j \; \forall m \in \{0, ..., M-1\}, \tag{7}$$

where the dynamic attention α^m for the tuple (i, j), denoted as $\alpha_{i,j}^m$, is computed by:

$$\alpha_{i,j}^m = \frac{\exp\left(\mathbf{a}^\top \text{LeakyReLU}\left(\mathbf{W}[\mathbf{x}_i \parallel \mathbf{x}_j]\right)\right)}{\sum_{k \in \mathcal{N}_i(t) \cup \{i\}} \exp\left(\mathbf{a}^\top \text{LeakyReLU}\left(\mathbf{W}[\mathbf{x}_i \parallel \mathbf{x}_k]\right)\right)}, \tag{8}$$

where $\mathbf{a}$ and $\mathbf{W}$ are learned. We denote $\hat{\mathbf{X}}_i = \mathbf{x}_i^0 \| \mathbf{x}_i^1 \| ... \| \mathbf{x}_i^{M-1}$, where $\|$ is the concatenation operator, as the concatenation of every attention output. Through message passing, each agent i receives $\hat{\mathbf{X}}_j, \forall j \in \mathcal{N}_i(t)$. These new representations are fed to the second GAT layer, where the computation follows the same logic seen in Eq. 7 and 8, producing the embedding $\hat{\mathbf{Z}}_i$. Finally, the output of each encoding stage is concatenated in a final latent representation $\mathbf{H}_i$:

$$\mathbf{H}_i = \mathbf{x}_i \| \hat{\mathbf{X}}_i \| \hat{\mathbf{Z}}_i. \tag{9}$$

At this point, $\mathbf{H}_i$ is fed to the two separate streams of the dueling network, namely the value network V and the advantage network A, parameterized by two separate MLPs with parameters α and β, respectively. Let us denote the parameterization previous to the dueling network, which produced the final latent representation $\mathbf{H}_i$ given o, as δ. The predicted Q values are then obtained as:

$$Q(o, a; \delta, \alpha, \beta) = V(o; \delta, \alpha) + \left(A(o, a; \delta, \beta) - \frac{1}{|\mathcal{A}|} \sum_{a' \in \mathcal{A}} A(o, a'; \delta, \beta) \right). \tag{10}$$

We note that the encoding process described above harmoniously integrates with the communication mechanisms present in protocols deployed in the real world, such as OLSR. We envision every node (agent) in the network, feeding its neighborhood structure through the encoding process described above. Subsequently, every agent shares their intermediate representation $\hat{\mathbf{X}}_i$ with their one-hop neighbors, in a similar way to how nodes communicate their MPR sets in OLSR. Once the representations are collected, agents feed them to the second GAT layer, obtaining $\mathbf{H}_i$. However, the process we just described generates a communication overhead of size proportional to $\hat{\mathbf{X}}_i$, an aspect which might need to be further minimized in networks where bandwidth is particularly constrained [30,31]. This observation leads us to our second approach.

Hyperlocal-DyAN. Intending to generate less communication overhead, we design HL-DyAN, which resembles L-DyAN in its form. We replace the three encoding stages with a single GAT layer with dynamic attention. Within agent i's observation, we apply the GAT encoding process to every node, followed by a ReLU activation function. Finally, a global max-pooling layer is applied to summarize the most salient neighborhood characteristics, as shown in Fig. 2.

The rationale for this approach is that agents can make informed decisions by processing their one-hop neighborhood dynamics from each neighbor's perspective, eliminating the need to share their latent representations, as seen in L-DyAN.

In detail, agent i's observation at time t is fed to the GAT layer and, as opposed to L-DyAN, such an operation is repeated for every node within the local observation of agent i, producing

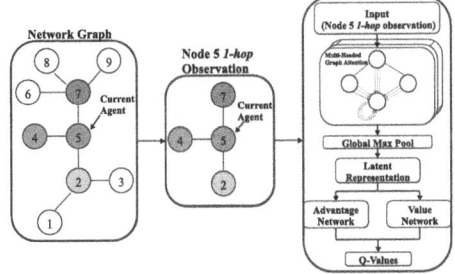

Fig. 2. The HL-DyAN architecture.

a set of latent representations comprising $\hat{\mathbf{Y}}_i$ and $\hat{\mathbf{Y}}_j, \forall j \in \mathcal{N}_i(t)$. We then perform global max pooling, obtained through a feature-wise max operation:

$$\mathbf{H}_i = \max_{j \in \mathcal{N}_i(t) \cup \{i\}} \hat{\mathbf{Y}}_j. \tag{11}$$

Finally, $\mathbf{H}_i$ is fed to the dueling network to compute the Q-Values (Eq. 10).

5 Experiments

Experimental Setup. We generated 50,000 connected graph topologies for training, each consisting of 50 nodes with a broadcasting range of 20 units and no constraints on the number of neighbors. For every learning algorithm, training was conducted five times adopting different random seeds (4, 9, 17, 42, 43) for 1 million agent steps. In each training episode, the environment randomly selected a graph as the initial graph and a node n_s as the source. To mitigate strong mobility pattern biases, a random state generator determined the nodes' directions, which was seeded anew at the beginning of every episode. The nodes' speed is defined in terms of distance units per step ($\frac{\text{units}}{\text{step}}$) and during training it was set to 6, generating random dynamic scenarios in which neighborhoods quickly evolve.

Furthermore, we utilized 4 distinct sets of connected starting graph topologies, not seen during training, for testing purposes. These sets comprised 50 nodes per graph with various constraints on the maximum node degree allowed (5, 10, 25, and 49). Our evaluation process involved testing each graph 50 times and selecting a different node as the source n_s in each iteration. To promote reproducibility and ensure the coherence of results, the same random state generator was used to control the nodes' movements across iterations for the same graph. Additional evaluations were conducted on the impact of the nodes' velocity, setting their speed to 1, 6, and 10 $\frac{\text{units}}{\text{step}}$. This allowed to simulate scenarios with different dynamics, where the agents' neighborhoods are more likely to suddenly change with higher velocity values, while this effect is reduced with lower values. Our analysis compared L-DyAN and HL-DyAN with the MPR heuristic and DGN. The DGN methodology excluded Temporal Relation Regularization, as it was unnecessary in our setting where agent interaction is temporally bounded by a short local horizon. To ensure a fair evaluation, we maintained consistent hyperparameters across all models.

Table 1. Evaluation of L-DyAN, HL-DyAN, MPR, and DGN - Coverage and Messages forwarded.

Initial Node Degree	Nodes Speed	Metric	L-DyAN	HL-DyAN	MPR	DGN
5 Neighbors	$1\frac{\text{unit}}{\text{step}}$	Messages	28.22	34.81	25.64	3.88
		Coverage	**93.78%**	90.02%	86.24%	24.34%
5 Neighbors	$6\frac{\text{units}}{\text{step}}$	Messages	24.05	33.40	8.21	6.69
		Coverage	79.34%	**83.64%**	34.02%	35.77%
5 Neighbors	$10\frac{\text{units}}{\text{step}}$	Messages	22.10	32.50	4.49	9.34
		Coverage	71.96%	**79.78%**	21.54%	45.32%
10 Neighbors	$1\frac{\text{unit}}{\text{step}}$	Messages	23.26	33.06	22.95	15.91
		Coverage	88.51%	**89.59%**	85.73%	24.35%
10 Neighbors	$6\frac{\text{units}}{\text{step}}$	Messages	24.31	37.01	8.10	7.03
		Coverage	86.33%	**91.69%**	37.98%	42.74%
10 Neighbors	$10\frac{\text{units}}{\text{step}}$	Messages	24.66	35.98	5.05	9.72
		Coverage	82.32%	**88.06%**	26.61%	52.15%
25 Neighbors	$1\frac{\text{unit}}{\text{step}}$	Messages	23.74	34.15	24.93	3.28
		Coverage	90.44%	**93.84%**	92.86%	26.33%
25 Neighbors	$6\frac{\text{units}}{\text{step}}$	Messages	24.29	36.35	10.03	6.96
		Coverage	88.19%	**92.23%**	46.80%	44.02%
25 Neighbors	$10\frac{\text{units}}{\text{step}}$	Messages	24.46	35.78	5.75	9.50
		Coverage	84.79%	**89.87%**	30.32%	51.89%
49 Neighbors	$1\frac{\text{unit}}{\text{step}}$	Messages	22.99	34.34	23.92	3.47
		Coverage	89.73%	**91.96%**	88.93%	27.16%
49 Neighbors	$6\frac{\text{units}}{\text{step}}$	Messages	24.81	36.61	9.39	6.81
		Coverage	88.86%	**92.42%**	43.49%	41.33%
49 Neighbors	$10\frac{\text{units}}{\text{step}}$	Messages	23.88	35.55	4.81	9.81
		Coverage	82.35%	**88.83%**	26.68%	52.69%

Results. Table 1 shows the results of our experiments in terms of coverage and messages required to achieve it, presenting the means and standard deviations. Our proposed methods, L-DyAN and HL-DyAN, consistently demonstrate higher coverage across different scenarios compared to MPR and DGN. In particular, HL-DyAN achieves the highest coverage with a mean of 90.37%. However, this comes at the cost of a higher number of messages, with HL-DyAN generating an average of 34.85 messages per

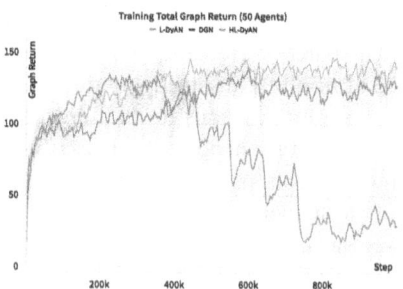

Fig. 3. Training performance in terms of average return achieved by the agents.

episode. DyAN, while slightly less successful in coverage (87.70%), requires significantly fewer messages (24.32), indicating its suitability in scenarios where message efficiency is prioritized over coverage.

As the nodes' speed increases, the performance gap between our proposed methods and MPR widens, with L-DyAN and HL-DyAN maintaining superior coverage across all tests with increased speed. This is further supported by additional tests we conducted with the nodes' speed set to 10, obtaining in average a coverage of 82.35%, resp. 88.83% with 23.88 resp. 36.61 messages, by L-DyAN and HL-DyAN. MPR, instead, struggled to reach 27% of coverage. Additionally, in the more dynamic scenarios with speed set to be greater than 1, the maximum node degree negatively influences the performance of all methods, but L-DyAN and HL-DyAN consistently outperform the other algorithms, which fail to reach 50% coverage. This indicates the robustness of L-DyAN and HL-DyAN to dynamic scenarios with less dense neighborhoods. In more static scenarios where the node speed is set to 1, L-DyAN and MPR reveal to be the more efficient reaching, respectively, 14.84 and 14.54% of coverage per message.

Figure 3 illustrates the training progress of L-DyAN, HL-DyAN, and DGN over multiple cycles using five distinct random seeds. Our proposed methodologies, L-DyAN and HL-DyAN, demonstrated an average of the total graph returns (sum of all the agents returns) of 141.47 ± 15.63 and 127.50 ± 13.57, respectively. In contrast, the training trajectory of DGN indicates a more brittle learning progress, unable to learn an effective multi-agent strategy for this task.

The results underscore the efficiency of L-DyAN and HL-DyAN in learning effective multi-agent strategies balancing message efficiency with coverage consistently across various scenarios. MPR falls short in both slightly and very dynamic and/or sparsely connected environments, with its performance worsening as nodes are faster and their starting neighborhood more sparse. The results also highlight the adaptability of L-DyAN and HL-DyAN in varying network densities and node velocities, making them suitable for a wide range of dynamic network environments. Additionally, the training behavior of the learning algorithms highlights how can tackle networked environments where the presence of the agents and the structure of their neighborhoods change dynamically.

6 Conclusion and Future Work

In this work, we captured the problem of information dissemination in dynamic broadcast networks in a novel POSG formulation and proposed two MARL methods to solve the task, namely L-DyAN and HL-DyAN. Our experiments showed how these methods outperform in terms of coverage and message efficiency both DGN and a popular heuristic employed in real-world networking scenarios.

Our future research agenda includes investigating more structured group communication tasks, where, for example, coverage is desired only for a subset of nodes or nodes with higher priority. We will also study methods to enable more controlled trade-offs between coverage and forwarded messages, as well as their application in deployed protocols for physical computer networks. Orthogonally, we will investigate the application of our approach to the dissemination of information in other domains, such as social networks and computational social choice.

References

1. Tonguz, O., Wisitpongphan, N., Bai, F., Mudalige, P., Sadekar, V.: Broadcasting in VANET. In: 2007 Mobile Networking for Vehicular Environments, pp. 7–12 (2007). https://doi.org/10.1109/MOVE.2007.4300825
2. Ibrahim, B.F., Toycan, M., Mawlood, H.A.: A comprehensive survey on VANET broadcast protocols. In: 2020 International Conference on Computation, Automation and Knowledge Management (ICCAKM), pp. 298–302 (2020). https://doi.org/10.1109/ICCAKM46823.2020.9051462
3. Suri, N., et al.: Comparing performance of group communications protocols over SCB versus routed manet networks. In: 2022 IEEE Military Communications Conference (MILCOM), MILCOM 2022, pp. 1011–1017 (2022). https://doi.org/10.1109/MILCOM55135.2022.10017772
4. Foerster, J.N., Assael, Y.M., de Freitas, N., Whiteson, S.: Learning to communicate with deep multi-agent reinforcement learning. In: Proceedings of the 30th International Conference on Neural Information Processing Systems, NIPS 2016, Red Hook, NY, USA, pp. 2145–2153. Curran Associates Inc. (2016). ISBN 9781510838819
5. Buşoniu, L., Babuška, R., De Schutter, B.: Multi-agent reinforcement learning: an overview. In: Srinivasan, D., Jain, L.C. (eds.) Innovations in Multi-Agent Systems and Applications - 1, vol. 310, pp. 183–221. Springer, Heidelberg (2010). https://doi.org/10.1007/978-3-642-14435-6_7
6. Sukhbaatar, S., Szlam, A., Fergus, R.: Learning multiagent communication with backpropagation. In: Proceedings of the 30th International Conference on Neural Information Processing Systems, NIPS 2016, Red Hook, NY, USA, pp. 2252–2260. Curran Associates Inc. (2016). ISBN 9781510838819
7. Peng, P., et al.: Multiagent bidirectionally-coordinated nets: emergence of human-level coordination in learning to play starcraft combat games (2017)
8. Jiang, J., Lu, Z.: Learning attentional communication for multi-agent cooperation. In: Proceedings of the 32nd International Conference on Neural Information Processing Systems, NIPS 2018, Red Hook, NY, USA, pp. 7265–7275. Curran Associates Inc. (2018)
9. Das, A., et al.: TarMAC: targeted multi-agent communication. In: Chaudhuri, K., Salakhutdinov, R. (eds.) Proceedings of the 36th International Conference on Machine Learning. Proceedings of Machine Learning Research, vol. 97, pp. 1538–1546. PMLR (2019). https://proceedings.mlr.press/v97/das19a.html
10. Jiang, J., Dun, C., Huang, T., Lu, Z.: Graph convolutional reinforcement learning. In: International Conference on Learning Representations (2020). https://openreview.net/forum?id=HkxdQkSYDB
11. Brody, S., Alon, U., Yahav, E.: How attentive are graph attention networks? In: International Conference on Learning Representations (Poster) (2022). https://openreview.net/forum?id=F72ximsx7C1
12. Wang, Z., Schaul, T., Hessel, M., Van Hasselt, H., Lanctot, M., De Freitas, N.: Dueling network architectures for deep reinforcement learning. In: Proceedings of the 33rd International Conference on International Conference on Machine Learning, ICML 2016, vol. 48, pp. 1995–2003. JMLR.org (2016)
13. Dearlove, C., Clausen, T.H.: Optimized Link State Routing Protocol Version 2 (OLSRv2) and MANET Neighborhood Discovery Protocol (NHDP) Extension TLVs. RFC 7188 (2014). https://www.rfc-editor.org/info/rfc7188
14. Guille, A., Hacid, H., Favre, C., Zighed, D.A.: Information diffusion in online social networks: a survey. SIGMOD Rec. **42**(2), 17–28 (2013). https://doi.org/10.1145/2503792.2503797. ISSN 0163-5808

15. Ye, Z., Zhou, Q.: Performance evaluation indicators of space dynamic networks under broadcast mechanism. Space: Sci. Technol. **2021** (2021). https://doi.org/10.34133/2021/9826517
16. Ma, X., Zhang, J., Yin, X., Trivedi, K.S.: Design and analysis of a robust broadcast scheme for VANET safety-related services. IEEE Trans. Veh. Technol. **61**(1), 46–61 (2012). https://doi.org/10.1109/TVT.2011.2177675
17. Sutton, R.S., Barto, A.G.: Reinforcement Learning: An Introduction. A Bradford Book, Cambridge (2018). ISBN 0262039249
18. Puterman, M.L.: Markov Decision Processes: Discrete Stochastic Dynamic Programming, 1st edn. Wiley (1994). ISBN 0471619779
19. Hausknecht, M.J., Stone, P.: Deep recurrent q-learning for partially observable MDPs. In: 2015 AAAI Fall Symposia, Arlington, Virginia, USA, 12–14 November 2015, pp. 29–37. AAAI Press (2015)
20. Albrecht, S.V., Christianos, F., Schäfer, L.: Multi-Agent Reinforcement Learning: Foundations and Modern Approaches. MIT Press (2023). https://www.marl-book.com
21. Ahmed, I.H., et al.: Deep reinforcement learning for multi-agent interaction. AI Commun. **35**(4), 357–368 (2022)
22. Qayyum, A., Viennot, L., Laouiti, A.: Multipoint relaying for flooding broadcast messages in mobile wireless networks. In: Proceedings of the 35th Annual Hawaii International Conference on System Sciences, pp. 3866–3875 (2002). https://doi.org/10.1109/HICSS.2002.994521
23. Garey, M.R., Johnson, D.S.: Computers and Intractability: A Guide to the Theory of NP-Completeness. W. H. Freeman (1979). ISBN 0-7167-1044-7
24. Macker, J.: RFC 6621: simplified multicast forwarding (2012)
25. Yahja, A., Kaviani, S., Ryu, B., Kim, J.H., Larson, K.A.: DeepADMR: a deep learning based anomaly detection for MANET routing. In: IEEE Military Communications Conference, MILCOM 2022, Rockville, MD, USA, 28 November–2 December 2022, pp. 412–417. IEEE (2022)
26. Kaviani, S., et al.: DeepCQ+: robust and scalable routing with multi-agent deep reinforcement learning for highly dynamic networks. In: 2021 IEEE Military Communications Conference, MILCOM 2021, San Diego, CA, USA, 29 November–2 December 2021, pp. 31–36. IEEE (2021)
27. Kaviani, S., et al.: DeepMPR: enhancing opportunistic routing in wireless networks through multi-agent deep reinforcement learning (2023)
28. Schulman, J., Wolski, F., Dhariwal, P., Radford, A., Klimov, O.: Proximal policy optimization algorithms (2017)
29. Veličković, P., Cucurull, G., Casanova, A., Romero, A., Liò, P., Bengio, Y.: Graph attention networks. In: International Conference on Learning Representations (2018). https://openreview.net/forum?id=rJXMpikCZ
30. Suri, N., et al.: Adaptive information dissemination over tactical edge networks. In: 2023 International Conference on Military Communications and Information Systems (ICMCIS), pp. 1–7 (2023). https://doi.org/10.1109/ICMCIS59922.2023.10253585
31. Galliera, R., et al.: Learning to sail dynamic networks: the marlin reinforcement learning framework for congestion control in tactical environments. In: 2023 IEEE Military Communications Conference (MILCOM), MILCOM 2023, pp. 424–429 (2023). https://doi.org/10.1109/MILCOM58377.2023.10356270

Toward Fair and Strategyproof Tournament Rules for Tournaments with Partially Transferable Utilities

David Pennock[1], Ariel Schvartzman[2], and Eric Xue[3]($\boxtimes$)

[1] DIMACS, Rutgers University, New Brunswick, NJ 08901, USA
dpennock@dimacs.rutgers.edu
[2] Google Research, Mountain View, CA 94043, USA
aschvartzman@google.com
[3] Princeton University, Princeton, NJ 08544, USA
ex3782@princeton.edu

Abstract. A tournament on n agents is a complete oriented graph with the agents as vertices and edges that describe the win-loss outcomes of the $\binom{n}{2}$ matches played between each pair of agents. The winner of a tournament is determined by a *tournament rule* that maps tournaments to probability distributions over the agents. We want these rules to be fair (choose a high-quality agent) and robust to strategic manipulation. Prior work has shown that under minimally fair rules, manipulations between two agents can be prevented when utility is nontransferable but not when utility is completely transferable. We introduce a partially transferable utility model that interpolates between these two extremes using a selfishness parameter λ. Our model is that an agent may be willing to lose on purpose, sacrificing some of her own chance of winning, but only if the colluding pair's joint gain is more than λ times the individual's sacrifice.

We show that no fair tournament rule can prevent manipulations when $\lambda < 1$. We computationally solve for fair and manipulation-resistant tournament rules for $\lambda = 1$ for up to 6 agents. We conjecture and leave as a major open problem that such a tournament rule exists for all n. We analyze the trade-offs between "relative" and "absolute" approximate strategyproofness for previously studied rules and derive as a corollary that all of these rules require $\lambda \geq \Omega(n)$ to be robust to manipulation. We show that for stronger notions of fairness, non-manipulable tournament rules are closely related to tournament rules that witness decreasing gains from manipulation as the number of agents increases.

Keywords: Tournaments · Computational Social Choice

Supplementary Information The online version contains supplementary material available at https://doi.org/10.1007/978-3-031-73903-3_12.

© The Author(s), under exclusive license to Springer Nature Switzerland AG 2025
R. Freeman and N. Mattei (Eds.): ADT 2024, LNAI 15248, pp. 174–188, 2025.
https://doi.org/10.1007/978-3-031-73903-3_12

1 Introduction

A tournament on n agents is a complete oriented graph in which the agents are vertices and an edge from agent i to agent j means "agent i defeats agent j". These structures frequently arise in sports as the outcome of $\binom{n}{2}$ pairwise matches between n agents or teams. However, tournaments can arise whenever the performance of every two agents is comparable (e.g., agents are candidates in an election and edges are pairwise majority votes).

A tournament rule maps a tournament to a probability distribution over the agents. These probabilities encode the likelihood that each agent is declared the tournament winner, or prescribe how to divide up a monetary reward [7]. While a tournament rule should be fair in that it chooses some qualified agent who beats many other agents, it also should not reward manipulations: for example, losing a match on purpose should not improve an agent's or their co-conspirator's chances of winning the tournament. If the rule is manipulable, then agents may act in ways that undermine the primary goal of choosing a highly qualified winner. In fact, instances of these actions are not unheard of in sports. At the London 2012 Olympic Games, four women's doubles teams were disqualified for attempting to throw their final matches in the round-robin group stage in order to earn a more favorable seed in the knockout stage of the tournament.

Unfortunately, prior work has shown that fairness and non-manipulability are largely incompatible. A prevailing notion of fairness studied by prior work [1, 2, 6, 10, 11] is Condorcet consistency. A tournament rule is Condorcet consistent if, whenever one agent beats all other agents, the undefeated agent wins the tournament with certainty.

Altman, Procaccia, and Tennenholtz [2] showed that any deterministic rule that satisfies this notion is susceptible to pairwise manipulations: for any Condorcet consistent rule, there exist tournaments in which two agents can influence the choice of winner by colluding to reverse the outcome of their match.

Altman and Kleinberg [1] extended this work to randomized rules that map tournaments to probability distributions over agents. They showed that there exist Condorcet consistent and pairwise non-manipulable rules when two agents collude only if one of them can strictly improve her probability of winning at no cost to the other. Rules that are pairwise non-manipulable under this assumption are said to be 2-Pareto non-manipulable (2-PNM). However, no Condorcet consistent rule exists when utility is completely transferable—that is, when two agents only care about the probability that at least one of them wins the tournament. Instead, the authors demonstrated rules that are approximately Condorcet consistent and pairwise non-manipulable in this setting, which the authors term 2-strongly non-manipulable (2-SNM). Another line of work [6, 10, 11] sought rules that were fair and approximately 2-SNM.

Motivated by the fact that collusion and the deliberate throwing of matches in sports occur less frequently than the negative results of prior work imply, we extend prior work to the setting in which utility is partially transferable. These settings are natural. For example, consider a setting in which a tournament rule is used to prescribe a division of monetary reward among the participants.

Because the reward is divisible, if two agents could improve their share of the reward by fixing the outcome of their match, then they may choose to do so and redistribute their winnings later so that the collusion is mutually beneficial. But collusions are rarely so frictionless in reality. There could be uncertainty as to whether the agent that benefits from fixing the outcome of the match will follow through with the redistribution. There could be penalties for agents found to have thrown their matches. Or there could be factors beyond the outcome of the tournament that matter, such as an agent's reputation, that a loss would negatively affect. With these frictions, agents would not be completely altruistic to their partner nor completely selfish. Instead, agents would care more about winning themselves but may be willing to sacrifice their own probability if it achieves a significant proportional gain for their partner.

We model each agent's values for her own probability of winning *and* for her collusion partner's probability of winning as being in some ratio and extend prior notions of non-manipulability by introducing a term that accounts for the range of selfishness of agents. More specifically, we say a rule is 2-NM$_\lambda$ if no agent can collude with another to improve her probability of winning by at least a $\lambda + 1$ factor of the decrease in probability witnessed by her colluding partner. Stated another way, under a 2-NM$_\lambda$ rule, no pairwise collusions would occur if we assume that each agent would not sacrifice her own chances of winning unless her partner gains at least $\lambda + 1$ times the amount that she loses.

We show that this model connects the notions of Pareto and strong non-manipulability by varying λ. Moreover, we conjecture that there exists a tournament rule that is monotone, Condorcet consistent, and 2-NM$_1$, implying that it is possible to prevent deliberate loss and collusion, as long as each agent weighs her own probability of winning twice as much as her opponents'. However, we show that none of the rules proposed in five previous papers [1,3,6,10,11] satisfy this combination of conditions by demonstrating how these rules trade-off between λ, our notion of relative approximate strategyproofness, and the established notion of absolute approximate strategyproofness [10].

In a separate direction, we introduce another notion of fairness, termed dominant sub-tournament consistency (DSTC), and show that several natural rules satisfy this condition. Intuitively, a rule is DSTC if the addition of an agent that loses to the original agents does not affect their probabilities. A closely related notion is top cycle consistency (TCC), which requires the winner to come from the top cycle with certainty. We show that within these notions of fairness, finding a rule that is 2-NM$_\lambda$ reduces to finding a rule that witnesses gains from manipulation that vanishes as the number of agents increases.

1.1 Related Work

For a broad discussion of recent developments on tournament design, see Suksompong's excellent survey [12]. We discuss work closely related to ours.

Altman and Kleinberg [1] and Altman et al. [2] were the first to consider the question of strategic manipulations of tournaments by agents. Their main conclusion is that Condorcet consistency and strong non-manipulability are directly at odds: no tournament rule, even randomized ones, can satisfy both properties. Later, Schneider et al. [10] considered a relaxation of the problem: they sought tournament rules that are Condorcet consistent and are minimally manipulable. Their main result is that the Randomized Single Bracket Elimination (RSEB) rule is 2-SNM-1/3, meaning that the most probability that any pair can gain is 1/3, and this is optimal among all Condorcet-consistent rules. This result was later strengthened to show that the Randomized King of the Hill (RKotH) rule is also 2-SNM-1/3 and cover consistent, a notion strictly stronger than Condorcet consistent [11]. Recent discoveries include a rule that is 3-SNM-31/60, meaning that the most probability that any coalition of three agents can gain is 31/60, the first explicit rule that is 3-SNM-α for $\alpha < 1$ [5], and a different rule that is 3-SNM-1/2 [9]. Parallel lines of work have considered variations on this problem, including probabilistic tournaments [6] and tournaments with prize vectors for multiple places rather than only one prize for the winner [4].

2 Preliminaries

Definition 1 (Tournament). *A tournament $T = (A, \succ_T)$ is a pair where A is a finite set of agents and $\succ_T$ is a complete asymmetric binary relation over A that describes the outcomes of the $\binom{|A|}{2}$ matches played between each pair of distinct agents. For agents $i \neq j \in A$, we write $i \succ_T j$ if i dominates j in T. Let $\mathcal{T}_n$ denote the set of tournaments where $[n]$ is the set of agents.*

Definition 2 (Tournament rule). *A tournament rule on n agents $r^{(n)} : \mathcal{T}_n \to \Delta^n$ maps a tournament $T \in \mathcal{T}_n$ to a probability distribution over the agents. A tournament rule r is a family of tournament rules on n agents $\{r^{(n)}\}_{n=1}^{\infty}$. For all $n \in \mathbb{N}$ and $T \in \mathcal{T}_n$, we write $r(T) := r^{(n)}(T)$, and for $i \in [n]$, we write $r_i(T)$ to denote the probability that i wins T under r.*

2.1 Fairness Properties

A desirable tournament rule should choose the most qualified agent as the winner. In line with this reasoning, we want a rule to choose an undefeated agent with probability 1 since this agent is clearly better than her opponents.

Definition 3 (Condorcet consistency). *A tournament rule on n agents $r^{(n)}$ is Condorcet consistent (CC) if for all $T \in \mathcal{T}_n$, $r_i^{(n)}(T) = 1$ when there exists i such that $i \succ_T j$ for all $j \neq i$. A tournament rule r is CC if $r^{(n)}$ is CC for all n.*

Note that Condorcet consistency is quite a minimal notion of fairness since it is binding only when there is an agent that is clearly superior than the others. Unfortunately, it is often the case that no such agent exists. The following notions of fairness seek to restrict the subset of agents that should be named the winner in such cases by eliminating those who are in some sense worse than her opponents.

Definition 4 (Top cycle consistency). *A subset of agents S is the top cycle in tournament T if it is the minimal subset of agents such that $i \succ_T j$ for all $i \in S, j \in [n] \setminus S$. The top cycle of a tournament always exists and is unique. Let $TC(T)$ denote the top cycle of T. A tournament rule on n agents $r^{(n)}$ is top cycle consistent (TCC) if for all $T \in \mathcal{T}_n$, $r_i^{(n)}(T) = 0$ for all $i \in [n] \setminus TC(T)$. A tournament rule r is TCC if $r^{(n)}$ is TCC for all n.*

Top cycle consistency extends Condorcet consistency quite naturally: Condorcet consistency requires that an undefeated agent be declared the winner, while top cycle consistency requires this winner to come from the smallest undefeated subset. Moreover, since the agents in the top cycle are undefeated by those outside of the top cycle, they are in some sense better. On the other hand, no agent in the top cycle is clearly superior than the others since every agent in the top cycle is defeated by another in the top cycle.

Definition 5 (Cover consistency). *For $i \neq j$, we say i covers j if $i \succ_T j$ and $j \succ_T k \implies i \succ_T k$ for all $k \in [n] \setminus \{i, j\}$. Moreover, we say j is covered if there exists $i \in [n]$ such that i covers j. A tournament rule on n agents $r^{(n)}$ is cover consistent if for all $T \in \mathcal{T}_n$, $r_j^{(n)}(T) = 0$ whenever j is covered. A tournament rule r is cover consistent if $r^{(n)}$ is cover consistent for all n.*

Cover consistency refines top cycle consistency by further restricting the set of potential winners. If i covers j, then not only did i defeat j, but i also defeated everyone that j defeated. Thus, covered agents are worse than the agents that cover them in some sense.

Definition 6 (Dominant sub-tournament consistency). *For a subset of agents $S \subseteq [n]$ and a tournament $T \in \mathcal{T}_n$, let $T|_S = (S, \{(i,j) \in S \times S : i \succ_T j\})$ denote the subgraph induced by S. $T|_S$ is a dominant sub-tournament in tournament T if $i \succ_T j$ for all $i \in S, j \in [n] \setminus S$. A tournament rule r is dominant sub-tournament consistent (DSTC) if $r_i(T|_S) = r_i(T)$ for all $i \in S$.*

Dominant sub-tournament consistency strengthens top cycle consistency in a different direction than cover consistency. Rather than narrow down the set of potential winners, dominant sub-tournament consistency requires that the probability of choosing a certain member of the top cycle as the winner is the same as the probability of choosing her if the agents outside the top cycle were removed. DSTC ensures that adding a totally inferior team, like a high-school team joining a professional league, will not affect who wins the tournament, similar to the Independence of Irrelevant Alternatives axiom in Arrow's Impossibility Theorem. To the best of our knowledge, dominant sub-tournament consistency has not been considered before in the tournament literature.

The following result formalizes the hierarchy of fairness conditions.

Proposition 1 (Fairness hierarchy). *Any tournament rule that satisfies either cover consistency or DSTC satisfies TCC. Moreover, any TCC rule is CC.*

Proof. Let $T \in \mathcal{T}_n$. Suppose r is a cover consistent tournament rule, and consider any $j \notin TC(T)$. Observe that any $i \in TC(T)$ covers j since by definition of the top cycle, we have that $i \succ_T j$, and for any $k \in [n]$ such that $j \succ_T k$, we have that $k \notin TC(T)$, so $i \succ_T k$. Since r is cover consistent and j is covered, we have that $r_j(T) = 0$. Thus, r is TCC.

Now, suppose r is a DSTC tournament rule. By definition, $TC(T)$ is a dominant sub-tournament of T. Thus, $\sum_{i \in TC(T)} r_i(T) = \sum_{i \in TC(T)} r_i(T|_{TC(T)}) = 1$. It follows that $r_i(T) = 0$ for all $i \in [n] \setminus TC(T)$, so r is TCC.

Now, suppose r satisfies TCC, and note that whenever some agent i is undefeated in T, i is the only member of $TC(T)$: i dominates every $j \neq i$, and no proper subset of $\{i\}$ satisfies this property. Thus, $r_i(T) = 1$ and r is CC. □

2.2 Non-manipulability Properties

In addition to satisfying some notion of fairness, tournament rules should be robust to manipulation. In this work, we consider manipulations where a single agent purposefully loses her match against one of her opponents and manipulations where two agents collude to reverse the outcome of their match.

Definition 7 (S-adjacent). $T, T' \in \mathcal{T}_n$ are *S-adjacent* where $S \subseteq [n]$ if $i \succ_T j \iff i \succ_{T'} j$ for $i \neq j \in [n] \setminus S$. In other words, T and T' are S-adjacent if they coincide on every match except possibly those between agents in S.

When utilities are nontransferable, two agents are willing to collude only if one of them can strictly improve her probability of winning at no cost to the other. Formally, distinct agents $i, j \in [n]$ collude from tournament T to tournament T' only if $\max\{r_i(T') - r_i(T), r_j(T') - r_j(T)\} > 0$ and $\min\{r_i(T') - r_i(T), r_j(T') - r_j(T)\} \geq 0$. Thus, to incentivize agents against such manipulations, a tournament rule must satisfy the following notion of non-manipulability.

Definition 8 (2-Pareto non-manipulability). *A tournament rule r is 2-Pareto non-manipulable (2-PNM) if for all $i \neq j \in [n]$ and $\{i, j\}$-adjacent tournaments $T \neq T' \in \mathcal{T}_n$, either (1) $\min\{r_i(T') - r_i(T), r_j(T') - r_j(T)\} < 0$ or (2) $\max\{r_i(T') - r_i(T), r_j(T') - r_j(T)\} \leq 0$.*

Altman and Kleinberg [1] give a rule that is monotone, TCC, and 2-PNM. The barrier to pairwise manipulation is much lower when utilities are completely transferable since two agents only care about the probability that at least one of them wins the tournament. In other words, i and j collude from T to T' only if $r_i(T') + r_j(T') > r_i(T) + r_j(T)$. Under this utility model, an agent may be willing to sacrifice and shift a significant portion of her probability to her partner in crime. Thus, tournament rules must satisfy a stronger notion of non-manipulability in this setting.

Definition 9 (2-strong non-manipulability). *A tournament rule r is 2-strongly non-manipulable (2-SNM) if $r_i(T') + r_j(T') \leq r_i(T) + r_j(T)$ for all $i \neq j \in [n]$ and $\{i, j\}$-adjacent tournaments $T \neq T' \in \mathcal{T}_n$.*

Prior work has shown that no Condorcet consistent tournament rule is 2-SNM. However, despite this strong impossibility result, instances of collusion are relatively infrequent in the real world, suggesting that settings where utilities are completely transferable are uncommon. On the other hand, instances of collusion are not unheard of, suggesting that utility is neither always nontransferable.

In this paper, we consider a third utility model in which utilities are partially transferable: distinct agents i and j collude from tournament T to tournament T' only if $r_i(T') + r_j(T') > r_i(T) + r_j(T) + \lambda \max\{r_i(T) - r_i(T'), r_j(T) - r_j(T')\}$. In this model, two agents always collude if both of them improve their chances of winning and never collude if both of their chances decrease. The interesting case is when one agent improves her chances at the expense of the other. One interpretation of this necessary condition is that agents would rather win the tournament themselves but are still willing to collude if the gain in probability is significantly larger than each agent's loss. Here, λ is a parameter that measures how transferable utility is. Note that when λ is low, utilities are more transferable. We will later see how λ can be interpreted as agents' level of selfishness. We now define a notion of non-manipulability for this model.

Definition 10 (2-non-manipulability for λ). *A tournament rule r is 2-non-manipulable for $\lambda \geq 0$ (2-NM$_\lambda$) if $r_i(T') + r_j(T') \leq r_i(T) + r_j(T) + \lambda \max\{r_i(T) - r_i(T'), r_j(T) - r_j(T')\}$ for all $i \neq j \in [n]$ and $\{i,j\}$-adjacent tournaments $T \neq T' \in \mathcal{T}_n$. We say r is 2-NM$_\infty$ if $r_i(T') + r_j(T') \leq r_i(T) + r_j(T) + \lim_{\lambda \to \infty} \lambda \max\{r_i(T) - r_i(T'), r_j(T) - r_j(T')\}$ (where the limit is taken in the extended reals).*

Observe that when $\lambda = 0$, our notion of non-manipulability coincides with strong non-manipulability. Moreover, we show that our notion coincides with Pareto non-manipulability when $\lambda = +\infty$. We remark that we do not interpret 2-NM$_\lambda$ as an approximation to 2-SNM. Unlike approximation algorithms, a tournament designer who finds herself faced with e.g., agents who value their opponents chances of winning as much as their own (completely transferable utility) may not find it in her best interest to use a 2-NM$_\lambda$ tournament rule for some $\lambda > 1$. Rather, λ is meant to model the behavior of the agents.

Proposition 2. *A tournament rule is 2-PNM if and only if it is 2-NM$_\infty$.*

By Proposition 2, our notion of non-manipulability generalizes strong and Pareto non-manipulability while connecting the two. As in previous work [10, 11], we are interested in approximately non-manipulable tournament rules: rules under which no two agents can collude to gain in joint probability more than α more than each agent's loss (weighted by λ). We will see that there is a range of λ for which fair and non-manipulable tournament rules do not exist. For λ in this range, it may be better to design approximately non-manipulable tournament rules tailored to λ than use a 2-NM$_{\lambda'}$ tournament rule for some $\lambda' > \lambda$.

Definition 11 (2-non-manipulability up to α for λ). *A tournament rule r is 2-non-manipulable up to α for $\lambda \geq 0$ (2-NM$_\lambda$-α) if $r_i(T') + r_j(T') \leq r_i(T) +*

$r_j(T) + \lambda \max\{r_i(T) - r_i(T'), r_j(T) - r_j(T')\} + \alpha$ for all $i \neq j \in [n]$ and $\{i,j\}$-adjacent tournaments $T \neq T' \in \mathcal{T}_n$.

In addition to being robust against pairwise manipulations, a tournament rule should be robust to the intentional throwing of matches.

Definition 12 (Monotonicity). *A tournament rule is monotone if $r_i(T) \geq r_i(T')$ for all $\{i,j\}$-adjacent tournaments $T \neq T' \in \mathcal{T}_n$ such that $i \succ_T j$.*

Intuitively, monotonicity says that no agent should be able to improve her chances of winning by deliberately losing one of her matches. Thus, agents have an incentive to win each of their matches under monotone rules. Violations of this property should be seen as quite severe.

Proposition 3. *Let r be a 2-NM_λ tournament rule for some $\lambda > 0$, then the following two statements are equivalent.*

1. *r is monotone*
2. *For all $i \neq j \in [n]$ and $\{i,j\}$-adjacent tournaments $T \neq T' \in \mathcal{T}_n$ such that $i \prec_T j$, $r_i(T') - r_i(T) \leq (\lambda + 1)(r_j(T) - r_j(T'))$*

Proposition 3 offers a natural interpretation of the parameter λ and the 2-NM_λ property for monotone tournament rules: λ is how much each agent weighs her own probability of winning over others' probabilities of winning and a tournament rule is 2-NM_λ if switching the outcome of a match does not increase the probability of winning for the new winner by more than a $\lambda + 1$ factor over the loss of the new loser. Note that Proposition 3 does not hold for $\lambda = 0$. Indeed, monotonicity and 2-SNM are independent properties: neither implies the other.

2.3 Tournament Rules

In this section, we define several tournament rules. See Table 1 for a summary of what was known about them prior to this work (to the best of our knowledge).

1. The **Iterative Condorcet Rule** (ICR) chooses the undefeated agent if one exists. Otherwise, eliminate an agent uniformly at random and repeat.
2. The **Randomized Voting Caterpillar** rule (RVC) begins by choosing a permutation of the agents uniformly at random. In the first iteration, RVC eliminates the loser between the first and second agents in the permutation. In each subsequent iteration until only one agent remains, RVC eliminates the loser between the previous winner and the next agent in the permutation.
3. The **Top Cycle Rule** (TCR) chooses an agent uniformly at random from the top cycle and declares her the winner.
4. A single elimination bracket is a complete binary tree whose leaves are labeled by a permutation of the agents. Each node is labeled by the winner of the match between its two children. The winner of the bracket is the agent labeling the root node. The **Randomized Single Elimination Bracket** rule (RSEB) introduces $2^{\lceil \log n \rceil} - n$ dummy agents who lose to the existing agents, chooses a bracket uniformly at random, and declares the winner of this bracket the winner of the tournament.

Table 1. Summary of relevant prior results. Unless stated otherwise, results come from the paper that proposed the tournament rule.

Rule	Monotone?	Fairness	2-PNM?	2-SNM-α
ICR [1]	Yes	TCC	Yes	$\alpha \geq \frac{1}{2} - \frac{1}{n(n-1)}$ [10]
RVC [1]	Yes	TCC	Yes	$\alpha \geq \frac{1}{2} - \frac{n-3}{n(n-1)}$ [10]
TCR [1]	Yes	TCC	Yes	$\alpha \geq 1 - 2/n$ [10]
RSEB [10]	Yes	CC	?	$\alpha = 1/3$
RKotH [11]	Yes	cover	?	$\alpha = 1/3$
RDM [6]	?	CC	?	$\alpha = 1/3$
PR [3]	?	?	?	?
PRSL [3]	?	?	?	?

5. The **Randomized King of the Hill** rule (RKotH) chooses the undefeated agent if one exists. Otherwise, choose an agent uniformly at random, eliminate her and the agents she dominates, and repeat.
6. The **Randomized Death Match** rule (RDM) chooses a pair of agents uniformly at random, eliminates the loser, and repeats.
7. The **PageRank** (PR) (**with Self-Loops** (PRSL)) rule chooses agent i as the winner with probability

$$r_i(T) = \begin{cases} \mathrm{pr}_i\left(T|_{TC(T)}\right) & i \in TC(T) \\ 0 & i \notin TC(T) \end{cases}$$

where for a strongly connected (sub)tournament S, $\mathrm{pr}(S)$ is the unique solution to the following linear system of equations:

$$\forall i, \mathrm{pr}_i(S) = \sum_{j:i \succ_S j} \frac{1}{|\{k : j \prec_S k\}| + \mathbb{1}(PRSL)} \mathrm{pr}_j(S)$$
$$\sum_i \mathrm{pr}_i(S) = 1$$

Note that both PR and PRSL are well-defined since the top cycle is strongly connected, so the stationary distribution is indeed unique. PageRank's recursive definition is natural for tournaments: an agent has high PageRank if she beats many other agents with high PageRank.

We show that many previously studied tournament rules actually satisfy stronger notions of fairness than previously demonstrated. We particularly highlight that many of them satisfy our proposed notion of DSTC.

Theorem 1. *RSEB satisfies TCC but neither DSTC nor cover consistency. ICR, RVC, TCR, RDM, PR and PRSL satisfy DSTC but not cover consistency. RKotH satisfies both DSTC and cover consistency. ICR, RVC, TCR, RSEB, RKotH, and RDM are monotone.*

Proof. Schvartzman et al. [11] showed that RKotH satisfies cover consistency. The authors of [1,10,11] proved the monotonicity of ICR, RVC, TCR, RSEB, and RKotH. RDM is monotone since for any sequence of matches, an agent gets at least as far as she did in the original tournament if she wins one more match.

RSEB satisfies TCC because in order for an agent outside of the top cycle to win, she must eventually defeat an agent in the top cycle. To see how RSEB violates DSTC and cover consistency, consider the 8-agent tournament T where

$$1 \succ_T 2, 2 \succ_T 3, 3 \succ_T 4, 4 \succ_T 1, 1 \succ_T 3, 2 \succ_T 4$$

and $i \succ_T j$ for all $i \in \{1,2,3,4\}, j \in \{5,6,7,8\}$. The relations between $\{5,6,7,8\}$ can be arbitrary. Note that 2 covers 3 in T yet 3 can win e.g., the bracket whose leaves are labeled by the permutation $(1,2,4,5,3,6,7,8)$. Moreover, the sub-tournament induced by the first four agents is a dominant sub-tournament. Observe that 4 never wins a bracket in $T|_{[4]}$, but in T, 4 can win e.g., the bracket whose leaves are labeled by the permutation $(1,2,3,5,4,6,7,8)$.

TCR satisfies DSTC because the addition of an agent that loses to all existing agents (a Condorcet loser) does not change the top cycle. ICR, RVC, and RKotH satisfy DSTC because inserting a Condorcet loser into a permutation does not change the winner, so each agent wins the same proportion of permutations they did before. Similarly, RDM satisfies DSTC because inserting a match involving a Condorcet loser into a sequence of matches does not change the winner. PR and PRSL are DSTC by definition.

To see why ICR, RVC, TCR, RDM, PR, and PRSL fail to satisfy cover consistency, it suffices to consider $T|_{[4]}$. 3 is in the top cycle, so she can win with positive probability under TCR, PR, and PRSL. 3 wins ICR and RVC if the chosen permutation is $(2,1,4,3)$. 3 wins RDM if $(1,2)$ is the first pair chosen, $(1,4)$ is the second, and $(3,4)$ is the last. □

3 Lower Bounds

Schneider et al. [10] showed that no Condorcet consistent tournament rule is 2-NM_0-α for $\alpha < 1/3$. The same lower bound construction yields Theorem 2.

Theorem 2. *No CC tournament rule is 2-NM_λ-α for $\lambda < 1 - 3\alpha$.*

Proof. We prove the theorem for monotone rules, but with some additional casework, one can extend the result to non-monotone rules.

Suppose tournament rule r is CC and 2-NM_λ-α, and consider any tournament T on $[n]$ in which 1 dominates 2, 2 dominates 3, and 3 in turn dominates 1. Note that any two agents among $\{1,2,3\}$ can collude so that one of them becomes undefeated. Since r is monotone, CC, and 2-NM_λ-α,

$$1 - r_2(T) \leq (\lambda + 1)r_1(T) + \alpha$$
$$1 - r_3(T) \leq (\lambda + 1)r_2(T) + \alpha$$
$$1 - r_1(T) \leq (\lambda + 1)r_3(T) + \alpha$$

Adding these three inequalities together and isolating λ yields

$$\frac{3(1-\alpha)}{r_1(T)+r_2(T)+r_3(T)} - 2 \leq \lambda$$

Since $r_1(T) + r_2(T) + r_3(T) \leq 1$, this inequality implies that $\lambda \geq 1 - 3\alpha$. □

Corollary 1. *No Condorcet consistent tournament rule is 2-NM$_\lambda$ for $\lambda < 1$.*

We believe that there exists a monotone, Condorcet consistent tournament rule that is 2-NM$_1$. Thus, pairwise collusion can be prevented without sacrificing fairness as long as agents prefer not to collude if their sacrifice in probability is greater than the joint gain. Figure 1 shows such an r for 4 agents. Expressing and computationally solving the problem as a feasibility linear program show that such rules exist for tournaments of up to 6 agents. Unfortunately, as the number of tournaments on n agents grows exponentially with n, it became computationally difficult to check if such rules exist for tournaments of larger size.

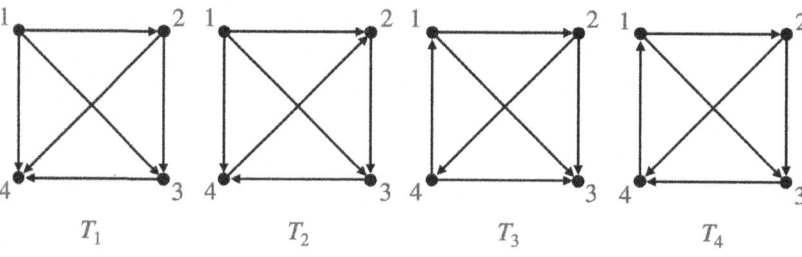

Fig. 1. All non-isomorphic tournaments on 4 agents. The following conditions are necessary and sufficient for a tournament rule on 4 agents r to be Condorcet-consistent and 2-NM$_1$. In T_1 and T_2, r chooses 1 as the winner with probability 1. In T_3, r chooses the winner uniformly at random among 1, 2, and 4. In T_4, r chooses the winner according to a distribution that is a convex combination of $\left(\frac{4}{9}, \frac{2}{9}, 0, \frac{3}{9}\right)$, $\left(\frac{5}{9}, \frac{1}{9}, 0, \frac{3}{9}\right)$, $\left(\frac{13}{33}, \frac{8}{33}, \frac{2}{33}, \frac{10}{33}\right)$, $\left(\frac{5}{12}, \frac{13}{48}, \frac{1}{48}, \frac{7}{24}\right)$, $\left(\frac{11}{21}, \frac{4}{21}, \frac{1}{21}, \frac{5}{21}\right)$, and $\left(\frac{17}{39}, \frac{7}{39}, \frac{4}{39}, \frac{11}{39}\right)$.

Conjecture 1. A monotone, Condorcet consistent, 2-NM$_1$ tournament rule exists.

We now consider several tournament rules and examine their trade-offs between α and λ. Table 2 provides a summary of our findings. Interestingly, the superman-kryptonite tournament identified by Schneider et al. [10] (and its variants) is responsible for all our lower bounds, suggesting that it is especially problematic. We note that Iglesias et al. [8] identified a general class of tournaments termed perfect manipulator tournaments that contains the superman-kryptonite tournament while studying a different problem.

Definition 13 (Superman-kryptonite tournament). *The superman kryptonite tournament on $[n]$ has $i \succ_T j$ whenever $i < j$, except $n \succ_T 1$. In particular, superman 1 dominates all agents but kryptonite n, and n is dominated by all agents except 1.*

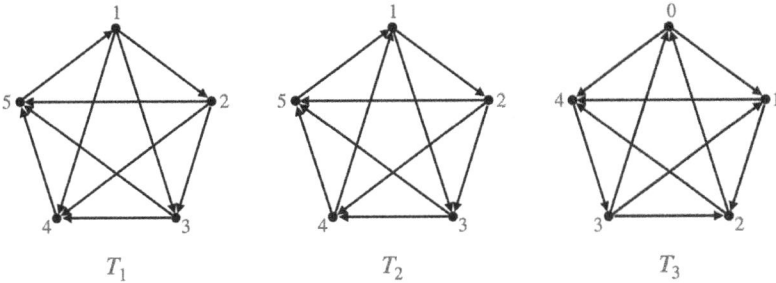

Fig. 2. The tournaments that lead to our lower bounds. T_1 is the superman-kryptonite tournament (here, on 5 players) that leads to most of our lower bounds. T_2 is the tournament that leads to the $\alpha \geq 1/10$ lower bound for RKotH. T_3 is the tournament (here, on 5 players) that leads to our lower bound for PRSL.

Theorem 3. *Let $\lambda \geq 0$. If RSEB satisfies 2-NM_λ-α, then $\alpha \geq 1/n$. If RKotH or PR satisfy this property, then $\alpha \geq \Omega(1)$. That is, RSEB, RKotH, and PR are always pairwise manipulable regardless of λ (Fig. 2).*

Proof (Sketch). The lower bound for RSEB is due to the superman kryptonite tournament on n agents. If the superman and kryptonite collude to make the superman the Condorcet winner, then the superman gains $1/n$ in probability, while the kryptonite's probability of winning remains the same, so $\alpha \geq 1/n$.

The lower bound for RKotH is due to the tournament $T \in \mathcal{T}_5$ in which $i \succ_T j$ whenever $i < j$, except both $4, 5 \succ_T 1$. If agents 1 and 5 collude to make agent 1 the Condorcet winner, then agent 1 gains $1/10$ in probability, while agent 5's probability of winning remains the same. Since RKotH is DSTC, this problematic tournament remains problematic for all $n \geq 5$, so $\alpha \geq 1/10$.

The lower bound for PR is due to the superman kryptonite tournament on four agents. If the superman colludes with agent 3, then agent 3 gains $1/13$ in probability, while the superman's probability of winning remains the same. Since PR is DSTC, this problematic tournament remains problematic even when there are more agents, so $\alpha \geq 1/13$. □

Theorem 4. *If ICR satisfies 2-NM_λ-α for some $\lambda \geq 0$, then $\lambda \geq (1 - O(\alpha))\Omega(n^2)$. If RDM satisfies this property, then $\lambda \geq (1 - O(n\alpha))\Omega(n)$. If RVC, TCR, or PRSL satisfy this property, then $\lambda \geq (1 - O(\alpha))\Omega(n)$.*

Proof (Sketch). The lower bounds for ICR, RDM, RVC, and TCR are all due to the superman kryptonite tournament on n agents. Under ICR, if the superman and kryptonite collude to make the superman the Condorcet winner, then the superman gains $\Omega(1)$ in probability, while the kryptonite loses $O(n^{-2})$ in probability. Thus, if ICR satisfies 2-NM_λ-α, then $\lambda \geq \frac{\Omega(1)-\alpha}{O(n^{-2})} - 1 = (1 - O(\alpha))\Omega(n^2)$.

Under RDM, the same collusion causes the superman to gain $\Omega(n^{-1})$ in probability, while the kryptonite loses $O(n^{-2})$ in probability. Thus, if RDM satisfies 2-NM_λ-α, then $\lambda \geq \frac{\Omega(n^{-1})-\alpha}{O(n^{-2})} - 1 = (1 - O(n\alpha))\Omega(n)$.

Under both RVC and TCR, if the superman and kryptonite collude to make the superman the Condorcet winner, then the superman gains $\Omega(1)$ in probability, while the kryptonite loses $O(n^{-1})$ in probability. Thus, if either satisfies 2-NM$_\lambda$-α, then $\lambda \geq \frac{\Omega(1)-\alpha}{O(n^{-1})} - 1 = (1 - O(\alpha))\Omega(n)$.

The lower bound for PRSL is due to the following tournament on $n = 2k+1$ agents, denoted $0, 1, \ldots, n-1$. Have team $n-1$ defeat team $n-2$. Have teams $0, 1, \ldots, n-3$ lose to team $n-2$ and defeat team $n-1$. For $i = 0, \ldots, n-3$, have team i defeat teams $(i+1) \mod (n-2), \ldots, (i+k-1) \mod (n-2)$ and lose to teams $(i-1) \mod (n-2), \ldots, (i-k+1) \mod (n-2)$. If agents $n-2$ and $n-1$ were to manipulate, then agent $n-2$ would become the Condorcet winner in the resulting tournament and gain $\Omega(1)$ in probability, while agent $n-1$ would lose $O(n^{-1})$ in probability. Thus, if PRSL satisfies 2-NM$_\lambda$-α, then $\lambda \geq \frac{\Omega(1)-\alpha}{O(n^{-1})} - 1 = (1 - O(\alpha))\Omega(n)$. □

Table 2. Performance summary. Only the strongest fairness properties satisfied by each tournament rule are listed. If a stronger or incomparable fairness property is not listed, then the tournament rule does not satisfy it.

Rule	Monotone?	Fairness	2-NM$_\lambda$-α	2-NM$_\lambda$-$f(n)$
ICR [1]	Yes [1]	DSTC	$\lambda \geq \left(\frac{1}{2} - \alpha\right)\Omega(n^2)$	$f(n) \geq 1/2$
RVC [1]	Yes [1]	DSTC	$\lambda \geq \left(\frac{1}{2} - \alpha\right)\Omega(n)$	$f(n) \geq 1/2$
TCR [1]	Yes [1]	DSTC	$\lambda \geq (1-\alpha)\Omega(n)$	$f(n) \geq 1$
RSEB [10]	Yes [10]	TCC	$\alpha \geq 1/n$	$f(n) \geq \varepsilon(\lambda) > 0$
RKotH [11]	Yes [11]	cover [11], DSTC	$\alpha \geq 1/10$	$f(n) \geq 1/10$
RDM [6]	Yes	DSTC	$\lambda \geq \left(1 - \frac{n\alpha}{2}\right)\Omega(n)$	$f(n) \geq \frac{\lambda-1}{\lambda(2\lambda+1)}$
PR [3]	?	DSTC	$\alpha \geq 1/13$	$f(n) \geq 1/13$
PRSL [3]	?	DSTC	$\lambda \geq (1-\alpha)\Omega(n)$	$f(n) \geq 1$

4 Reductions

In a separate direction, we consider fair tournament rules that for fixed λ become increasingly non-manipulable with the number of agents. Formally, we sought rules that satisfy 2-NM$_\lambda$-$f(n)$ where n is the number of agents and f is a non-negative, non-increasing function. Under notions of fairness stronger than Condorcet consistency, it turns out this problem is just as hard as finding rules that satisfy 2-NM$_\lambda$-$(\lim_{n \to \infty} f(n))$.

Theorem 5. *Let $\lambda \geq 0$ and $f \geq 0$ be a non-increasing function such that $f(n) \to \alpha$. A DSTC tournament rule is 2-NM$_\lambda$-$f(n)$ if and only if it is 2-NM$_\lambda$-α.*

The idea behind the proof is as follows: by DSTC, the gains from manipulation in a tournament T on n agents are exactly the same as the gains from manipulation among these agents in a larger tournament on $n' > n$ agents in which T is a dominant subtournament. Thus, the gains from manipulation in T are in fact at most $f(n')$ for all $n' > n$ and hence, at most $\lim_{n' \to \infty} f(n')$.

A similar but weaker result holds for top cycle consistent rules.

Theorem 6. *Let $\lambda \geq 1$ and $f \geq 0$ be a non-increasing function such that $f(n) \to \alpha$. There exist a TCC tournament rule satisfying 2-NM_λ-$f(n)$ if and only if there exists a TCC tournament rule satisfying 2-NM_λ-α.*

The proof is similar to that of Theorem 5. However, because we do not have DSTC, we cannot directly relate the gains from manipulation in tournaments on n agents to those in tournaments on $n' > n$ agents. Nonetheless, we can define a tournament rule on n agents as the limit point of a sequence of tournament rules on n' agents for all $n' > n$. The gains from manipulation under this limit point will then be at most $\lim_{n' \to \infty} f(n')$.

Theorem 7. *If ICR, RVC, TCR, RKotH, PR, or PRSL satisfy 2-NM_λ-$f(n)$ for some fixed $\lambda \geq 1$ and some non-increasing function $f \geq 0$, then $f \geq \Omega(1)$. If RDM satisfies this property, then $f \geq \Omega(1/\lambda)$. If RSEB satisfies this property, then $f \geq \varepsilon(\lambda)$ where $\varepsilon(\lambda)$ is some strictly positive function of λ.*

5 Discussion

In this work, we introduced a partially transferable utility model to study the tension between fairness and strategic robustness in the design of tournaments. In our model, two agents are willing to fix the outcome of their match only if their joint gain is greater than λ times any of their losses. Theorem 2 demonstrates that tournament designers cannot prevent manipulations while maintaining some degree of fairness if agents care about their chances of winning less than twice as much as their opponents' chances. However, it is possible that caring twice as much is sufficient for the existence of fair and non-manipulable tournament rules. Unfortunately, we do not know of any tournament rule that achieves this, and Theorems 3 and 4 show that the tournament rules previously studied in this line of work require agents to care at least $\Omega(n)$ times more about their own chances of winning than their opponents' in order to be non-manipulable. We leave finding a Condorcet consistent tournament rule that is non-pairwise-manipulable when $\lambda = 1$ as a major open problem.

Theorems 5 and 6 may help in resolving this question. If proving that a DSTC tournament rule witnesses vanishing gains from manipulation is easier than proving that it is non-manipulable (e.g., one can only get upper bounds that approach 0), then Theorem 5 would imply that the rule is in fact non-manipulable. On the other hand, if one finds a TCC tournament rule that witnesses vanishing gains from manipulation, then this rule, together with Theorem 6, would yield a non-constructive proof that a non-manipulable rule exists.

Acknowledgments. This work was carried out while one of the authors, Eric Xue, was a participant in the 2021 DIMACS REU program at Rutgers University, supported by NSF grant CCF-1852215, under the supervision of Ariel Schvartzman (who at the time was affiliated with DIMACS) and David Pennock.

Disclosure of Interests. The authors have no competing interests to declare that are relevant to the content of this article.

References

1. Altman, A., Kleinberg, R.: Nonmanipulable randomized tournament selections. In: Proceedings of the Twenty-Fourth AAAI Conference on Artificial Intelligence, AAAI 2010, pp. 686–690. AAAI Press (2010)
2. Altman, A., Procaccia, A.D., Tennenholtz, M.: Nonmanipulable selections from a tournament. In: Proceedings of the 21st International Joint Conference on Artificial Intelligence, IJCAI 2009, pp. 27–32 (2009)
3. Brandt, F., Fischer, F.: PageRank as a weak tournament solution. In: Deng, X., Graham, F.C. (eds.) WINE 2007. LNCS, vol. 4858, pp. 300–305. Springer, Heidelberg (2007). https://doi.org/10.1007/978-3-540-77105-0_30
4. Dale, E., Fielding, J., Ramakrishnan, H., Sathyanarayanan, S., Weinberg, S.M.: Approximately strategyproof tournament rules with multiple prizes. In: The 23rd ACM Conference on Economics and Computation, EC 2022, Boulder, CO, USA, 11–15 July 2022, pp. 1082–1100. ACM (2022)
5. Dinev, A., Weinberg, S.M.: Tight bounds on 3-team manipulations in randomized death match. In: Hansen, K.A., Liu, T.X., Malekian, A. (eds.) WINE 2022. LNCS, vol. 13778, pp. 273–291. Springer, Cham (2022). https://doi.org/10.1007/978-3-031-22832-2_16
6. Ding, K., Weinberg, S.M.: Approximately strategyproof tournament rules in the probabilistic setting. In: 12th Innovations in Theoretical Computer Science Conference (ITCS 2021) (2021)
7. Felsenthal, D.S., Machover, M.: After two centuries, should condorcet's voting procedure be implemented? Behav. Sci. **37**(4), 250–274 (1992)
8. Iglesias, J., Ince, N., Loh, P.S.: Computing with voting trees. SIAM J. Discret. Math. **28**(2), 673–684 (2014). https://doi.org/10.1137/130906726
9. Mikšaník, D., Schvartzman, A., Soukup, J.: On approximately strategy-proof tournament rules for collusions of size at least three (2024). https://arxiv.org/abs/2407.17569
10. Schneider, J., Schvartzman, A., Weinberg, S.M.: Condorcet-consistent and approximately strategyproof tournament rules. In: 8th Innovations in Theoretical Computer Science Conference (ITCS 2017) (2017)
11. Schvartzman, A., Weinberg, S.M., Zlatin, E., Zuo, A.: Approximately strategyproof tournament rules: on large manipulating sets and cover-consistence. In: 11th Innovations in Theoretical Computer Science Conference (ITCS 2020) (2020)
12. Suksompong, W.: Tournaments in computational social choice: recent developments. In: Proceedings of the Thirtieth International Joint Conference on Artificial Intelligence, IJCAI 2021, pp. 4611–4618. International Joint Conferences on Artificial Intelligence Organization (2021)

Preference Theory

Noise-Tolerant Active Preference Learning for Multicriteria Choice Problems

Margot Herin[1]([✉]), Patrice Perny[1], and Nataliya Sokolovska[2]

[1] LIP6, Sorbonne University, Paris, France
margot.herin@lip6.fr
[2] LCQB, Sorbonne University, Paris, France

Abstract. To make a choice in the presence of multiple criteria, we generally use an aggregation function which determines, for each alternative, the balance of its strengths and weaknesses and its overall evaluation. The aggregation function uses weights to adapt the model to the decision-maker's value system, by specifying the importance of the criteria and possibly their interactions. In this paper, we propose a noise-tolerant active learning method for these parameters, which not only effectively reduces the indeterminacy of the weights to identify an optimal or near-optimal decision among a given set of alternatives, but also simultaneously determines a predictive model of preferences capable of making relevant choices for the decision-maker on new instances. These outcomes are achieved by leveraging a general disagreement-based active learning approach that is theoretically guaranteed to be tolerant to noisy answers. The proposed method applies to various weighted aggregation functions, linear or not, classically used in decision theory.

Keywords: Multicriteria Decision Making · Active Preference Learning · Weighted Sum · Choquet integral · Weighted norms

1 Introduction

In multicriteria choice problems, it is commonly accepted that the exploration of admissible trade-offs should be restricted to Pareto-optimal solutions, i.e. solutions that cannot be improved on one criterion without having to be degraded on another. These solutions are, however, potentially very numerous, and it is necessary to collect additional preference information to define how the evaluations from the different criteria combine to define the overall preference. Decision theory provides numerous mathematical models to account for agents' preferences, in particular scalarizing functions which associate with any vector of partial evaluations a global evaluation defining the overall utility of the solution for the decision-maker (DM). The multicriteria choice problem can then be reformulated as a problem of maximizing the scalarizing function over all feasible performance vectors, i.e., the evaluation vectors associated with the alternatives in the problem under consideration.

In order to explore Pareto-optimal solutions and identify the optimal compromise for the DM, the scalarizing function f must be monotonically increasing with respect to Pareto dominance. That is, if a solution x is at least as good as x' on all criteria, we must have $f(x) \geq f(x')$. A wide range of such aggregation functions is available, from the simplest such as the weighted sum, to more sophisticated and expressive models such as the multilinear model and non-additive integrals (Choquet or Sugeno integrals), which can be used to model interactions between criteria, without forgetting weighted norms, which have proved effective for interactively exploring the set of non-dominated solutions in the Pareto sense. The weights used in these aggregators specify the relative importance of criteria and/or sometimes also the positive or negative interactions between criteria. They enable us to adapt a generic model to the DM's particular value system. We therefore optimize a function f_w parameterized by the weighting vector w. The phase of eliciting preferences and learning the w weighting vector is absolutely crucial, as it completely determines the nature of the compromise that will be found by optimizing f_w and the recommendation that will follow.

Various approaches have been proposed for specifying the parameters of a decision model in different contexts and thus determining an optimal choice. Here, we distinguish three types of approaches.

The Local and Interactive Judgment Approach: we choose an initial vector of parameters w, calculate an optimal solution for f_w, then let w evolve according to user feedback until we arrive at a solution that satisfies the DM. This approach, widely used in interactive multicriteria optimization [18,19], allows a user-driven exploration of the Pareto set, alternating phases of calculation of the current optimal solution and phases of dialogue with the user. It may require numerous interactions, and the quality of the solution chosen at the end of the process is only validated by the decision-maker's instant sense of satisfaction.

Incremental Preference Elicitation: a first approach to incremental elicitation consists in progressively reducing the space of admissible parameters. Iteratively, a preference query is chosen, the answer to which induces a new constraint on the parameter space. The set of parameters compatible with the constraints induced by the preference judgments expressed is progressively reduced until the point where an alternative proves optimal for all remaining parameters (necessarily optimal solution). This approach is introduced in the ISMAUT method [21]. A principle of active question selection is often used, based on the minimization of maximum regret, to choose the most informative question [3,5,20] and derive a robust recommendation. Another approach, more tolerant to noisy responses, is to manage a probability distribution (or other uncertainty model [1]) over the parameter space and revise it according to the answers to questions, to choose a decision having the maximum expected value [6] or minimizing the expectation of regret [4]. These methods are question-saving, as they direct the questionnaire towards the resolution of a particular instance. On the other hand, they do not produce a learned model and are generally not sufficient to solve a choice problem involving a new set of alternatives.

Complete Learning of the Decision Model: we use a base of preference examples and perform a regression (on the values or the order induced by the values) to determine the parameter w that best fits the example base [11,16]. To determine the parameters of the aggregation function accurately and reliably, the model must be trained on a large base of examples, and requires a much larger number of preference queries than in incremental approaches. On the other hand, the learned model can be reused to treat a new choice problem with the same decision-maker, on new alternatives.

In this paper we propose a hybrid active learning approach that combines the objectives of the last two items, namely to quickly identify the optimal choice on the instance to be solved while providing a model capable of explaining the decision-maker's preferences and predicting his choices or formulating recommendations adapted to his preferences on new instances of choice problems.

Another aim of the work presented here is to improve the elicitation method's tolerance to DM's noisy answers. Indeed, the aim of minimizing the elicitation effort and the number of questions asked or examples used to determine an optimal alternative often leads to taking each answer as valid information, likely to definitively constrain the space of admissible parameters, without any further possibility of questioning or checking. This approach, which aims for questionnaire efficiency, is of course rather risky, as it omits any validation operation through partial redundancy of questions, nor any compromise between partially contradictory answers within the framework of a given decision model. The pitfalls of incremental elicitation by progressive and definitive reduction of possible parameters are well illustrated by the following example.

Example 1. Consider a set $X = \{a^0, \ldots, a^q\}$ of $q+1$ alternatives evaluated on two criteria and represented by performance vectors $a^i = (i, q-i)$ for $i = 0, \ldots, q$. Suppose the DM has expressed a first preference $a^r \succ a^t$ for two indices $r, t \in \{0, \ldots, q\}$ such that $r > t$. Suppose we want to learn the weights of a weighted sum model of the form $f_w(x) = wx_1 + (1-w)x_2$ for an unknown parameter $w \in [0, 1]$. The preference $a^r \succ a^t$ implies $wa_1^r + (1-w)a_2^r > wa_1^t + (1-w)a_2^t$ and therefore $wr + (1-w)(q-r) > wt + (1-w)(q-t)$, or equivalently $w(r-t) > (1-w)(r-t)$, hence $w > 1-w$ and thus $w > 1/2$. Under this constraint, it's easy to see that $f_w(a^q) > f_w(a^i)$ for all $i < q$. We indeed have $f_w(a^q) - f_w(a^i) = wq - (wi + (1-w)(q-i)) = (2w-1)(q-i) > 0$ since $w > 1/2$ and $q > i$. Hence a^q is necessarily an optimal solution in X. Note, however, that if the decision-maker was mistaken in the first answer ($a^t \succ a^r$ being the actual preference), then the same reasoning would have led to the choice of a^0. In this case, the recommendation a^q is in fact the worst possible recommendation given the actual DM's preferences.

Although this example is a bit of a caricature, it does illustrate that a concern for efficiency in the active choice of a question to ask can lead to choices that are not robust to noisy responses. In this paper, we will propose a non-Bayesian approach to active learning of decision-maker preferences, which is more robust to noisy responses than usual methods based on regret minimization and enables us to identify or approximate a necessary winner in a given set of alternatives, as well as to build an explanatory model of decision-maker preferences. The paper is organized as follows. Section 2 introduces background and notations. In Sect. 3, we introduce an algorithm for active preference learning in a noisy setting, and in Sect. 4, we demonstrate its benefits on synthetic preference data. Concluding remarks close the paper.

2 Background and Notations

Multicriteria decision making problems are characterized by alternatives evaluated with respect to n dimensions representing various points of view (criteria evaluation or individual opinions) possibly conflicting with each other. In the sequel, $N = \{1, \ldots, n\}$ denotes a set of criteria. Any alternative $x \in X$ is represented by an evaluation vector $x = (x_1, \ldots, x_n)$ where $x_i \in [0, 1]$ represents the value of x with respect to criterion i for $i = 1, \ldots, n$. Thus, the set $\mathcal{X} = [0,1]^n$ represents the criteria space. Let us consider a choice problem over a finite set $X \subseteq \mathcal{X}$ representing all feasible evaluation vectors. The subset of Pareto optimal alternatives in X is denoted by X_P.

Let $f_w : \mathcal{X} \to \mathbb{R}$ be the scalarizing function used to represent the DM's preference $\succ$, i.e., :

$$x \succ x' \iff f_w(x) > f_w(x'), \quad \forall x, x' \in X \quad (1)$$

and let W be the associated definition domain of parameter w. A basic example is the weighted sum used in Example 1 (formally defined as $f_w(x) = \sum_{i=1}^n w_i x_i$ with $W = \{w \in [0,1]^n | \sum_{i=1}^n w_i = 1\}$). Beyond its simplicity and intuitive appeal, this linear aggregation function may suffer from well-known limits in view of exploring Pareto-optimal tradeoffs when the set of alternatives is discrete (or more generally non-convex). Some Pareto-optimal tradeoffs are indeed impossible to obtain by maximizing a weighted sum (e.g., solutions that are interior points of the convex hull of feasible points). Such solutions are referred to as unsupported Pareto-optimal solutions. By limiting ourselves to the supported solutions we deprive the decision-maker of the possibility of examining various interesting trade-offs that could possibly better correspond to its value system. We give below two major examples of more sophisticated scalarizing functions able to cover a wider subset of Pareto-optimal tradeoffs.

The *Choquet integral* [13], used in multicriteria decision-making to model preferences in the presence of interacting criteria, employs a weighting system $w : 2^N \to \mathbb{R}$ that attaches a weight $w(S)$ to any possible set of criteria $S \subseteq N$, to aggregate the evaluation vector's values as follows:

$$f_w(x) = \sum_{i=1}^n \left[w(X_{(i)}) - w(X_{(i+1)}) \right] x_{(i)} \quad (2)$$

where $(.)$ is any permutation of N such that $x_{(i-1)} \leq x_{(i)}$, $i = 1, \ldots, n$, $X_{(i)} = \{(i), \ldots, (n)\}$ and $X_{(n+1)} = \emptyset$. The set function w is called a capacity and is supposed to be normalized, i.e., $w(\emptyset) = 0$ and $w(N) = 1$. Also, to guarantee the monotonicity of f_w w.r.t. Pareto dominance, the capacity is supposed to be monotonic w.r.t. set inclusion, i.e., for any $A \subseteq B$, $w(A) \leq w(B)$ and thus $W = \{w : 2^N \to [0,1] \mid w(A) \leq w(B)$ for all A, B such that $A \subseteq B\}$. An alternative representation of any capacity w is given by its Möbius transform m_w defined as follows:

$$m_w(S) = \sum_{T \subseteq S} (-1)^{|S \setminus T|} w(T) \text{ with } w(S) = \sum_{T \subseteq S} m_w(T) \quad (3)$$

The values $m_w(S)$ are called Möbius masses. We remark that we necessarily have $\sum_{T \subseteq N} m_w(T) = 1$, since $w(N) = 1$. Let us also recall that a Choquet integral admits a simple reformulation from m_w [7]:

$$f_w(x) = \sum_{S \subseteq N} m_w(S) \min_{i \in S} \{x_i\} \quad (4)$$

This formulation can be further simplified when considering k-additive capacities, i.e., capacities whose Möbius masses are zero for subsets larger than k and non-zero for at least one subset of size k. Whenever w is 1-additive, which is equivalent to $w(A) = \sum_{i \in A} w(\{i\})$, $A \subseteq N$ according to Eq. 3, the Choquet integral boils down to a weighted sum. Whenever w is 2 additive, the weighted sum is augmented with a linear combination of pairwise minimum of type $\min\{x_i, x_j\}$ allowing the representation of positive or negative synergies for every pair of criteria. Obviously, Choquet integrals used with k-additive capacities ($2 \leq k \leq n$) involve larger criteria interactions as k increases, with enhanced descriptive possibilities. In particular, their maximization can lead to unsupported solutions.

Despite its enhanced descriptive power compared to the weighted sum, some non-supported Pareto-optimal solutions may not be accessible by maximizing a Choquet integral. In contexts where no prior information about the preference system of the DM is available, it may be the case that any Pareto-optimal solution is of possible interest and must be accessible by the scalarizing function. In this case, the standard approach is to minimize a weighted Chebyshev norm measuring the distance to ideal point [22]. This is equivalent to maximize the following scalarizing function:

$$f_w(x) = -\max_{i \in N}\{w_i | x_i - u_i |\} \quad (5)$$

where $u = (u_1, \ldots, u_n)$ is the ideal point, defined by $u_i = \max_{x \in X_p}\{x_i\}$ and the associated weight definition domain is $W = \{w \in [0,1]^n \mid \sum_{i=1}^n w_i = 1\}$.

The relative ability of the three families of scalarizing functions mentioned above to describe observed choices is illustrated on the simple instance of a biobjective problem involving 12 Pareto-optimal solutions depicted in Fig. 1. We can see the supported solutions (accessible by maximization of a weighted sum) (blue round) that lie on the upper part of the convex hull of the alternative set, then those unsupported solutions that are accessible by maximization of a Choquet integral (green diamond), and finally those unsupported solutions that are only accessible by maximization of a Chebyshev norm (red square).

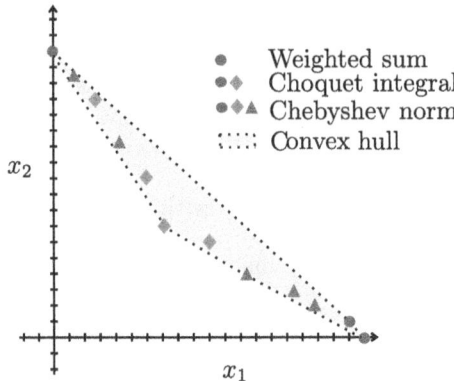

Fig. 1. Accessibility of various Pareto-optimal tradeoffs (Color figure online)

3 Noise-Tolerant Active Preference Learning

Disagreement-based active learning [9,15] refers to a branch of theoretically grounded active learning algorithms for binary classification, that share with incremental preference elicitation methods the common objective of reducing the space of admissible models as fast as possible (in terms of the number of asked questions). In the following, we establish a link between both worlds and propose a disagreement-based active preference learning for choice problems, which we illustrate on the toy case of Example 1.

3.1 Disagreement-Based Active Learning

In binary classification, the input consists of data points z^i belonging to an input space $\mathcal{Z}$ and their labels $y^i \in \mathcal{Y} = \{-1, +1\}$ distributed according to a joint distribution over $\mathcal{Z} \times \mathcal{Y}$ denoted by $\mathbb{P}_{ZY}$. In this setting, based on data samples $(z^i, y^i)_{i=1}^m$, the goal is to find the classifier h^* in a hypothesis class $\mathcal{H}$ that allows to best predict the target value of a point z, i.e., that minimizes the expectation of making a bad prediction. More formally, $h^* = \arg\min_{h \in \mathcal{H}} \ell(h)$ where $\ell(h) = \mathbb{E}_{(z,y) \sim \mathbb{P}_{ZY}}[\mathbb{1}_{\{h(z) \neq y\}}] = \mathbb{P}_{(z,y) \sim \mathbb{P}_{ZY}}(h(z) \neq y)$, for any $h \in \mathcal{H}$.

In this setting, the general idea of disagreement-based active learning is embodied by the CAL algorithm [8] (called after the authors' names Cohn, Atla and Ladner). Starting with $\mathcal{H}$ as the space of admissible models (also called the version space), CAL iteratively proceeds a sequence of unlabeled points $(z^i)_{i=1}^m$, and at each iteration asks for the label y^i of the point z^i if and only if there exist two classifiers h_1, h_2 in the current space of admissible model that disagree on z^i, i.e., such that $h_1(z^i) \neq h_2(z^i)$. The portion of the input space $\mathcal{Z}$ in which this condition holds defines the disagreement region. If asked, the newly obtained label y^i provides the additional constraint $h(z^i) = y^i$ on the set of admissible models, which now excludes the models classifying z^i differently. By doing so, the algorithm asks for a label if and only if the new constraint $h(z^i) = y^i$ surely reduces the space of admissible models, allowing to narrow down the version space around h^* with minimal labeling effort.

Obviously, this algorithm works to identify h^* under the separability hypothesis, i.e., $h^*(z^i) = y^i, i = 1, \ldots, m$. In the more realistic noisy case where $\ell(h^*) > 0$ (referred to as the agnostic setting), the hard constraints $h(z^i) = y^i$ will eventually exclude h^* from the set of admissible models. Extensions of the CAL algorithm in the noisy case [2,10] bypass this issue by defining the set of admissible models as the set of models that proved to yield small errors on the learning examples $(z^i, y^i)_{i=1}^m$. Among them, the DMH algorithm [10] (also called after the authors' names Dasgupta, Hsu, and Monteleoni), which relies on supervised learning sub-tasks, provides a simple way of cautiously excluding models associated with significantly high errors. Here we propose to exploit and extend it to the multicriteria choice problem under noisy answers.

In the multicriteria choice problem setting, the determination of the weight vector w in the preference model (1) from pairwise preference examples $x^i \succ x'^i$, $x^i, x'^i \in X_P$ can be indeed formulated as a binary classification problem where $z^i = (x^i, x'^i) \in \mathcal{Z} = X_P^2$ and $y^i = 1$ if $x^i \succ x'^i$ and $y^i = -1$ otherwise $(x^i \precsim x'^i)$. In this case, the hypothesis class can be defined as $\mathcal{H} = \{h : X_P^2 \to \mathbb{R} | \exists w \in W, h((x, x')) = \text{sign}(f_w(x) - f_w(x'))\}$ where $\text{sign}(t) = 1$ if $t > 0$ and $\text{sign}(t) = -1$ otherwise, and thus can be identified to the set of admissible weights W. In this set, let w^* denote the weight vector that best represents the DM's preferences, i.e., $w^* = \arg\min_{w \in W} \ell(f_w)$. The supervised learning sub-tasks of the DMH algorithm involve minimizations over W of the empirical error on preference examples $S = (z^i, y^i)_{i=1}^m$ defined as $\ell_S(f_w) = \frac{1}{|S|} \sum_{(x^i, x'^i, y^i) \in S} \mathbb{1}_{\{\text{sign}(f_w(x^i) - f_w(x'^i)) \neq y^i\}}$. However, the 0–1 loss $\ell_S(f_w)$ is non-convex and discontinuous and its optimization on a set W is known to be NP-hard in the case of noisy answers, even for linear models f_w [12]. To bypass this issue, we use a randomly generated finite approximation $W_0 \subseteq W$ of W and solve the optimization tasks by exhaustive search.

In the next subsection, we introduce an algorithm designed to achieve a twofold objective: on the one hand, finding a near-optimal solution within X_P, and on the other hand, assessing parameter w to have a predictive model f_w of DM's preferences. To this end, we propose a tailoring of the DMH algorithm to the aforementioned preference learning setting, extended with a mechanism to control the regret of the recommended solutions.

3.2 A Disagreement-Based Active Preference Learning Algorithm

Let us introduce Algorithm 1 that takes as input a stream of pairs of alternatives $z^k = (x^k, x'^k)$ drawn randomly and uniformly from the set of alternatives X_P and an initial set of admissible weight vectors W_0, the definition domain of weight w used in function f_w. This algorithm sequentially proceeds the pairs of alternatives and at iteration k, only when necessary, asks for the DM to provide an answer y^k to the pairwise comparison query $x^k \succ x'^k$?. Since the DM possibly provides answers in contradiction with the preferences induced by f_{w^*}, the obtained preference examples are not exploited as hard constraints to reduce W_0, but used to construct a growing learning dataset of preference examples T_k.

Algorithm 1:

Inputs: W_0, α, ρ

1. draw a pair (x^0, x'^0) uniformly in X_P^2;
2. $y^0 \leftarrow$ answer to the query "$x^0 \succ x'^0$?";
3. $T_0 \leftarrow \{(x^0, x'^0, y^0)\}$, $\text{MR}_0 \leftarrow 1$, $k \leftarrow 1$;
4. **while** $\text{MR}_{k-1}/\text{MR}_0 > \rho$ **do**
5.   draw a pair (x^k, x'^k) uniformly in X_P^2;
6. $W_k^+, W_k^- \leftarrow \{w \in W_{k-1} | f_w(x^k) > f_w(x'^k)\}, \{w \in W_{k-1} | f_w(x^k) \leq f_w(x'^k)\}$;
7.   **if** $W_k^+ \neq \emptyset$ **and** $W_k^- \neq \emptyset$ **then**
8. $w^+, w^- \leftarrow \arg\min_{w \in W_k^+} \ell_{T_{k-1}}(f_w), \arg\min_{w \in W_k^-} \ell_{T_{k-1}}(f_w)$;
9. $\beta_k \leftarrow \sqrt{(4/k) \ln(8(k^2 + k)|W_0|^2/\delta)}$;
10. $\Delta_k \leftarrow \beta_k^2 + \beta_k \left(\sqrt{\ell_{T_{k-1}}(f_{w^+})} + \sqrt{\ell_{T_{k-1}}(f_{w^-})}\right)$
11.   **end**
12.   **if** $W_k^- \neq \emptyset$ **and** $(W_k^+ = \emptyset$ **or** $\ell_{T_{k-1}}(f_w^+) - \ell_{T_{k-1}}(f_w^-) > \alpha \Delta_k)$ **then**
13. $W_k \leftarrow W_k^-$, $T_k \leftarrow T_{k-1}$
14.   **else if** $W_k^+ \neq \emptyset$ **and** $(W_k^- = \emptyset$ **or** $\ell_{T_{k-1}}(f_w^-) - \ell_{T_{k-1}}(f_w^+) > \alpha \Delta_k)$ **then**
15. $W_k \leftarrow W_k^+$, $T_k \leftarrow T_{k-1}$;
16.   **else**
17. $y^k \leftarrow$ answer to the query "$x^k \succ x'^k$?";
18. $W_k \leftarrow W_{k-1}, T_k \leftarrow \{(x^k, x'^k, y^k)\}$;
19.   **end**
20. $\hat{w}_k \leftarrow \arg\min_{w \in W_k} \ell_{T_k}(f_w)$;
21. $\hat{x}_k, \text{MR}_k \leftarrow \arg\max_{x \in X_P} f_{\hat{w}_k}(x), \max_{x' \in X_P} \max_{w \in W_k}(f_w(x') - f_w(\hat{x}_k))$;
22. $k \leftarrow k + 1$;
23. **end**

Outputs: $\hat{w}_{k-1}, \hat{x}_{k-1}, \text{MR}_{k-1}$

More precisely, at iteration k, to determine whether query "$x^k \succ x'^k$?" is worth asking (i.e., whether the answer y^k is likely to provide new information), the algorithm assesses the level of disagreement of the current set of admissible models W_{k-1} on the pair (x^k, x'^k). To this end, W_{k-1} is partitioned into $W_k^+ = \{w \in W_{k-1} | f_w(x^k) > f_w(x'^k)\}$ (set of weights that verify $x^k \succ x'^k$) and $W_k^- = \{w \in W_{k-1} | f_w(x^k) \leq f_w(x'^k)\}$ (set of weights that verify $x^k \precsim x'^k$) and the minimal empirical errors on the learning database T_{k-1} are computed on both set W_k^+ and W_k^-, and compared. If w^+, w^- respectively denote the best elements in W_k^+ and W_k^-, i.e., $w^+ = \arg\min_{w \in W_k^+} \ell_{T_{k-1}}(f_w)$ and $w^+ = \arg\min_{w \in W_k^-} \ell_{T_{k-1}}(f_w)$, this amounts to assessing the gap $\ell_{T_{k-1}}(f_{w^+}) - \ell_{T_{k-1}}(f_{w^-})$.

If the magnitude of the error difference $|\ell_{T_{k-1}}(f_{w^+}) - \ell_{T_{k-1}}(f_{w^-})|$ does not exceed a certain threshold Δ_k, the elements of W_{k-1} somehow "disagree" on whether $x^k \succ x'^k$ or $x^k \precsim x'^k$. Indeed, the weight vectors verifying $x^k \succ x'^k$

(W_k^+) and the weight vectors verifying $x^k \precsim x'^k$ (W_k^-) are attached to similar minimal errors on the learning database T_{k-1}. Therefore, the answer y^k is likely to provide new information, and thus the query "$x^k \succ x'^k$?" is asked to the DM. Then, the answer y^k is stored as a new preference example (x^k, x'^k, y^k) in the learning database, i.e., $T_k = T_{k-1} \cup \{(x^k, x'^k, y^k)\}$.

However, if $|\ell_{T_{k-1}}(f_{w^+}) - \ell_{T_{k-1}}(f_{w^-})| > \Delta_k$, the elements of W_{k-1} somehow "agree" on whether $x^k \succ x'^k$ or $x^k \precsim x'^k$, since one of the two sets W_k^+, W_k^- yields significantly higher errors on the learning database T_{k-1} than the other, and thus is not likely to contain the best predictor w^*. Then, the query is not worth asking and the algorithm exploits the agreement of W_{k-1} on the instance (x^k, x'^k) to reduce, with high confidence, the set of admissible models by adding the preference as a hard constraint, i.e., $W_k = W_k^+$, if the constraint $x^k \succ x'^k$ yields the smallest error, and $W_k = W_k^-$ otherwise. In this case, no example is added to the learning database, i.e., $T_k = T_{k-1}$. Note that cases $W_k^+ = \emptyset$ or $W_k^- = \emptyset$ are omitted here for clarity but are included in Algorithm 1.

This cautious reduction of the set of admissible weights makes it possible to incrementally control the remaining level of uncertainty on the DM's best alternative in X_P, as in incremental preference elicitation methods based on the notion of maximum regret [3,5,20]. Indeed, at the end of iteration k, the learned model is $\hat{w}_k = \arg\min_{w \in W_k} \ell_{T_k}(f_w)$ and naturally the recommended solution is $\hat{x}_k = \arg\max_{x \in X_P} f_{\hat{w}_k}(x)$. Then, we propose to assess the remaining level of uncertainty on the DM's best alternative by computing the maximum regret attached to the recommendation of $\hat{x}_k$ knowing that the current set of admissible weights is W_k, i.e.,: $\mathrm{MR}_k = \max_{x \in X_P} \max_{w \in W_k} \{f_w(x) - f_w(\hat{x}_k)\}$. Once MR_k is sufficiently reduced (ratio $\mathrm{MR}_k/\mathrm{MR}_1$ below a configurable threshold $\rho \in [0,1)$), the algorithm stops and outputs a recommended solution $\hat{x}_k$.

We now establish Proposition 1 showing how threshold values Δ_k must be set to make sure that $W_k, k \geq 1$ contain with high probability weight vector w^* that best fits to DM's preferences. This result relies on the known guarantee attached to the DMH algorithm [10] and shows that MR_k upper bounds the real regret $\max_{x \in X_P} \{f_{w^*}(x) - f_{w^*}(\hat{x}_k)\}$, with high probability.

Proposition 1. *For $\delta > 0$, $\beta_k = \sqrt{(4/k)\ln(8(k^2+k)|W_0|^2/\delta)}$, and $\Delta_k = \beta_k^2 + \beta_k \left(\sqrt{\ell_{T_{k-1}}(f_{w^+})} + \sqrt{\ell_{T_{k-1}}(f_{w^-})}\right)$, with probability at least $1-\delta$, we have $w^* \in W_k$ and MR_k upper bounds the real regret, i.e., $\max_{x \in X_P}\{f_{w^*}(x) - f_{w^*}(\hat{x}_k)\} \leq \mathrm{MR}_k$, for any $k \geq 1$, in Algorithm 1.*

Proof. Algorithm 1 is a specification of the DMH algorithm [10] (with a modified stopping criterion) for the input space $\mathcal{Z} = X_P^2$ associated to the uniform distribution, and the hypothesis class $\mathcal{H} = \{h : X_P^2 \to \mathbb{R} | \exists w \in W_0, h((x,x')) = \mathrm{sign}(f_w(x) - f_w(x'))\}$. Let $\mathcal{H}_k$ denote the hypothesis class at iteration k, i.e., $\mathcal{H}_k = \{h : X_P^2 \to \mathbb{R} | \exists w \in W_k, h((x,x')) = \mathrm{sign}(f_w(x) - f_w(x'))\}$. Then, for $\delta > 0$ and $\gamma_k = \sqrt{(4/k)\ln(8(k^2+k)\mathcal{S}(\mathcal{H},2k)^2/\delta)}$ where $\mathcal{S}(\mathcal{H},2k)$ is the k-th shatter coefficient of $\mathcal{H}$, with probability at least $1-\delta$, and all $(h,h') \in \mathcal{H}_k \times \mathcal{H}_k$, we have for any $k \geq 1$ ([10], Corollary 1):

$$\ell_{T_{k-1}}(h) - \ell_{T_{k-1}}(h') \leq \ell(h) - \ell(h') + \gamma_k^2 + \gamma_k \left(\sqrt{\ell_{T_{k-1}}(h)} + \sqrt{\ell_{T_{k-1}}(h')}\right)$$

Recall that the k-th shatter coefficient is defined as the maximal number of ways $\mathcal{H}$ can classify k input points. More formally, $\mathcal{S}(\mathcal{H}, k)$ is the maximal number of different target vector $(h(z_1), \ldots, h(z_k))$ that can be associated to k input points $(z_1, \ldots, z_k)$ by $\mathcal{H}$, i.e., $\mathcal{S}(\mathcal{H}, k) = \max_{z \in \mathcal{Z}} |\{(h(z_1), \ldots, h(z_k))| h \in \mathcal{H}\}|$. Then, by definition of the shatter coefficient, when $\mathcal{H}$ is finite, i.e., $|\mathcal{H}| < \infty$, for any $k \geq 1$, $\mathcal{S}(\mathcal{H}, k) \leq |\mathcal{H}|$. Here, $|\mathcal{H}| \leq |W_0| < \infty$ and thus $\mathcal{S}(\mathcal{H}, k) \leq |W_0|$ for any $k \geq 1$. Therefore $\beta_k = \sqrt{(4/k)\ln(8(k^2+k)|W_0|^2/\delta)} \geq \gamma_k$, and thus we have, with probability at least $1 - \delta$, for all $(h, h') \in \mathcal{H}_k \times \mathcal{H}_k$ and all $k \geq 1$:
$$\ell_{T_{k-1}}(h) - \ell_{T_{k-1}}(h') \leq \ell(h) - \ell(h') + \beta_k^2 + \beta_k \left(\sqrt{\ell_{T_{k-1}}(h)} + \sqrt{\ell_{T_{k-1}}(h')} \right)$$

Therefore, for $\Delta_k = \beta_k^2 + \beta_k \left(\sqrt{\ell_{T_{k-1}}(f_{w+})} + \sqrt{\ell_{T_{k-1}}(f_{w-})} \right)$, with probability at least $1 - \delta$, we have $w^* \in W_k$, for any $k \geq 1$ ([10], Lemma 3). Thus, with probability at least $1 - \delta$, $\max_{x \in X_P}\{f_{w^*}(x) - f_{w^*}(\hat{x}_k)\} \leq \max_{w \in W_k} \max_{x \in X_P}\{f_w(x) - f_w(\hat{x}_k)\} = \mathrm{MR}_k$. □

In the next Section dedicated to numerical tests, we will see that Δ_k is an over-cautious threshold, that may induce many preference queries. In practice, a more aggressive threshold $\alpha \Delta_k$ where $\alpha \in [0, 1]$ can be used to reduce the number of preference queries without too much sacrificing the recommendation quality. Parameter α can be set using cross-validation on a small test set of preference examples $\{(x^i, x'^i, y^i)\}$. On the other side, parameter δ is set to 0.95.

3.3 Illustration on a Toy Example

To illustrate the benefit of Algorithm 1 for preference elicitation with noisy answers, we exploit the easy-to-grasp toy case of Example 1 on the choice set $X = \{a^0, \ldots, a^q\}$. The first pair under consideration is $(x^0, x'^0) = (a^r, a^t)$ with $r > t$. Since the first DM's preference statement is $a^r \succ a^t$, Algorithm 1 starts with the initial learning database $T_0 = \{(x^0, x'^0, 1)\}$. Then, examples of pairs $(x^k, x'^k) = (a^{r_k}, a^{t_k})$ are repeatedly drawn uniformly from $X_P^2 = X^2$ (where $r_k > t_k$) and Algorithm 1 either asks the query $x^k \succ x'^k$? or if confident enough to predict the DM's answer, does not ask the query and reduces the current set of admissible weights accordingly (either $(\frac{1}{2}; 1]$ or $[0; \frac{1}{2}]$). In this case, the recommended alternative $\hat{x}_k = \arg\max_{x \in X_P} f_{\hat{w}_k}(x)$ with $\hat{w}_k = \arg\min_{w \in W_k} \ell_{T_k}(f_w)$ is necessarily a^q if $W_k = (\frac{1}{2}; 1]$ or a^0 if $W_k = [0; \frac{1}{2}]$. Finally, the recommended alternative being associated with a null maximum regret (for instance if $W_k = (\frac{1}{2}; 1]$, $\max_{r \in \{0, \ldots, q\}} \max_{w \in (1/2, 1]}(f_w(a^r) - f_w(a^q)) = \max_{r \in \{0, \ldots, q\}} \max_{w \in (1/2, 1]}\{(r-q)(2w-1)\} = 0$)), the algorithm stops. Also, at each iteration k, one can easily see that $\ell_{T_{k-1}}(f_{w+}) = \frac{1}{|T_{k-1}|} \sum_{i=0}^{k-1} \mathbb{1}_{\{y^i = 1\}}$ and $\ell_{T_{k-1}}(f_{w-}) = \frac{1}{|T_{k-1}|} \sum_{i=0}^{k-1} \mathbb{1}_{\{y^i = -1\}}$ and thus $\ell_{T_{k-1}}(f_{w+})$ and $\ell_{T_{k-1}}(f_{w-})$ are respectively the frequencies of occurrences of preferences of type $x^i \succ x'^i$ and of type $x^i \precsim x'^i$ in the sequence of the past DM's preference statements. Thus, Algorithm 1 recommends a solution when one of the two frequencies becomes significantly higher than the other, i.e., with a difference higher than $\alpha \Delta_k$ which, according to Proposition 1, is decreasing with k. This choice process is more robust than taking for granted the very first DM's preference statement, as illustrated below.

In Table 1, we present numerical results obtained for the described toy case with $q = 20$. To assess the benefit of Algorithm 1 in the context of noisy answers we introduce a random noise in the simulated DM's answers which swaps the answers with probability $p = 0.1$ and $p = 0.3$. For the two noise levels, we compare Algorithm 1 to an incremental preference elicitation method based on the minimization of the maximal regret and the current solution strategy (CSS) [3,5,20]. CSS consists of asking to compare solution $x \in X_P$ that minimizes the maximum regret $\text{MR}(x) = \max_{x' \in X_P} \max_{w \in W} \{f_w(x') - f_w(x)\}$ and its best opponent $x' = \arg\max_{x' \in X_P} \max_{w \in W} \{f_w(x') - f_w(x)\}$. Then the set of admissible models W is reduced using the obtained preference as a hard constraint, before iterating on this new set of admissible models. Parameters α and δ of Algorithm 1 are respectively set to $\alpha = 0.05$ and $\rho = 0$. In Table 1, we compare the results of both methods on 500 simulations in terms of number of queries and accuracy of the recommendation (number of simulations where the recommendation was correct). Looking at the results, we can see that CSS, while always terminating after one query, suffers from noisy answers and does recommend the optimal alternative in only 89% of the time for the low noise level, and 74% of the time for the high noise level. On the contrary, Algorithm 1 recommends the optimal alternative in nearly 100% of the time for the low noise level while asking only 3.5 questions on average, and in 90% of the time for the high noise level with about 14 questions on average. Further tests are conducted in the next section.

Table 1. Comparison of Algorithm 1 and CSS over 100 simulations.

	$p = 0.1$		$p = 0.3$	
	Number of query	Rec. accuracy	Number of query	Rec. accuracy
CSS	1.0	89%	1.0	74%
Algorithm 1	3.5	99%	14.0	90%

4 Numerical Tests

In this section, we present the results of numerical tests performed on synthetic preference data. We test the ability of Algorithm 1 to provide accurate recommendations while receiving noisy answers from the DM, and when possible, we compare those results to CSS. The tests are conducted with f_w taken as the weighted sum, the 2-additive Choquet integral and the Chebyshev distance. We consider random finite sets of admissible models $W_0 \subseteq W$ of size $|W_0| = 5000$ for each experiment. For the weighted sum and the Chebyshev distance, W_0 is obtained by uniform sampling of the simplex $W = \{w \in [0,1]^n | \sum_{i=1}^{n} w_i = 1\}$.

For the 2-additive Choquet integrals, the set of admissible weights is the set of capacities $W = \{w : 2^N \to \mathbb{R} | A \subseteq N, w(A) \leq w(B), w(\emptyset) = 0, w(N) = 1\}$, restricted to 2-additive capacities. The set of 2-additive capacities is a polyhedron admitting a polynomial number of extreme points [14] (Theorem 2.65), namely, the *unanimity games* defined for any $i \in [\![1, \frac{n(n+1)}{2}]\!]$ by $v_i(S) = 1$ if $Y_i \subseteq S$ and 0 otherwise, where $Y_i \subseteq N$ is any nonempty subset of size at most 2, and the *conjugates of unanimity games* defined for any $i \in [\![\frac{n(n+1)}{2} + 1, n^2]\!]$, by $v_i(S) = 1$ if $Y_i \cap S \neq \emptyset$ and 0 otherwise, for $i \in [\![\frac{n(n+1)}{2} + 1, n^2]\!]$. Hence, any 2-additive capacity w can be generated by a convex combination $w = \sum_{i=1}^{q} \beta_i v_i$ with $q = n^2, \beta_i \in [0,1], \sum_{i=1}^{q} \beta_i = 1$. Thus, we obtain samples W_0 of the set of 2-additive capacities by uniform sampling of the simplex $\{\beta \in [0,1]^q | \sum_{i=1}^{q} \beta_i = 1\}$.

For all the experiments, the set X_P of Pareto optimal solutions is randomly drawn within $\mathcal{X}$ as follows: vectors μ of size $n-1$ are uniformly drawn in $[0,1]^{n-1}$, then performance vectors $x \in \mathcal{X}$ are obtained by setting $x_i = \mu_{(i)} - \mu_{(i-1)}$ for $i = 1, \ldots, n$, where $\mu_{(0)} = 0$ and $\mu_{(p)} = 1$. To avoid that all generated alternatives share the same hyperplane $\sum_i x_i = 1$ in the utility space, the square root function is applied on all components x_i of each performance vector x. Also, the DM's answers are simulated according to a ground truth model $f_{w_{gt}}$ with a random weight vector w_{gt} generated in the same way as the elements of W_0. The answers are disturbed with random noises ϵ such that $y^k = \text{sign}(f_{w_{gt}}(x) - f_{w_{gt}}(x') + \epsilon_k)$ and ϵ_k is uniformly distributed within $[-\sigma, \sigma]$ with noise level $\sigma > 0$, for any k.

In the first experiment, we compare the noise tolerance of Algorithm 1 and CSS when f_w is the Choquet integral, $n = 5$, and $X_P = 100$. Parameter α of Algorithm 1 is set to $\alpha = 1.7 \times 10^{-2}$. Figures 2a and 2b respectively show, for Algorithm 1 and CSS, the average regret (left) and rank in the DM's hidden ranking (right) of the recommended solution over 100 simulations w.r.t. the number of query. More precisely, 2a(left) (resp. 2b(left)) shows the real regret of the recommended solution in percentage w.r.t. the initial level of uncertainty MR_1 (resp. the initial min-max regret $\min_{x \in X_P} \max_{x' \in X_P} \max_{w \in W} \{f_w(x') - f_w(x)\}$) along with the upper bound of the real regret MR_k (resp. the min-max regret of the reduced set of admissible models after k queries). In Fig. 2b, we observe that while the min-max regret quickly reduces with CSS, it does not induce a reduction of the real regret and yields recommended solutions with increasing real ranks. On the contrary, in Fig. 2a, we observe that while decreasing more slowly, the bound MR_k of Algorithm 1 decreases accordingly with the real regret and rank of the recommended solution $\hat{x}_k$. After 30 queries, the real rank of the recommended solution is about 8 for Algorithm 1 and 16 for CSS.

In the second experiment, we show the different tradeoffs between quality of the recommendation and number of asked queries that can be achieved with Algorithm 1 by varying the α parameter, which controls the threshold value $\alpha \Delta_k$. The tests are conducted for the Chebyshev distance for $n = 10$, $X_P = 100$ and α varying in a uniform grid within $[5 \times 10^{-3}, 5 \times 10^{-2}]$. The results are averaged over 100 simulations, and for this experiment, the DM's answers are disturbed with a random noise which swaps the answers with probability $p = 0.1$ and $p = 0.2$. For both noise levels, Fig. 3a (left) represents the average real regret of the

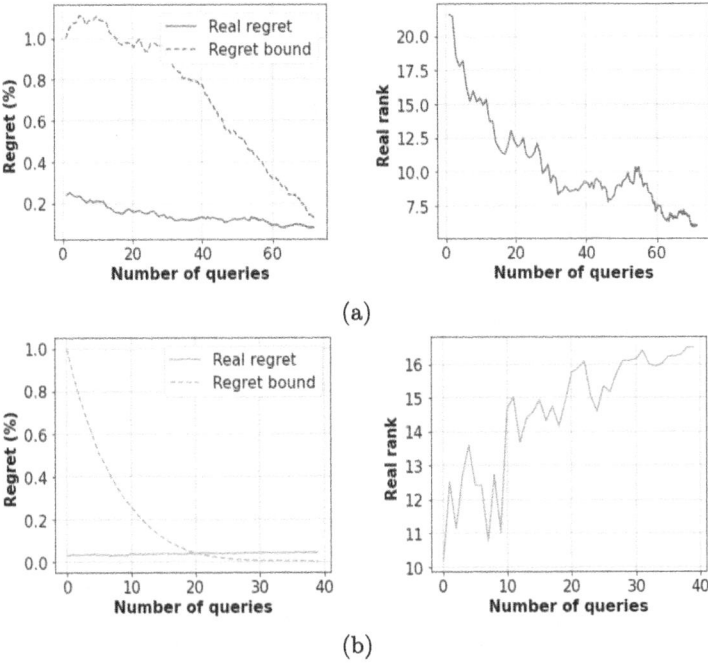

Fig. 2. Real regret and rank w.r.t. query number for Algorithm 1 (a) and CSS (b).

recommended solutions (in percentage w.r.t. MR_1) versus the average number of asked queries and Fig. 3a (right) shows the average real rank of the recommended solution, again versus the average number of queries. For all figures, the higher the α value, the higher the caution level of Algorithm 1, and thus the higher the number of asked queries. For $p = 0.1$ (red), asking 7 queries yields a real regret of 20% in average with an average real rank equal to 21 and asking 50 queries reduces the real regret to 10% and the average rank to 8. When the noise level increases, the performances weaken. For instance, for $p = 0.2$, 7 questions yield an average real rank equal to 26.

In the third experiment, we compare Algorithm 1 to another non-Bayesian active learning method recently proposed for linear models [17]. This method also exploits the idea of minimizing the 0–1 loss error on the set of admissible models W instead of irreversibly reducing the set of admissible models such as in CSS. However, while being effective at solving choice problems with small number of queries, the used querying strategy focuses only on the most plausible best element of X_P, and thus, the learned model is further from the hidden model w_{gt} and shows lower generalization performances on $\mathcal{X}$ than Algorithm 1. This can be seen in Table 2 where both methods are compared in terms of query number and real rank of the recommended solutions; we also give the average absolute distance to w_{gt} of the learned preference model and the test accuracy defined as the percentage of preference inversion on a test set of pairwise

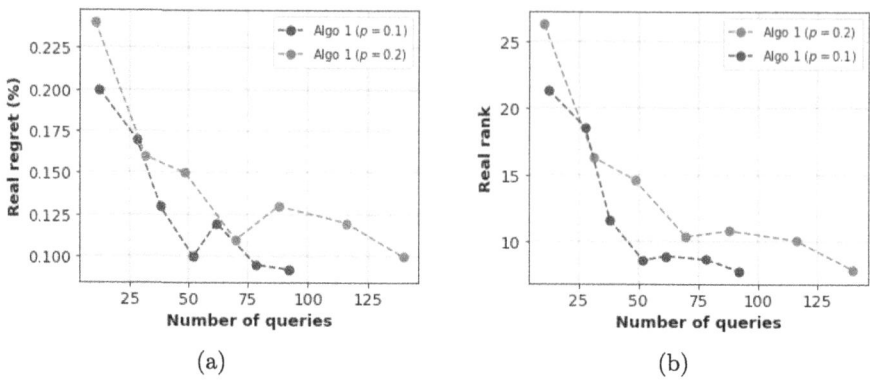

Fig. 3. Real regret (a) and rank (b) w.r.t. query number (Chebyshev distance).

comparison in $\mathcal{X}$. The tests are conducted with f_w taken as the weighted sum, $n = 10$, $\sigma = 0.05$, $|X_P| = 1000$ and for Algorithm 1 the parameter are set to $\alpha, \rho = (2 \times 10^{-10}, 0.5)$ which allow yielding similar query numbers for both methods. We observe that while yielding similar results in terms of query number and real rank, Algorithm 1 better recovers preference model w_{gt} and achieves a higher test accuracy. Computation times are comparable for both methods (3.35 sec. for Algorithm 1 and 1.33 sec for [Pourkhajouei, 23] on average).

Table 2. Comparison with [Pourkhajouei, 23] over 100 simulations.

	Query number	Real rank	Distance to w_{gt}	Test accuracy
Algo 1	68.0 ± 22.7	$40.1/1000 \pm 59.8$	0.03 ± 0.01	$88.5\% \pm 2.5\%$
[Pourkhajouei, 23]	60.9 ± 17.3	$44.75/1000 \pm 81.13$	0.07 ± 0.02	$78.0\% \pm 5.0\%$

5 Conclusion

We have presented a new approach for determining an optimal solution in a given set, by actively learning the parameters of an aggregation function describing the DM's preferences. This approach is a cautious version of the standard CSS based on the minimax regret criterion that progressively reduces the set of admissible model parameters, until a zero-regret (or near-zero-regret) solution appears as a necessary winner. In our view, our approach offers three significant advantages.

Firstly, it is more error-tolerant, since the DM's responses are not systematically interpreted as hard constraints on the parameter space. The numerical tests carried out in Sect. 4 clearly demonstrated the gain in robustness in the face of noisy responses. The second advantage is that, beyond the identification of an optimal choice, the method provides a learned model that can be used to explain decisions and make choices on new instances. Finally, it does not require the scalarizing function to be linear in its parameters and thus applies to a wider class of aggregators, including the weighted Chebyshev norm, or the Sugeno integral that is generally not learned by regret minimization.

Algorithm 1 also brings some advantages compared to recently proposed approaches for preference learning with noisy DM's answers, whether Bayesian [4,6] or non-Bayesian [17]. On the one hand, being non-Bayesian, the proposed approach does not require knowledge of a prior distribution on the model parameters, a strong assumption often necessary to initiate Bayesian learning. On the other hand, concerning non-Bayesian approaches, the numerical tests presented at the end of Sect. 4 show that Algorithm 1, while exhibiting comparable performance to recent alternative proposals [17] in terms of robustness to noisy responses, achieves significantly better generalization performance and thus is likely to make better decision on new instances of choice problems.

References

1. Adam, L., Destercke, S.: Handling inconsistency in (numerical) preferences using possibility theory. Inf. Fusion **103**, 102089 (2024)
2. Balcan, M.F., Beygelzimer, A., Langford, J.: Agnostic active learning. In: Proceedings of the 23rd International Conference on Machine Learning, pp. 65–72 (2006)
3. Benabbou, N., Perny, P., Viappiani, P.: Incremental elicitation of Choquet capacities for multicriteria choice, ranking and sorting problems. Artif. Intell. **246**, 152–180 (2017)
4. Bourdache, N., Perny, P., Spanjaard, O.: Incremental elicitation of rank-dependent aggregation functions based on Bayesian linear regression. In: Proceedings of IJCAI-19, pp. 2023–2029 (2019)
5. Boutilier, C., Patrascu, R., Poupart, P., Schuurmans, D.: Constraint-based optimization and utility elicitation using the minimax decision criterion. Artif. Intell. **170**(8–9), 686–713 (2006)
6. Chajewska, U., Koller, D., Parr, R.: Making rational decisions using adaptive utility elicitation. In: AAAI/IAAI, pp. 363–369 (2000)
7. Chateauneuf, A., Jaffray, J.Y.: Some characterizations of lower probabilities and other monotone capacities through the use of Möbius inversion. Math. Soc. Sci. **17**(3), 263–283 (1989)
8. Cohn, D., Atlas, L., Ladner, R.: Improving generalization with active learning. Mach. Learn. **15**, 201–221 (1994)
9. Dasgupta, S.: Two faces of active learning. Theoret. Comput. Sci. **412**(19), 1767–1781 (2011)
10. Dasgupta, S., Hsu, D.J., Monteleoni, C.: A general agnostic active learning algorithm. Adv. Neural Inf. Process. Syst. **20** (2007)
11. Domshlak, C., Joachims, T.: Unstructuring user preferences: efficient non-parametric utility revelation. arXiv preprint arXiv:1207.1390 (2012)
12. Feldman, V., Guruswami, V., Raghavendra, P., Wu, Y.: Agnostic learning of monomials by halfspaces is hard. SIAM J. Comput. **41**(6), 1558–1590 (2012)
13. Grabisch, M., Marichal, J.L., Mesiar, R., Pap, E.: Aggregation Functions, vol. 127. Cambridge University Press, Cambridge (2009)

14. Grabisch, M., et al.: Set Functions, Games and Capacities in Decision Making, vol. 46. Springer, Cham (2016). https://doi.org/10.1007/978-3-319-30690-2
15. Hanneke, S., et al.: Theory of disagreement-based active learning. Found. Trends® Mach. Learn. **7**(2-3), 131–309 (2014)
16. Herin, M., Perny, P., Sokolovska, N.: Learning preference models with sparse interactions of criteria. In: Proceedings of the of IJCAI (2023)
17. Pourkhajouei, S., Toffano, F., Viappiani, P., Wilson, N.: An efficient non-Bayesian approach for interactive preference elicitation under noisy preference models. In: Bouraoui, Z., Vesic, S. (eds.) ECSQARU 2023. LNCS, vol. 14294, pp. 308–321. Springer, Cham (2023). https://doi.org/10.1007/978-3-031-45608-4_23
18. Steuer, R.E.: Multiple Criteria Optimization. Theory, Computation, and Application (1986)
19. Vanderpooten, D., Vincke, P.: Description and analysis of some representative interactive multicriteria procedures. In: Models and Methods in Multiple Criteria Decision Making, pp. 1221–1238. Elsevier (1989)
20. Wang, T., Boutilier, C.: Incremental utility elicitation with the minimax regret decision criterion. In: IJCAI, vol. 3, pp. 309–316 (2003)
21. White, C.C., Sage, A.P., Dozono, S.: A model of multiattribute decisionmaking and trade-off weight determination under uncertainty. IEEE Trans. Syst. Man Cybern. **2**, 223–229 (1984)
22. Wierzbicki, A.P.: On the completeness and constructiveness of parametric characterizations to vector optimization problems. Oper.-Res.-Spektr. **8**(2), 73–87 (1986)

Learning Multiple Multicriteria Additive Models from Heterogeneous Preferences

Vincent Auriau[1,2](✉) [iD], Khaled Belahcène[1] [iD], Emmanuel Malherbe[2] [iD], and Vincent Mousseau[1] [iD]

[1] MICS, CentraleSupélec, Université Paris Saclay, Gif sur Yvette, France
{vincent.auriau,khaled.belahcene,vincent.mousseau}@centralesupelec.fr
[2] Artefact Research Center, Paris, France
emmanuel.malherbe@artefact.com

Abstract. Additive preference representation is standard in Multiple Criteria Decision Analysis, and learning such a preference model dates back from the UTA method [11]. In this seminal work, an additive piecewise linear model is inferred from a learning set composed of pairwise comparisons. In this setting, the learning set is provided by a single Decision-Maker (DM), and an additive model is inferred to match the learning set. We extend this framework to the case where *(i)* multiple DMs with heterogeneous preferences provide part of the learning set, and *(ii)* the learning set is provided as a whole without knowing which DM expressed each pairwise comparison. Hence, the problem amounts to inferring a preference model for each DM and simultaneously "discovering" the segmentation of the learning set. In this paper, we show that this problem is computationally difficult. We propose a mathematical programming based resolution approach to solve this *Preference Learning and Segmentation* problem (PLS). We also propose a heuristic to deal with large datasets. We study the performance of both algorithms through experiments using synthetic and real data.

Keywords: Additive preference model · preference learning · heterogeneous preferences

1 Introduction and Motivation

Multiple Criteria Decision Analysis (MCDA) aims at supporting a Decision-Maker (DM) and at providing him/her recommendations concerning the decision tackled. These recommendations are based on a preference model that faithfully represents the DM's judgments. We consider, in this paper, an additive value-based representation of preferences [13]. For the additive model to faithfully represent the DM's preferences, adequate values should be assigned to the model parameters. This is usually done through a preference learning phase, which takes as input observed preferences such as pairwise comparisons [22].

A standard value-based setting for preference learning in multicriteria ranking problems with an additive representation of preferences dates back from the

UTA method [11]. This seminal work considers a multicriteria ranking problem in which an additive piece-wise linear model is inferred from preferences. In this setting, a single DM provides the learning set, and the UTA method infers, with linear programming, the additive piece-wise linear model that best matches the learning set, even with noisy preferences (Fig. 1).

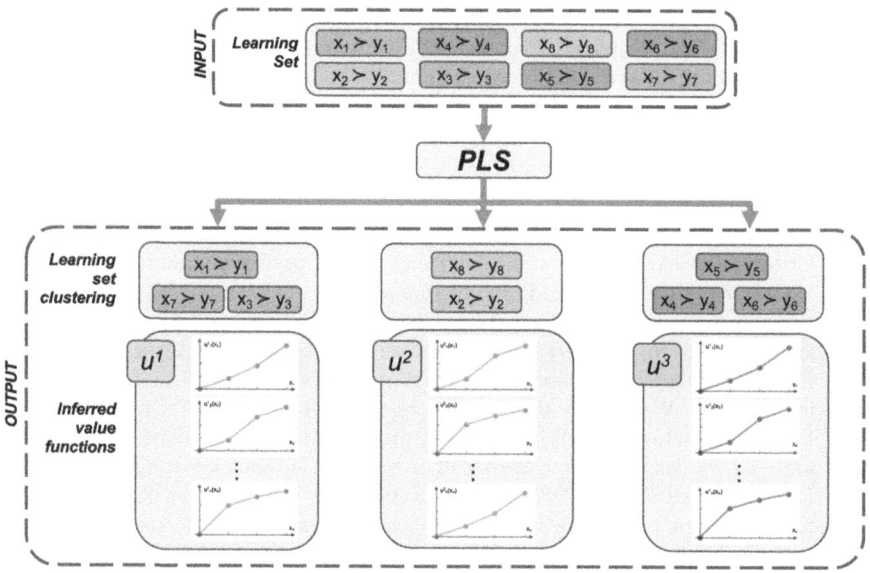

Fig. 1. Illustration of Preference Learning and Segmentation with 3 clusters

Example 1 (Clients preferences in a supermarket): A supermarket is willing to adapt the list of products to its customer base. The products selected should align with the client's preferences. Yet, not all customers necessarily have the same preferences, and it is standard to consider a market segmentation in which each segment represents a group of clients with homogeneous preferences. Each store can rely on actual sales to identify the segmentation and learn the clients' preferences. For each product sale x, one can derive a comparison $x \succcurlyeq y$ for each product y sold in the store and substitutable to x. Hence, the problem can be formulated as clustering the clients and learning a preference model for each cluster.

Example 2 (Films comparisons on social media): A website about cinema presents several films and requests user feedback. Users provide film scores from which pairwise comparisons can be derived. Users are not requested to have an account and are hence unidentifiable. For marketing purposes, the website wants to identify, from the collected data, k clusters of users, with each a specific preference model.

Example 3 (Wine tasting): A group of oenologist experts is performing a blind tasting evaluation of various wines. These wines have been previously evaluated from a chemical point of view, and the concentration of several chemical components known to characterize the quality of the wines are measured. Each expert is required to attach to wines a number of "stars" (from ★ to ★★★★★). All experts in the panel obviously are not fully aligned in the way they assess wines. We would like to identify groups of experts with similar tastes and how each group accounts for wine characteristics in their preferences.

These examples correspond to instances of a *Preference Learning and Segmentation* (PLS) problem. Such PLS problem takes as input a learning set composed of a list of pairwise comparisons and should determine a given number of preference models so that each comparison in the learning set is correctly represented by at least one of the models. A key issue in PLS problems is that the partition of the learning set is not known and should be computed along the inferred preference models. Hence, the outputs of the PLS problem are both *(i)* the inferred preference models, and *(ii)* the corresponding partition of the learning set.

In this paper, we propose a formal setting for the PLS problem, usually not considered in preference learning for MCDA, in which decisions from heterogeneous sources are observed. We present two algorithms for preference learning segmentation: a MILO formulation and a heuristic method. Finally, we have conducted experiments to establish our model's performances using synthetic and real data.

The paper is organized as follows. In the next section, we review related works that are of interest with respect to our contribution. Section 3 is devoted to the formalization of the PLS problem, its resolution via a mathematical programming formulation, and the study of its computational complexity. A heuristic method specifically developed for the PLS problem is presented in Sect. 4. The experiments performed on synthetic data, in Sect. 5, allows to analyze the performance of both proposed algorithms. In Sect. 6, we show the benefit of our contribution on a a-posteriori experiment based on real data. Final section groups conclusions and avenues for further research.

2 Related Work and Previous Research

2.1 MCDA and the UTA Model

In multi-criteria decision analysis, alternatives are evaluated on several conflicting criteria. Formally, we consider a set $\mathcal{N} = \{1, \cdots, n\}$ of n criteria, where the evaluation scale on criterion i, $i \in \mathcal{N}$, is denoted $X_i \subset \mathbb{R}$. Hence an alternative $x = (x_1, ..., x_n)$ is an element of $X = \prod_{i \in \mathcal{N}} X_i$.

To compare such alternatives x and y, a value-based approach consists in defining a value function u such that $x \succcurlyeq y \Leftrightarrow u(x) \geq u(y)$, where $\succcurlyeq$ is a weak preference relation defined on X ($\succ$ and $\sim$ its symmetric and asymetric part).

In the following we consider w.l.o.g. that criteria are to be maximized, i.e., $x_i \succcurlyeq_i y_i \Leftrightarrow x_i \geq y_i$, where $\succcurlyeq_i$ denotes the weak preference relation on criterion i.

The UTA [11] method defines a value function $u(\cdot)$ using additive marginal value functions $u_i(\cdot)$, defined on X_i, $i \in \mathcal{N}$: $u(x) = \sum_{i \in \mathcal{N}} u_i(x_i)$, $x \in X$. On criterion $i \in \mathcal{N}$, the marginal value function $u_i(\cdot)$ is piecewise linear. More precisely, the scale of each criterion X_i is divided into L equal interval, $[x_i^0, x_i^1], [x_i^1, x_i^2], \ldots, [x_i^{L-1}, x_i^L]$, and the marginal value function $u_i(\cdot)$ is linear in each interval. Hence, $u_i(\cdot)$ is defined by its values at each breakpoint, denoted $u_i(x_i^l)$, $i \in \{1..n\}$, $l \in \{0..L\}$. The piecewise marginal functions $u_i(\cdot)$ when $x_i \in [x_i^l, x_i^{l+1}]$ can be written as:

$$u_i(x_i) = u_i(x_i^l) + \frac{x_i - x_i^l}{x_i^{l+1} - x_i^l} \cdot \left(u_i(x_i^{l+1}) - u_i(x_i^l)\right) \tag{1}$$

The parameters of the value model, $u_i(x_i^l) \in \mathbb{R}^+$, $i \in \mathcal{N}$, $l = 0..L$ are to be estimated using a learning set $\mathcal{P}$ provided by the DM. This learning set is composed of P pairwise comparisons:

$$\mathcal{P} = \{(x^{(1)}, y^{(1)}), (x^{(2)}, y^{(2)}), \cdots, (x^{(P)}, y^{(P)})\}$$

where $x^{(j)} = (x_1^{(j)}, ..., x_n^{(j)})$ is preferred to $y^{(j)} = (y_1^{(j)}, ..., y_n^{(j)})$, $j = 1..P$.

In such settings, the objective is to obtain the functions u_i so that:

$$x^{(j)} \succ y^{(j)} \Leftrightarrow \sum_i u_i(x_i^{(j)}) > \sum_i u_i(y_i^{(j)}) \tag{2}$$

The model follows additional normalization and monotonicity constraints:

$$u_i(x_i^l) \leq u_i(x_i^{l+1}), \forall i \in \mathcal{N}, \forall l = 0..L - 1 \tag{3a}$$

$$u_i(x_i^0) = 0, \forall i \in \mathcal{N} \tag{3b}$$

$$\sum_{i=1}^{n} u_i(x_i^L) = 1 \tag{3c}$$

The model can be formalized as a Linear Optimization problem. It ensures an efficient and scalable computation using solvers. For this purpose, we introduce potential estimation errors, $\sigma^{(j)} \in \mathbb{R}^+$. They are used to add an additional preference constraint for all observed pairs $j \in \{1..P\}$:

$$u(x^{(j)}) - u(y^{(j)}) + \sigma^{(j)} \geq 0, \forall j \in \{1..P\} \tag{3d}$$

The optimization objective is formulated as the minimization of the estimation errors:

$$\min \sum_{j=1}^{P} \sigma^{(j)} \tag{UTA_{obj}}$$

Variants of the UTA model [24] exist for different types of data or optimality criteria. Along this adaptability and efficient computation, the method has been popularized by its interpretability potential.

2.2 Market Segmentation

Understanding its customer base is a critical stake for retailers [25]. It is used to adjust the marketing strategy as well as increasing satisfaction. A common approach is to delineate market segments: buyers with similar characteristics are grouped together and represent a segment of the customer base. The similarity between individuals is usually based on demographic, geographic or choice data. Natural tools to uncover these different segments are clustering algorithms such as k-means [28] or hierarchical clustering [16]. If using the customer characteristics is natural, as stated in [17], individual with similar socioeconomic attributes may have very different preferences. Introducing attitudinal information [17] helps taking into account the decision process of the customer. Such data can be collected through surveys which are difficult and expensive to carry out. Leveraging past transactions for modeling user choices has also been studied, in particular with the family of latent class models [27]. Recent work shows promising perspectives on large datasets [2], however this probabilistic modeling may not reveal distinct clusters of preferences. Engineering features [15] can also prove to be efficient but usually needs several data points for each customer so that it is possible to represent well its behavior.

In order to overcome these limitations, we propose to create segments based on customer decisions, taking into account the alternatives attributes. Specific information on the individuals is not needed although it could be integrated.

2.3 Clustering Rankings

When disagreements or incompatibilities are observed in the learning set, the UTA model is designed to find the best consensus. In order to integrate heterogeneity in the modelization, a preprocessing step can be carried out to partition the data into homogeneous sets. A standard method is to work with clustering algorithms based on the different decision makers attributes. However when few or no data is available or if decision makers have only been exposed to a subset of alternatives, an approach is to leverage their decision process with finer granularity. A common algorithm used with heterogeneous decision makers and ranking data is the collaborative filtering [8]. The algorithm is able to generalize from partial evaluations of alternatives by individuals. It leverages a proximity distance between individuals based on their common evaluations of the same alternatives. Several methods have proven their efficiency to create segments of the population based on a collaborative filtering output [26]. Another possibility is to define a similarity measure between two rankings that can be used to create clusters. [4] leverages the Mahalanobis distance to create groups of decisions within a defined dissensus threshold.

These methods do not describe the different alternatives on relevant criteria. An advantage is that the data collection process is simpler. However one might want to generalize a model to unknown alternatives or to take into account criteria that may vary over time such as the price. For this purpose, [5,7] propose to compute a ranking function by decision maker and then to clusterize these value functions. In order to use clustering algorithms, it is needed to define the appropriate distance measure in the value function space. [7] uses such approach for a search engine that generates recommendations based on a click history and [5] regroups spectators according to their movie tastes.

These methods suppose that enough preferences are collected for every user. In the context of a webpage or a supermarket, it is difficult to track customers. The database may be an agglomeration of anonymous contributions, that must therefore be handled differently. Latent class models have been developed for such cases. For instance, mixtures of Plackett-Luce models [9,19] or Mallows [3,18] have been employed to model elections. It is a typical case of an unidentified population that has extremely heteregenous tastes. However, these mixture models consider defined and fixed alternatives. For instance, candidates are well identified, the clusters are not based on the different elements of their programs but the rankings of the candidate themselves. Going back to our introductive *Example* 3, it would mean that the computed model would not be able to generalize to an untasted wine or to the next year when wines will be different. The experts decision process is not learned nor used. Furthermore some criteria such as the price can prove to be particularly important in the decision making of customers. In *Example* 1 it has a direct impact on customers preferences and can change. It is therefore compulsory to integrate such information. Moreover, obtaining parameters such as a sensitivity toward a criterion will not only better match the customer decisions but also provide meaningful insight.

Finally, [20] uses a hierarchical model to represent heteogenerous consumers that is estimated with Bayesian inference. Priors are defined and adjusted to represent each customer membership to a customer segment and then its preferences. While this model is able to work with large datasets, it mainly relies on the choice of the right priors. This prior selection is mainly subjective and lacks an efficient validation method.

3 Preference Learning and Segmentation (PLS) Problem

3.1 Problem Setting

We consider a heterogeneous set of P observed preferences. These preferences are expressed as a pair of alternatives $(x^{(j)}, y^{(j)})$, $j \in \{1..P\}$, where $x^{(j)}$ has been preferred to $y^{(j)}$ by an unknown decision maker. The alternatives are described on n criteria whose evaluation scale is noted X_i for criterion $i \in \{1..n\}$.

The Preference Learning and Segmentation problem aims to cluster the P pairs into K clusters and to learn a multicriteria preference model for each of these clusters. A particular case occurs when these preference models are represented by an additive value function, i.e., when we learn a value function u^k on each cluster. An obvious intent of the PLS is to represent the maximum number observed preferences. By that, we mean that for all $(x^{(j)}, y^{(j)})$, $j \in \{1..P\}$, at least one of the k value functions expresses the observed decisions or differently written:

$$x^{(j)} \succcurlyeq^K y^{(j)} \iff \exists k \in \{1..K\} \text{ such that } \sum_i u_i^k(x_i^{(j)}) \geq \sum_i u_i^k(y_i^{(j)}) \quad (4)$$

3.2 Mixed Integer Linear Optimization Resolution Approach

We propose a Mixed Integer Linear Optimization (MILO) approach to solve our Preference Learning and Segmentation problem. The main objective is to cover at best the observed preferences with K different UTA models. In the following, the models are noted $u^k(.), k \in \{1..K\}$ and their marginal values on criterion $i, u_i^k(.)$.

As similarly introduced in Sect. 2.1, the $u^k, k \in \{1..K\}$ value functions of the UTA model are represented by their marginal coefficients on criterion $i \in \{1..n\}$, $u_i^k(x_i^l), l \in \{1..L\}$, to be computed. They also follow the same normalization and monotonicity constraints defined in Eqs. 3a, 3b and 3c $\forall k \in \{1..K\}$:

$$\sum_{i=1}^{n} u_i^k(x_i^L) = 1 \tag{5a}$$

$$u_i^k(x_i^0) = 0, \forall i \in \{1..n\} \tag{5b}$$

$$u_i^k(x_i^l) \leq u_i^k(x_i^{l+1}), \forall i \in \{1..n\}, \forall l \in \{0..L-1\} \tag{5c}$$

Additionally to compute the marginal coefficients, our optimization formulation attributes each observed preference to one of the K models. For this purpose, we introduce the binary variables $z^{k,(j)}, j \in \{1..P\}$ and $k \in \{1..K\}$ as well as the following constraints $\forall j \in \{1..P\}$:

$$\sum_{k=1}^{K} z^{k,(j)} \geq 1 \tag{6a}$$

$$\sum_{i=1}^{n} u_i^k(x_i^{(j)}) - \sum_{i=1}^{n} u_i^k(y_i^{(j)}) + M \cdot (1 - z^{k,(j)}) + \sigma^{(j)} \geq 0, \forall k \in \{1..K\} \tag{6b}$$

with M a majorant of sufficient magnitude and $\sigma^{(j)} \geq 0$ estimation errors. The constraint in Eq. 6a makes sure that every pair is attributed to a cluster. The constraint in Eq. 6b ensures that when a pair is attributed to cluster k, the corresponding value function u^k respects the preference. The linear interpolation of $u_i^k(x_i)$ between $u_i^k(x_i^l)$ and $u_i^k(x_i^{l'})$ is defined as in (1).

Finally we formulate the following optimization objective, minimizing the total sum of estimation errors:

$$\min \sum_{j=1}^{P} \sigma^{(j)} \tag{PLS_{obj}}$$

This formulated MILO can be solved using optimization solvers. In terms of complexity, it handles $K \times P$ binary variables as well as $K \times n \times L$ continuous variables in $[0; 1]$. The complexity grows with the number of clusters to be estimated, but most of all with the number of observed preferences. In the experiments, we provide solving time insights on different instance sizes.

The presented model defines a set of binary variables $z^{k,(j)}$ for each of the observed pairwise comparisons. One can ensure that groups of pairs are clusterized together using the same binary variables in 6b. For instance, as further described in Sect. 6, a choice can be broken into several pairwise comparisons to be clusterized. The $z^{k,(j)}$ binary variables are set to be equal for these pairs.

3.3 Problem Difficulty

In this section, we study the problem of rationalizing a set of observed preferences by segmenting them and finding a model for each segment from a computational perspective. When the number of segments is one and the sought model is the additive value model with piecewise linear marginal utilities with fixed breakpoints, the UTA method reformulates an instance of the 1-DM UTA SEGMENTATION problem as an instance of LINEAR OPTIMIZATION (see Sect. 2.1) which can be solved in polynomial time [23]. When allowing for several DMs, the problem involves partitioning the data and seems more arduous. We indeed show that, even in the simplest configuration where each criterion is evaluated on a binary scale, the problem is NP-complete for any fixed value of $K \geq 2$.

We begin by giving a proper definition of this trimmed down version, where the observed preferences can be interpreted as subset comparisons. We name the problem by referring to the majority rule because when trying to represent preferences with an additive value function, each criterion is given a weight that can be interpreted as a voting power.

MULTIPLE DECISION-MAKER MAJORITY RULE SEGMENTATION (K-DM MRS)	
Input:	set N of criteria, set $\mathcal{P}$ of comparisons between subsets of N, positive integer $K < |N|$
Question:	is there a partition of $\mathcal{P}$ into K disjoint subsets $\mathcal{P}^1, \ldots, \mathcal{P}^K$ and K functions $W_1, \ldots W_K : N \Rightarrow \mathbb{Q}^+$ such that, for each statement $A \succ B$ in $\mathcal{P}^k$ we have $\sum_{a \in A} W_k(a) > \sum_{b \in B} W_k(b)$?

Membership of this problem to the class NP is obvious, as the partition and weights form a polynomial certificate by design. We detail the hardness results, differentiating between the case with 3 or more decision-makers, where we directly interpret the K-partition of preference statements as a K-coloring, from the case with 2 decision-makers, as obtaining a 2-coloring is polytime but we leverage the 2-partition of preference statements to mimic conditional statements allowing to represent the constraints of a vertex cover problem. Complete proofs can be found in the supplementary material.

Proposition 1. *for $K \geq 3$, K-DM MRS is NP-complete.*

Proof (sketch). reduction from K-COLORABILITY [6,12]. Given a graph $\mathcal{G}=(V,E)$, we produce an instance $(N, \mathcal{P})$ of K-DM MRS as follows.

- *Criteria:* let $N := V \cup \{\tau_0, \tau_1, \tau_2\} \cup \bigcup_{k \in [K]} \{\omega_k\}$. We note $\Omega_{-k} := \bigcup_{k' \in [K]: k' \neq k} \{\omega_k\}$ and $N_{-k} := N \setminus \{\omega_k\}$.
- *Preference statements:* let $\mathcal{P} := \bigcup_{v \in V} \pi_v \cup \bigcup_{k \in [K]} \mathcal{P}_k^E \cup \bigcup_{k \in [K]} \mathcal{P}_k^\tau \cup \bigcup_{k \in K} \pi_k$ where:
  - for all $v \in V$, π_v denotes the statement $\{v\} \succ \{\tau_1\}$;
  - for all $k \in [K]$, $\mathcal{P}_k^E$ and $\mathcal{P}_k^\tau$ denote the following collections of statements $\mathcal{P}_k^E := \bigcup_{\{u,v\} \in E} \{\{\tau_2\} \succ \{u,v\} \cup \Omega_{-k}\}\}$, $\mathcal{P}_k^\tau := \{\{\tau_1\} \succ \{\tau_0\} \cup \Omega_{-k},$ $\{\tau_0, \tau_1\} \succ \{\tau_2\} \cup \Omega_{-k}\}$
  - finally, for all $k \in [K]$, π_k^ω denotes the statement $\{\omega_k\} \succ N_{-k}$.
- *Claim:* $\mathcal{G}$ is a positive instance of K-COLORABILITY $\iff N, \mathcal{P}$ is a positive instance of K-DM MRS.
- *Hint:* observe the statements $\bigcup_{k \in [K]} \pi_k^\omega$ are pairwise inconsistent, so each segment contains exactly one of those, defining the color k associated to the vertices v whose statement π_v appear in the same bundle as π_k^ω. Statements in $\mathcal{P}_k^E$ or $\mathcal{P}_k^\tau$ are pairwise inconsistent with $\pi_{k'}^\omega$ with $k \neq k'$ thus appear in the same segment as π_k^ω and ensure the coloring is correct.

Proposition 2. *2-DM MRS is NP-complete.*

Proof (sketch). reduction from VERTEX COVER [6,12]. Given a graph $\mathcal{G} = (V, E)$ and a positive integer $K < |V|$, we produce an instance $(N, \mathcal{P})$ of 2-DM MRS as follows.

- *Criteria:* Let $N := V \cup \{\alpha, \beta, \omega_1, \omega_2\} \cup \{\gamma_1, \ldots, \gamma_K\}$, such that $|N| = |V| + K + 4 < 2|V| + 4$.
- *Preference statements:* let $\mathcal{P} := \bigcup_{v \in V} \{\pi_v^+, \pi_v^-\} \cup \bigcup_{\{u,v\} \in E} \pi_{\{u,v\}}^E \cup \{\pi_2^\alpha, \pi_1^\omega, \pi_2^\omega\} \cup \mathcal{P}_1^\alpha$ where:
  - for all vertices $v \in V$, $\pi_v^+ : \{v\} \succ \{\alpha\}$ and $\pi_v^- : \{\beta\} \succ \{v\}$;
  - for all edges $\{u, v\} \in E$, $\pi_{u,v}^E : \{u, v\} \succ \{\alpha, \beta, \omega_2\}$;
  - $\pi_1^\omega : \{\omega_1\} \succ V \cup \{\omega_2, \alpha, \beta\}$, $\pi_2^\omega : \{\omega_2\} \succ V \cup \{\omega_1, \alpha, \beta\}$ and $\pi_2^\alpha : \{\alpha\} \succ \{\beta, \omega_1\}$;
  - $\mathcal{P}_1^\alpha : \left\{\{\alpha\} \succ \{\beta, \omega_2\}, \{\alpha\} \cup \bigcup_{k \in [K]} \{\gamma_k\} \succ V \cup \{\omega_2\}\right\} \cup \bigcup_{k=1}^K \{\{\alpha\} \succ \{\gamma_k, \omega_2\}\}$.
- *Claim:* $\mathcal{G}$ has a VERTEX COVER (i.e. a subset $V' \subset V$ such that for each edge $\{u, v\} \in E$ at least one of u and v belongs to V') of size at most $K \iff N, \mathcal{P}$ is a positive instance of 2-DM MRS.
- *Hint:* Observe the statements π_1^ω and π_2^ω cannot appear in the same segment. Thus, comparative statements with ω_1 (resp. ω_2) at the RHS are bundled together. Meanwhile, for each vertex $v \in V$, statements π_v^+ and π_v^- are pairwise inconsistent and appear in distinct bundles. The set $\{v \in V : \pi_v^+$ appears in the same segment as $\pi_1^\omega\}$ is a vertex cover of size at most K.

Hence, finding an exact solution to the PLS problem is difficult; the MILO formulation presented in Sect. 3.2 is therefore expected to be relevant for instances up to a certain size. The detailed proofs are shared in the online appendices[1].

---

[1] Appendices available at github.com/artefactory/learning-heterogeneous-preferences.

4 Heuristic

In order to mitigate the difficulty to scale up the MILO formulation, we have developed a first heuristic that can work with large number of observed preferences. Similar to the K-Means algorithm [14], it consists of a succession of two steps: assignment and update.

Initialization:
K - value chosen by the user - clusters are initialized with the KMeans++ [1] procedure:

- Choose randomly one of the data points as the first cluster center
- Compute the Euclidian distance $D(x^{(j)}, y^{(j)}) = ||(x^{(j)} - y^{(j)}) - (x^* - y^*)||_2$ between each data point $(x^{(j)}, y^{(j)})$ and the closest cluster center (x^*, y^*)
- Choose randomly the next cluster center among the remaining data points with a probability proportional to $D(x)^2$
- Once K cluster centers are chosen, assign each data point to the cluster whose center is the closest

After the initialization, we proceed to an update step and then alternate assignment and update steps until convergence.

Assignment Step:
The assignment step attributes a cluster to each of the data point $(x^{(j)}, y^{(j)})$. Considering K UTA models computed in the update step, the data point is associated to the cluster $k = \arg\max_{k \in \{1..K\}} \left(u^k(x^{(j)}) - u^k(y^{(j)}) \right)$.

Update Step: Given a partitioning in K cluster of our learning set, we proceed to compute a UTA model u^k, as defined in Sect. 2.1, on each of the partitions.

Convergence: We consider that the algorithm has converged when the partition does not change between two assignment steps. A softer stopping criterion can be set through a maximum number of successive iterations with the number of data points attributed to different cluster under a negligible ratio of the learning set.

The heuristic consists on a succession of Linear Programming problems solving, ensuring better scalability than our MILO formulation.

5 Empirical Assessment of Algorithms

We present experimental results of our two algorithms on synthetic data. The different algorithms have been implemented in Python. We have used the Gurobi solver [10] to solve the different optimization problems. The codebase is shared at github.com/artefactory/learning-heterogeneous-preferences.

5.1 Experiments on Synthetic Datasets

Experimental Objectives. Synthetic data can be leveraged to better understand a model's behavior and performance. In particular, the synthetic experiments are designed toward three objectives:

- Assessing descriptive power of the algorithm, i.e., its ability to learn from heterogeneous preferences compared to a "single preference" formulation such as *UTA*.
- Assessing predictive power of the algorithm, i.e., its ability to generalize to unseen data.
- Assessing the scalability of the model with large datasets.

Experimental Design. The following protocol has been carried out for the synthetic experiments:

Data generation
- Select the number of ground truth clusters, $K \in \{2,3,4\}$, the number of criteria, $n \in \{4,8,10\}$, the number of pairs in the learning set $P \in \{128, 256, 512, 1024, 2048, 4096\}$, and the noise level $NL \in \{0, 0.05, 0.1, 0.2\}$,
- Draw uniformly at random K additive piecewise linear functions following *UTA*'s normalization constraints,
- Draw uniformly at random pairs $(x^{(j)}, y^{(j)}) \in \prod_{i \in \mathcal{N}} X_i \times \prod_{i \in \mathcal{N}} X_i$ ($\frac{P}{K}$ for learning and 8192 for testing),
- with probability $1 - NL$, place the correct preference in the dataset (i.e., $a > b$ when $u^k(a) > u^k(b)$ for any pair (a,b)), and the opposite preference with probability NL.

Model Computation. Knowing the right number of clusters, each instance is solved using the MILO formulation as well as using the heuristic.

Evaluation. For each instance, we evaluate the computed models on the learning set and test set. The *ExplainedPairs* metric estimates the ratio of pairwise comparisons that are correctly restored by at least one the models.

$$ExplainedPairs = \frac{1}{P} \sum_{j=1}^{P} \mathbb{1}\{\exists k \leq K \mid u^k(x^{(j)}) \geq u^k(y^{(j)})\}$$

The full procedure is repeated 100 times for each combination of parameters.

Results and Discussion. In Fig. 2, we look into the computing time of our two models for different learning set sizes. We observe that the MILO formulation, quickly takes significant amounts of time, particularly when noise is added to the dataset, in Fig. 3. On the contrary the heuristic solving time seems to only increase linearly with the data quantity and seems not to be impacted by the chosen number of clusters.

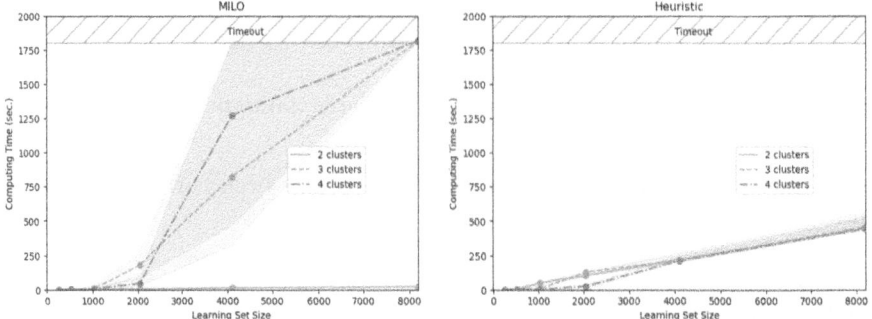

Fig. 2. Computing time for 8 criteria, 0% noise and different number of clusters. Shaded areas represent the values of first and 3rd quartiles over experiment repetitions.

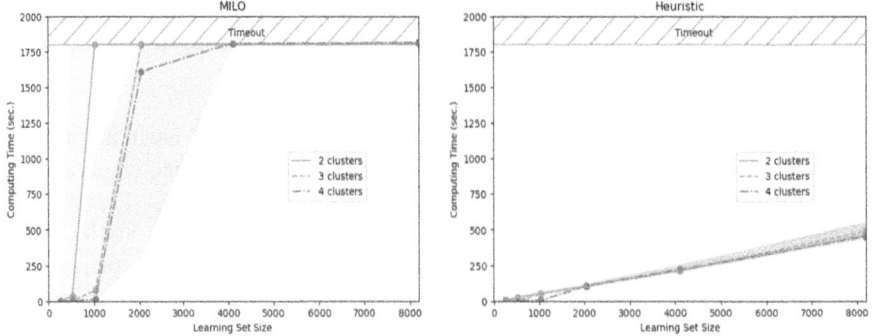

Fig. 3. Computing time for 8 criteria, 5% noise and different number of clusters. Shaded areas represent the first and 3rd quartiles over experiment repetitions.

We explore the ability of our MILO and heuristic to learn and generalize the heterogeneous preferences with the *ExplainedPairs* metric computed on the test Set. In Fig. 4, we observe the variations with different numbers of clusters and learning set sizes and in Fig. 5, the variations with different noise levels. While we can explain some performance drops with the fact that the MILO reaches the defined timeout, we can globally observe that the more data, the better.

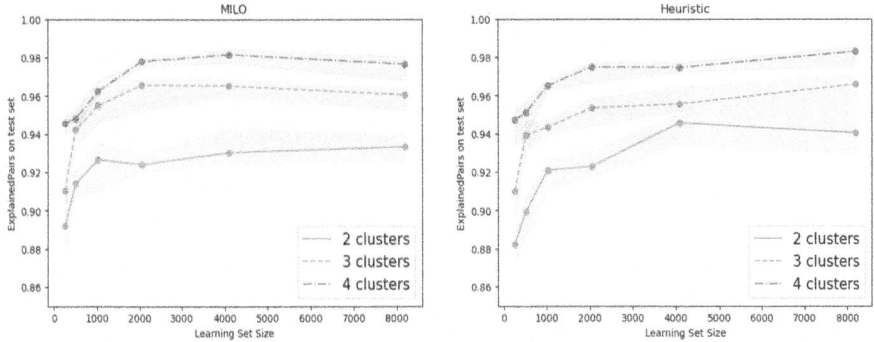

Fig. 4. Proportion of ExplainedPairs (%) on test set with 8 criteria, 5% noise and different cluster numbers. Shaded areas represent the first and 3rd quartiles over experiment repetitions.

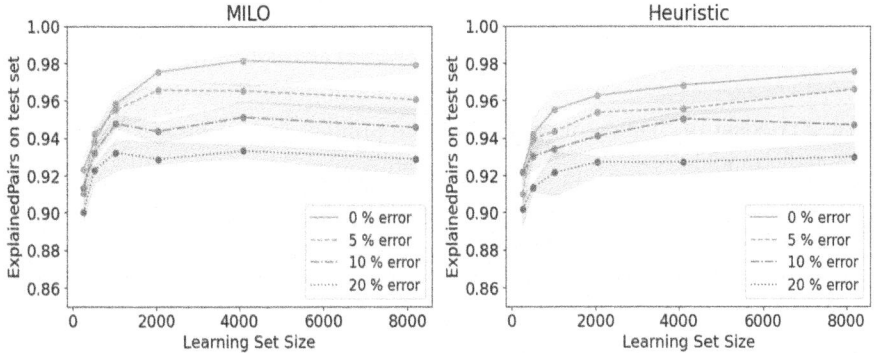

Fig. 5. Proportion of ExplainedPairs (%) on test set for 3 clusters, 8 criteria and different noise levels. Shaded areas represent the first and 3rd quartiles over experiment repetitions.

In a similar expected fashion, the less noise in the learning set, the better the performances are. As a general comment, we consider that our approaches fullfill their main task, namely learning heterogeneous preferences. Additional results and settings are shared in the online appendices[2].

6 Experiments on Real Datasets

We investigate the performances of our algorithms on the car preferences dataset[3] [21]. This dataset results from a study conducted in California where respondents were asked to choose a car among six alternatives. The cars are evaluated on eight monotonous criteria described in Table 1. The feasibility of

---
[2] Appendices available at github.com/artefactory/learning-heterogeneous-preferences.
[3] Dataset available at github.com/artefactory/choice-learn.

our approach is not guaranteed, in particular, the appropriate number of clusters needs to be questioned.

Table 1. Car preferences criteria description.

Criterion	Description	Objective
price	Price of vehicle divided by the logarithm of income	min
range	Hundreds of miles vehicle can travel between refuelings/rechargings	max
acc	Acceleration, tens of seconds required to reach 30 mph from stop	min
speed	Highest attainable speed in hundreds of mph	max
pollution	Tailpipe emissions as fraction of those for new gas vehicle	min
space	Fraction of luggage space in comparable new gas vehicle	max
cost	Cost per mile of travel (tens of cents)	min
station	Fraction of stations that can refuel/recharge vehicle	max

Data Processing. Different processing are applied to the dataset before estimating the model.

- alternatives with out-of-distribution criteria values are removed,
- preference pairs with dominance are removed from the dataset,
- criteria are normalized in $[0, 1]$ with 1 being the best value possible for each criterion.

The original dataset that includes 4654 choices, each involving 6 alternatives, leads to 18.498 pairwise comparisons.

Use Case Adaptation. The original dataset contains a set of choices made on sets of 6 alternatives. We exploit these choices as follows. The choice of x^1 in the set $\{x^1, x^2, \ldots, x^6\}$ results in the 5 following pairwise comparisons: $x^1 \succ x^2$, $x^1 \succ x^3$, $x^1 \succ x^4$, $x^1 \succ x^5$, and $x^1 \succ x^6$. Conversely, a computed model succeeds at restoring the choice of x^1 among $\{x^1, x^2, \ldots, x^6\}$ iff the 5 corresponding comparisons are simultaneously restored.

Experimentations. The dataset is randomly split in two: 10.000 randomly selected pairwise comparisons constitute a basis from which a number of learning sets of varying sizes are sampled. The remaining comparisons compose the test set. The comparisons derived from the same choice are grouped together. Both approaches are computed with $K \in \{1, 2, 3, 4, 5\}$ clusters on each of the learning sets. The experimental design is repeated 5 times with different random seeds.

6.1 Results

First, the resulting computing times lead to similar conclusions with regards to the synthetic experiments. The MILO formulation needs more than 6 h with a few hundreds of pairs while the heuristic finds satisfying results within an hour for the largest set size, 10.000 pairs.

Figure 6 showcases the heuristic performances on the test set with different learning set sizes. The performances are analyzed at the pair level, on the left, or at the choice level, on the right. First, we can observe that a single UTA model (1 cluster) cannot restore more than 64% of the preferences and 48% of the choices. The heterogeneity of the data cannot be well handled, showcasing the need for preference segmentation. A recurring problem in clustering is finding the right number of clusters representing the data. While several methods exist, such as using the silhouette score or the elbow method, it usually comes to a human-informed choice from the experimental observations. From Fig. 6, we can infer that 2 clusters are enough to clusterize the pairwise comparisons. However, at the choice level, 4 clusters seem like a better choice, with 90% of representation of the choices from the test set. The improvement brought by the fifth cluster is too marginal to be considered as the right value.

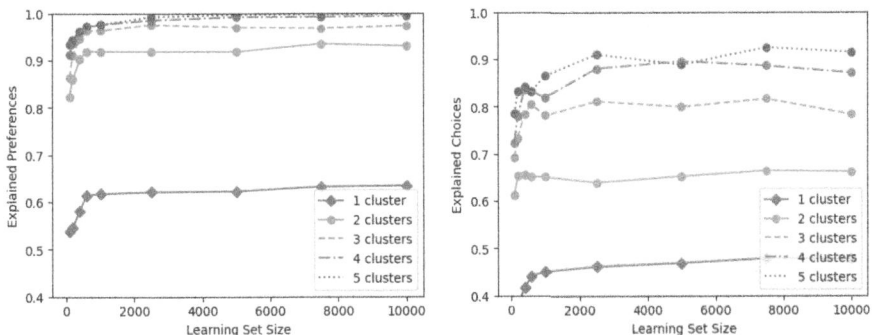

Fig. 6. Average *ExplainedPairs* and *ExplainedChoices* for the heuristic and different settings.

Figure 7 provides an example of resulting model coefficients for $K = 3$ clusters. A salient strength of our formulation is the direct interpretability of the compute models. We can observe how each criterion weighs on the decision with the coefficient values. For instance, the cluster 1 uses the price and pollution criteria to take a decision, while the cluster 2 mainly uses the speed criterion. Finally, the cluster 3, uses 4 criteria, with similar importance to represent the preferences.

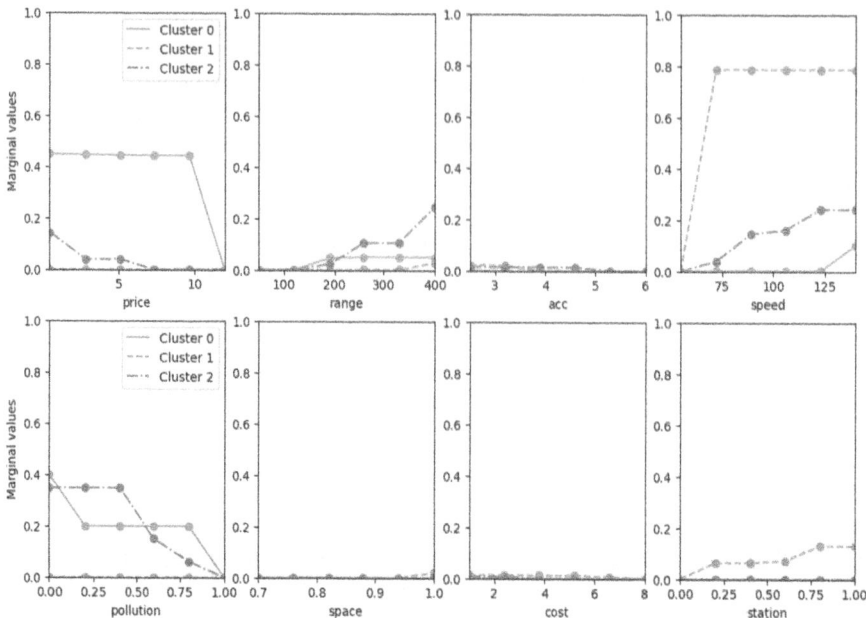

Fig. 7. Example of resulting UTA marginal values on the stated car preferences dataset.

7 Conclusion

In this paper we propose and study the *Preference Learning and Segmentation* (PLS) problem based on pairwise comparisons. In this problem, we want to clusterize the observed preferences and to compute a value function on each of the clusters. We show that the PLS problem is computationally difficult. We propose a mathematical programming as well as a heuristic resolution approach. We present experimental results on synthetic and real-world data.

These first results give rise to future research directions. First, we would like to use our approach in actual applications such as customer segmentation in supermarkets. Second, we plan to extend our approach to be able to specify PLS algorithms when replacing the additive preference model with alternative preference representations, such as pairwise comparison methods. Another interesting extension corresponds to a situation in which observed preferences that we consider as input are not comparisons but assignment examples, in which case we should move from choice problem formulation to sorting problems.

References

1. Arthur, D., Vassilvitskii, S., et al.: k-means++: the advantages of careful seeding. In: Soda, vol. 7, pp. 1027–1035 (2007)
2. Auriau, V., Aouad, A., Désir, A., Malherbe, E.: Choice-learn: large-scale choice modeling for operational contexts through the lens of machine learning. J. Open Sour. Softw. (2024)
3. Busse, L.M., Orbanz, P., Buhmann, J.M.: Cluster analysis of heterogeneous rank data. In: Proceedings of the 24th International Conference on Machine Learning, pp. 113–120 (2007)
4. Cascon, J., González-Arteaga, T., de Andres Calle, R.: A new preference classification approach: the λ-dissensus cluster algorithm. Omega **111**, 102663 (2022)
5. Díez, J., del Coz, J.J., Luaces, O., Bahamonde, A.: Clustering people according to their preference criteria. Expert Syst. Appl. **34**(2), 1274–1284 (2008)
6. Garey, M.R., Johnson, D.S.: Computers and Intractability, vol. 174. Freeman, San Francisco (1979)
7. Giannopoulos, G., Brefeld, U., Dalamagas, T., Sellis, T.: Learning to rank user intent. In: Proceedings of the 20th ACM International Conference on Information and Knowledge Management, pp. 195–200 (2011)
8. Goldberg, D., Nichols, D., Oki, B.M., Terry, D.: Using collaborative filtering to weave an information tapestry. Commun. ACM **35**(12), 61–70 (1992)
9. Gormley, I.C., Murphy, T.B.: Exploring voting blocs within the Irish electorate: a mixture modeling approach. J. Am. Stat. Assoc. **103**(483), 1014–1027 (2008)
10. Gurobi Optimization, LLC: Gurobi Optimizer Reference Manual (2023)
11. Jacquet-Lagreze, E., Siskos, J.: Assessing a set of additive utility functions for multicriteria decision-making, the UTA method. EJOR **10**(2), 151–164 (1982)
12. Karp, R.: Reducibility among combinatorial problems. Complexity Comput. Comput. 85–104 (1972)
13. Keeney, R.L., Raiffa, H.: Decisions with Multiple Objectives: Preferences and Value Trade-Offs. Cambridge University Press, Cambridge (1993)
14. Krishna, K., Murty, M.N.: Genetic k-means algorithm. IEEE Trans. Syst. Man Cybern. Part B (Cybern.) **29**(3), 433–439 (1999)
15. Lefait, G., Kechadi, T.: Customer segmentation architecture based on clustering techniques. In: 2010 Fourth International Conference on Digital Society, pp. 243–248 (2010)
16. Li, J., Wang, K., Xu, L.: Chameleon based on clustering feature tree and its application in customer segmentation. Ann. Oper. Res. **168**(1), 225–245 (2009)
17. Li, Z., Wang, W., Yang, C., Ragland, D.R.: Bicycle commuting market analysis using attitudinal market segmentation approach. Transp. Res. Part A: Policy Pract. **47**, 56–68 (2013)
18. Liu, A., Moitra, A.: Efficiently learning mixtures of Mallows models. In: 2018 IEEE 59th Annual Symposium on Foundations of Computer Science (FOCS), pp. 627–638 (2018)
19. Liu, A., Zhao, Z., Liao, C., Lu, P., Xia, L.: Learning Plackett-Luce mixtures from partial preferences. In: Proceedings of the AAAI Conference on Artificial Intelligence, vol. 33, no. 01, pp. 4328–4335 (2019)
20. Liu, J., Wang, Y., Kadziński, M., Mao, X., Rao, Y.: A multiple criteria Bayesian hierarchical model for analyzing heterogeneous consumer preferences. Omega 103113 (2024)

21. McFadden, D., Train, K.: Mixed MNL models for discrete response. J. Appl. Economet. **15**(5), 447–470 (2000)
22. Mousseau, V., Pirlot, M.: Preference elicitation and learning. EURO J. Decis. Process. **3**(1-3) (2015)
23. Nesterov, Y., Nemirovskii, A.: Interior-Point Polynomial Algorithms in Convex Programming. SIAM (1994)
24. Siskos, Y., Grigoroudis, E., Matsatsinis, N.F.: UTA methods. In: Multiple Criteria Decision Analysis: State of the Art Surveys, pp. 315–362 (2016)
25. Tynan, A.C., Drayton, J.: Market segmentation. J. Mark. Manag. **2**(3), 301–335 (1987)
26. Ungar, L.H., Foster, D.P.: Clustering methods for collaborative filtering. In: AAAI Workshop on Recommendation Systems, vol. 1, pp. 114–129. Menlo Park, CA (1998)
27. Wang, Q., Yang, X., Song, P., Sia, C.L.: Consumer segmentation analysis of multichannel and multistage consumption: a latent class MNL approach. J. Electron. Commer. Res. **15**(4), 339 (2014)
28. Zakrzewska, D., Murlewski, J.: Clustering algorithms for bank customer segmentation. In: 5th International Conference on Intelligent Systems Design and Applications, pp. 197–202 (2005)

Independent Relaxed Subproblems for Dominance Testing in CP-Nets

Liu Jiang and Thomas E. Allen[(✉)]

Centre College, Danville, KY 40422, USA
thomas.allen@centre.edu

Abstract. Conditional preference networks (CP-nets) have received significant attention for modeling preferences over combinations of features. However, dominance testing, the problem of inferring from a CP-net whether one outcome is always preferred over another, is NP-hard, requiring exponential time in practice. In this paper, we introduce the use of independent relaxed subproblems as a heuristic for faster inference from CP-nets. We show how to relax the constraints in the conditional preference tables and partition the network, yielding smaller subproblems that can be solved easily. The lengths of these solutions can then be added together to provide a heuristic for informed search algorithms such as A* and applied to dominance testing in CP-nets. We prove that the resulting additive heuristic function is admissible and consistent, guaranteeing optimality and completeness. We show from experiments on randomly generated binary and multivalued CP-nets that our method performs better than the state of the art, and can further be combined with other pruning techniques, resulting in greatly improved performance for dominance testing.

Keywords: CP-nets · dominance testing · relaxed problems · independent subproblems · heuristic search

1 Introduction

A conditional preference network (CP-net) [5] is a graphical representation used to model and reason about preferences over sets of attributes that must be chosen together. Each node in the network represents a specific attribute, and is associated with a conditional preference table that specifies the preferred value for that attribute given the values of its parent attributes.

To illustrate, consider the CP-net in Fig. 1 describing a colleague's preferences over poke bowls at a restaurant. The restaurant offers a choice of salmon, shrimp, or tofu for protein, white or brown sushi rice, and classic or spicy poke sauce. Assuming all the ingredients are available, there are a total of 12 different combinations from which to choose. The nodes correspond to the *features* or *variables* (protein, rice, and sauce). The tables of the CP-net, known as *conditional preference tables* (CPTs), specify the preference order for each feature

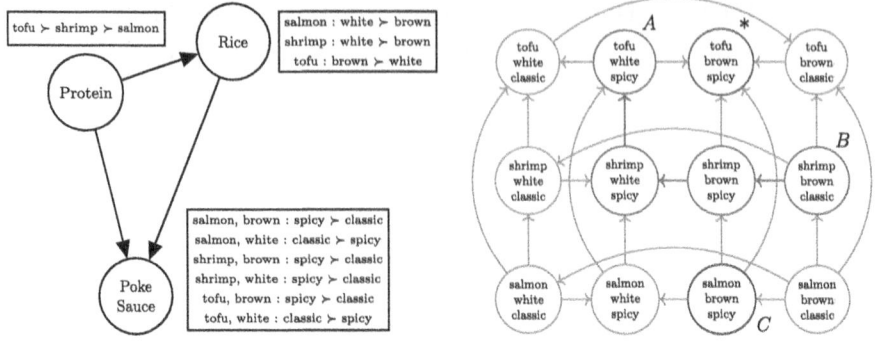

Fig. 1. CP-net and preference graph specifying poke bowl preferences

given dependencies. Looking closely, we can see that the colleague's most preferred *outcome* is a bowl with tofu, brown rice, and spicy sauce.

Suppose the colleague asked you to order a bowl for them. However, upon arriving at the restaurant, you discover that they have no tofu or spicy sauce that day. If you only knew your colleague's most preferred bowl, you would not know what to order. Thankfully, your colleague has specified their preferences as a CP-net, allowing you to make an informed decision. In this case, since tofu is unavailable, you know that your colleague prefers shrimp over salmon as a protein. Moreover, while they prefer the spicy sauce with shrimp, today they will have to settle for the classic sauce.

CP-nets have been proposed for a variety of important applications, such as product configuration [5], scheduling in grid computing environments [7], service selection in cloud computing [2], cybersecurity [4], preference aggregation and social choice [17], and dialogue systems [6]. More recently, CP-nets have been proposed as a tool to support the decisions of moral agents and to measure the distance between an individual's ethical principles and those of society [14,16].

CP-nets offer many advantages. They allow human beings to thoughtfully reason in advance about some potentially complex configuration problem, expressing their preference in the form of a set of rules. These rules, once elicited, can then be entrusted to a proxy, such as an automated system or buyer's agent, to make decisions on their behalf. CP-nets offer transparency and interpretability, making them well-suited for applications requiring explanations, such as ethical decision-making. Moreover, given a CP-net, the most preferred outcome can be decided efficiently, an advantage in real-time environments such as high-frequency securities trading or ticket selection for Taylor Swift concerts.

However, when the proxy must choose from a menu of preconfigured alternatives, the problem becomes more difficult. Consider a buyer's agent tasked with selecting an apartment in a city with limited housing, where the buyer's conditional preferences—proximity to metro, amenities, natural light exposure, noise level, etc.—are submitted in the form of a CP-net. The proxy must then infer from the CP-net which of two apartments the buyer prefers. Unfortunately, inferring from a CP-net the user's preference between two arbitrary alternatives—known as *dominance testing*—turns out to be NP-hard [5,9], requiring exponential time in practice. The difficulty arises because the CPTs only specify how to improve one feature at a time, assuming everything else remains the same (*ceteris paribus*). Comparing outcomes that differ in multiple features requires a search for a transitive sequence of feature changes known as a *flipping sequence*, where the validity of each flip is determined by a rule of the CP-net.

Flipping sequences can be visualized with the help of a *preference graph* like the one to the right of the CP-net in Fig. 1. The nodes of the preference graph have a different meaning from those of the CP-net. In the preference graph, nodes represent all possible outcomes, and arcs represent improving flips. Flipping sequences correspond to paths in this graph. Suppose you wanted to know whether your colleague prefers bowl A with tofu, white rice, and spicy sauce, or B with shrimp, brown rice, and classic sauce. The flipping sequence

(shrimp, brown, classic) $\rightarrow$ (shrimp, brown, **spicy**)

$\rightarrow$ (shrimp, **white**, spicy) $\rightarrow$ (**tofu**, white, spicy)

shows that your colleague prefers A to B. What if you were then asked to compare B with C (salmon, brown, spicy)? No flipping sequence exists from B to C or vice versa. Such outcomes are said to be *incomparable* under the network. In practice, we may think of such pairs as ties since they could occur in either relative order in a topological ordering of the outcomes.

Since the number of outcomes is exponential in the number of features, the preference graph is exponentially larger than the CP-net that induces it. Our toy example consists of only 12 outcomes. The real life restaurant on which the example is based offers a combination of 40 features, with 2^{40} outcomes. Understandably, threading a path through a trillion nodes can be difficult, requiring specialized techniques like heuristics to guide the search and enable more efficient exploration.

We propose a new heuristic to accelerate dominance testing in CP-nets. Inspired by Korf and Felner's approach to solving the sliding block 24-puzzle [11], we partition CP-nets into smaller networks, decomposing the problem into independent relaxed subproblems. These subproblems can be solved either on the fly or in advance, providing insight into potential intermediary outcomes encountered during the search. We prove that our heuristic function is both admissible and consistent, guaranteeing optimal solutions. Additionally, we introduce a powerful pruning method to reduce the search space. By combining our approach with existing pruning methods, we achieve significant speedups over the state of the art.

In the following sections, we formalize key notions (Sect. 2), review related work (Sect. 3), and introduce our method along with theoretical results (Sect. 4). We then present experimental evaluations (Sect. 5) and conclude with suggestions for future research (Sect. 6).

2 Preliminaries

Let $\mathbf{X} = \{X_1, \ldots, X_n\}$ be a set of variables, each with a finite domain $\mathrm{Dom}(X_i)$. An **outcome** o is a complete assignment to all variables in $\mathbf{X}$ from their domains, i.e., a specific configuration of these variables. We denote the set of all possible outcomes as $\mathcal{O}$, the Cartesian product of the domains of all variables. $\mathrm{Asst}(\mathbf{U})$ denotes the set of all (partial) assignments to variables $\mathbf{U} \subseteq \mathbf{X}$. When we wish to refer to the values of particular variables $\mathbf{U} \subset \mathbf{X}$ in an outcome o, we use $o[\mathbf{U}]$. When referencing specific assignments, we use string notation so that each element in the string corresponds to the value of a variable. For example, if $o = a_1 b_1 c_3 d_1$ then $o[A,C] = a_1 c_3$ denotes the partial assignment to A and C.

The expression $o \succ o'$ indicates that o is strictly preferred to o', and $o \succeq o'$ indicates that o is at least as good as o'. The use of $\prec$ and $\preceq$ is analogous. We respectively denote indifference and incomparability by $o \sim o'$ and $o \parallel o'$. We assume familiarity with basic concepts of order theory. A preorder $\succeq$ is a reflexive, transitive relation. A preorder is said to be *strict* if $o \succeq o'$ implies $o' \not\succeq o$, *total* if $o \succeq o'$ or $o' \succeq o$ for all o, o', and *partial* if incomparability is allowed. We assume a total preorder unless otherwise noted.

A **conditional preference network** (CP-net) N is a directed acyclic graph in which each node represents a variable $X_i \in \mathbf{X}$, and arcs (directed edges) encode dependencies: an arc from X_h to X_i means that the preference over $\mathrm{Dom}(X_i)$ depends on the value of X_h. We denote the parents and children of a node X_i respectively by $\mathrm{Pa}(X_i)$ and $\mathrm{Ch}(X_i)$. To avoid confusion with other graphs, such as the induced preference graph, the graph structure of a CP-net is referred to as the **dependency graph**.

Each node $X_i \in \mathbf{X}$ is associated with a **conditional preference table** $\mathrm{CPT}(X_i)$ that specifies a total preorder over $\mathrm{Dom}(X_i)$ for each assignment to its parents in the dependency graph. Formally, $\mathrm{CPT}(X_i)$ maps each assignment $\mathbf{z} \in \mathrm{Asst}(\mathbf{Z})$, $\mathbf{Z} = \mathrm{Pa}(X_i)$, to a total preorder $\succeq_i^{\mathbf{z}}$ over $\mathrm{Dom}(X_i)$. A conditional preference **rule** is a pair $(\mathbf{z}, \succeq_i^{\mathbf{z}})$ in $\mathrm{CPT}(X_i)$, which we write as $\mathrm{CPT}(X_i \mid \mathbf{z})$.

A CP-net N models the user's underlying **preference ranking**, assumed to be a total preorder $\succeq$ over $\mathcal{O}$. As with most useful models, CP-nets come with certain tradeoffs. An important limitation is that a CP-net induces only a partial order $\succeq_N$ over $\mathcal{O}$ that can be extended with many total orders. A preference ranking $\succeq$ is said to satisfy N if it is consistent with the induced partial order $\succeq_N$. If an outcome o is ranked before o' in *every* total order that extends $\succeq_N$, we say o **dominates** o', denoted $N \models o \succ o'$. However, if o and o' have conflicting rankings in different total orders that can extend $\succeq_N$, they are said to be incomparable.

Consider that a conditional preference rule only specifies how to compare outcomes that differ in just one variable, i.e., those with Hamming distance 1. This limitation is inherent in the *ceteris paribus* semantics of the CP-net as a compact model of the preference ranking. The order over $\mathrm{Dom}(X_i)$ specified by the CPT is guaranteed to hold as long as no other variable changes value. Thus, if two outcomes o_{t-1} and o_t differ in just one variable X_i, we can consult $\mathrm{CPT}(X_i \mid \mathbf{z})$ directly to look up the order $\succeq_i^{\mathbf{z}}$, $\mathbf{z} = o_i[\mathrm{Pa}(X_i)]$. If $o_{t-1} \prec o_t$, we can *flip* X_i to its more preferred value, improving the state of things. We assume CPTs are fully specified, with a rule for every assignment to parents.

To compare two outcomes that differ in more than one variable, we iteratively apply the rules of the CP-net, constructing transitive sequences of outcomes that differ by one variable at a time. Formally, a monotone improving **flipping sequence** from o' to o is a sequence of outcomes, $\pi = \langle o_0, o_1, \ldots, o_L \rangle$, $o' = o_0$, $o = o_L$, in which each pair o_{t-1}, o_t differs in the value of just one variable, and $o_{t-1} \preceq o_t$, $0 < t \leq L$. If the CP-net specifies strict orders, then the flipping sequence is strictly improving, meaning that $o_{t-1} \prec o_t$ for all t. If such a sequence can be shown to exist, consistent with the rules of the network, we can conclude that $N \models o \succeq o'$ (monotone improving) or $N \models o \succ o'$ (strictly improving). The existence of a flipping sequence establishes that o is at least as good as (or better than) o' in every preference ranking that satisfies N.

Flipping sequences correspond to paths in a graph over outcomes known as the **preference graph**. An induced preference graph $G = (\mathcal{O}, E)$ over outcomes is a directed graph in which an arc from $\mathbf{u}$ to $\mathbf{v}$ corresponds to a possible improving flip. That is, an arc from $\mathbf{u}$ to $\mathbf{v}$ indicates that the successor differs only in the value of X_i and that $\text{CPT}(X_i)$ specifies an improvement from $\mathbf{u}$ to $\mathbf{v}$. Formally, $(\mathbf{u}, \mathbf{v}) \in E$ if and only if x is ordered before x' in $\text{CPT}(X_i \mid \mathbf{z})$ where $x, x' \in \text{Dom}(X_i)$, $\mathbf{z} = \mathbf{u}[\text{Pa}(\mathbf{u})]$, $\mathbf{u} = \mathbf{w}x\mathbf{y}$, $\mathbf{v} = \mathbf{w}x'\mathbf{y}$, and $\mathbf{w}$ and $\mathbf{y}$ respectively denote the unchanged values of $X_1, \ldots, X_{i-1}$ and $X_{i+1}, \ldots, X_n$.

Dominance testing in the context of the induced preference graph can thus be understood as a path-finding problem: $N \models o \succ o'$ if and only if there exists a directed path from o' to o in G, i.e., $o' \rightsquigarrow o$. We denote by $c(o', o)$ the distance (or *cost*) of a shortest path between two outcomes, where distance is the number of arcs in the path, or equivalently, the number of flips. If no path exists from o' to o, then by definition $c(o', o) = \infty$.

Consider that if a CP-net has n nodes representing variables with domain size d, the induced preference graph will have d^n nodes. Since the preference graph grows exponentially with the size of the dependency graph that induces it, it is only feasible to construct G when n and d are small. Nevertheless, it serves as a useful conceptual tool for reasoning about heuristic search strategies, even when it is too large to construct in practice. Thus, although it is often described as a decision problem, here we frame dominance testing (DT) as a search problem.

Definition 1. *Let N be a CP-net over V and o and o' be two outcomes in $\mathcal{O}$. A **dominance testing problem** $\text{DT}(N, o', o)$ specifies a search for a flipping sequence $\pi = \langle \pi_0, \pi_1, \ldots, \pi_L \rangle$ in the preference graph G induced by N, from the start state o' to the goal state o, $o' = \pi_0$, $o = \pi_L$, proving that $N \models o \succeq o'$.*

*A path from o' to o is generated by **actions** $\alpha = (\alpha_1, \ldots, \alpha_L)$ in which each $\alpha_t = (X_j, x)$ specifies that variable X_j flips from its value x in π_{t-1} to $x' \in \text{Dom}(X_j)$, $x \neq x'$, in step t, generating the transition from π_{t-1} to π_t.*

*A **solution** to a DT problem is the action sequence α that generates a flipping sequence π from o' to o.*

If α is a shortest path from o' to o, i.e., $|\alpha| = c(o', o)$, we say it is *optimal*, and denote the flipping and action sequences by π^* and α^*, respectively.

3 Related Work

CP-nets were introduced by Boutilier et al. [5], who studied their properties and showed that dominance testing is NP-hard. They identified special cases in which DT problems can be solved in polynomial time, such as tree- and polytree-shaped CP-nets, and proposed methods for reducing the search space, including suffix fixing and forward pruning. They also introduced two incomplete DT methods: consistent orderability and a least variable pruning heuristic.

Effective search methods have been a fundamental problem in artificial intelligence for decades. Various approaches have been proposed, with A* and IDA* a staple of undergraduate AI classes. The literature is extensive, but we highlight the work of Korf and Felner [11], who studied independent subproblems as heuristics and introduced the concept of a pattern database, where relaxed subproblems are pregenerated, solved in advance, and stored for use as a heuristic during search. We applied many of their ideas to CP-nets in this paper.

Beyond Boutilier's work, various heuristics and pruning methods have been proposed for DT. Li et al. [13] introduced DT*, which involves using a penalty function that considers the number of variables out of position with respect to the goal state and their topological order within the dependency graph. The resulting heuristic function, although highly effective, is not admissible since it can lead to suboptimal solutions.

Laing et al. [12] took a similar approach, introducing a ranking function that considers the position of variable in the dependency graph. Their method requires arbitrary precision rational numbers and up-front computational work, resulting in a powerful pruning method that can eliminate many unpromising outcomes. They proposed using their rank function to guide best-first search that prioritizes nodes of higher rank. However, their function does not estimate the number of flips to the goal, making it less suitable for heuristic search, and their experiments only consider networks with up to 10 nodes.

Like our method, Ahmed and Mouhoub [1] also partition the dependency graph into subgraphs. However, whereas our approach relies on heuristic search, they introduce a recursive divide-and-conquer algorithm that reaches special base cases that can be solved efficiently, such as tree-shaped CP-nets. While their approach is theoretically interesting, it lacks experimental validation, limiting direct comparisons with our method.

4 Method

Consider the dependency graph in Fig. 2. If each variable has domain size 4, dominance testing (DT) requires searching a space of up to $4^8 = 65536$ outcomes. But what if the arc from C to A could somehow be removed? Then we would have two small problems, each with only $4^4 = 256$ outcomes. This motivates an idea: partition the dependency graph to create subproblems, using the lengths of the solutions to these as a search heuristic. But what do we do with the CPT when a child is separated from its parents? The problem with removing parents is that this leads to conflicting rankings for assignments to the remaining parents.

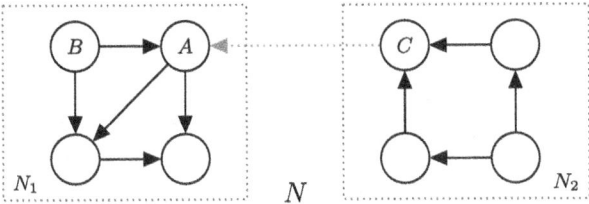

Fig. 2. If there were no arc from C to A, DT problems for this CP-net would decompose into two independent subproblems.

4.1 Creating Relaxed Problems with DT-RELAX

To address this, we propose DT-RELAX. As shown in Fig. 3, we construct a new CPT for X_i when parents $\mathbf{Y} \subseteq \text{Pa}(X_i)$ are removed, resolving conflicts with the help of a domain transition graph (DTG) [5] and Tarjan's algorithm for finding strongly connected components (SCCs) [19]. A **domain transition graph** $\text{DTG}(X_i)$ is a directed graph with nodes for each value $x \in \text{Dom}(X_i)$ and arcs indicating possible improvements.

function DT-RELAX(X_i, $\mathbf{Y}$):
   $C = $ **new** CPT()
   for all $\mathbf{u} \in \text{Asst}(\text{Pa}(X_i) \setminus \mathbf{Y})$ **do:**   ▷ assignments $\mathbf{u}$ to remaining parents
      $D = $ **new** DTG()
      for all $\mathbf{y} \in \text{Asst}(\mathbf{Y})$ **do:**   ▷ assignments $\mathbf{y}$ to removed parents
         $D = D \cup \text{DTG}(\text{CPT}(X_i \mid \mathbf{uy}))$
      $C(X_i \mid \mathbf{u}) = \text{Preorder}(\text{Tarjan}(D))$
   return C

Fig. 3. Pseudocode for DT-Relax

In the inner loop, we construct a DTG for each assignment $\mathbf{y}$ to the removed parents, while holding the assignment $\mathbf{u}$ to the remaining parents constant, adding an arc from x' to x for each consecutive pair $x \succ x'$ in the linear order $\succ_i^{\mathbf{uy}}$. The union operation ensures that no improving flip defined by the original CPT is lost in the relaxation. We then apply Tarjan's algorithm to the aggregated DTG to discover SCCs in a depth-first traversal, from best to worst, with each SCC representing a set of equivalent values. The resulting preorder is then added to the CPT. We repeat this for each assignment to the remaining parents $\mathbf{u}$. When DT-RELAX has been applied to a CP-net N, we call the result $R(N)$ a **relaxation** of N.

Example 1. Figure 4 illustrates the process when parent C is detached. The domain of A has 4 values, and B and C are binary. Thus, CPT(A) originally has 4 rows, one for each combination of parent values, and only 2 after C is removed.

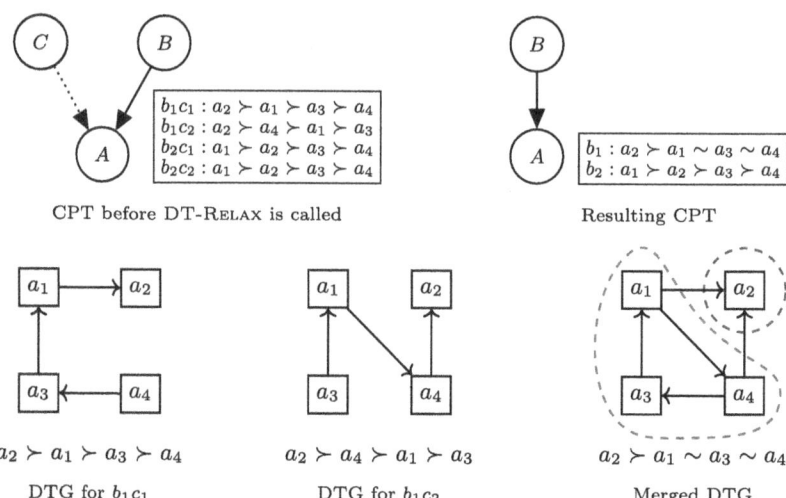

Fig. 4. The effect of DT-RELAX on CPT(A) when parent C is removed

Notice that for $B = b_1$, there are conflicting orders for A. The illustration shows how we resolve these. DTGs are constructed for b_1c_1 and b_1c_2 and then merged. The dotted ellipsoids show the SCCs that are discovered when Tarjan is called: $(\{a_2\}, \{a_1, a_3, a_4\})$. The arcs between components indicate the direction of improvement, resulting in the preorder $a_2 \succ a_1 \sim a_3 \sim a_4$.

When we come to $B = b_2$ next in the outer loop, the situation is less interesting, since the order for b_2c_1 and b_2c_2 is the same: $a_1 \succ a_2 \succ a_3 \succ a_4$.

Example 2. We now consider the effect of DT-RELAX on flipping sequences and equivalent paths in the induced preference graph. Figure 5 shows a CP-net N, its relaxation $R(N)$, and their respective preference graphs G and H. We make the following observations:

First, G is a *subgraph* of H. DT-RELAX adds arcs to accommodate influences of detached parents. As such, every arc (path) in G is an arc (path) in H.

Second, the new arcs create shortcuts. For example, G has a path of length 4 from u to v. The same path exists in H, but now there are two shorter paths due to the relaxation. Moreover, although no path connects y and z in G, in H there is now a path from y to z. Intuitively, we see that the distance between any two vertices in H must be a lower bound on their distance in G.

Third, H has cycles. These have the same interpretation as for DTGs: the SCCs create equivalence classes, and arcs that directly link them denote strict improvement. Thus, $a_1b_1c_1 \sim a_1b_1c_2$ in H, whereas $a_1b_1c_1 \prec a_1b_1c_2$ in G.

Remark 1. In the absence of information gained through search, we should be careful to interpret an arc as $\preceq$ rather than $\prec$ when the graph admits cycles.

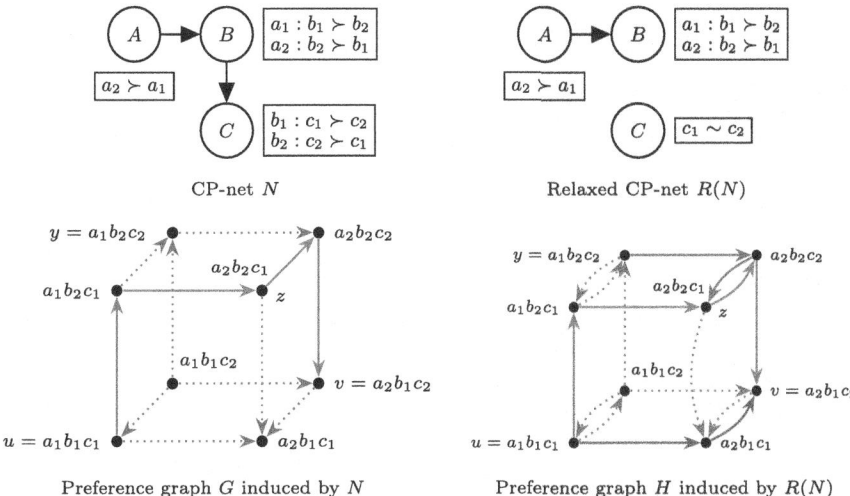

Fig. 5. A different example showing the effect of DT-RELAX on flipping sequences

4.2 Heuristic Search Guarantees

DT-RELAX is supported by properties that enable its use as the basis of an admissible and consistent DT heuristic. We formalize these in the following theorems, guaranteeing that our method will find an optimal solution if one exists.

Theorem 1. *If $G = (\mathcal{O}, E)$ and $H = (\mathcal{O}, F)$ are the preference graphs induced from N and $R(N)$, respectively, then G is a subgraph of H, with $E \subseteq F$, such that every flipping sequence under N is preserved under $R(N)$.*

Proof. Consider an arbitrary arc $(o, o') \in E$, corresponding to a flip of some variable X_i from x to x'. In the CP-net N, this flip is governed by $\text{CPT}(X_i \mid \mathbf{uy})$, where $\mathbf{U}$ denotes the parents of X_i in both N and $R(N)$, $\mathbf{Y}$ denotes the parents of X_i only in N, and $\mathbf{u}$ and $\mathbf{y}$ are assignments to $\mathbf{U}$ and $\mathbf{Y}$, respectively.

Let $E_{\mathbf{uy}}$ denote the arcs in the DTG of $\text{CPT}(X_i \mid \mathbf{uy})$ in N. Then, by construction of DT-RELAX, the arcs in the DTG of $\text{CPT}(X_i \mid \mathbf{u})$ in $R(N)$ are given by $E_{\mathbf{u}} = \bigcup_{\mathbf{y} \in \text{Asst}(\mathbf{Y})} E_{\mathbf{uy}}$. That is, each set of arcs $E_{\mathbf{uy}}$, corresponding to each specific assignment $\mathbf{y}$, is included in the aggregated set $E_{\mathbf{u}}$. Since $E_{\mathbf{uy}} \subseteq E_{\mathbf{u}}$, every flip $x \to x'$ under N is preserved under $R(N)$.

Since this holds for every arc $(o, o') \in E$, we conclude that G is a subgraph of H, and the theorem follows. □

Theorem 2. *Let $R(N)$ be the relaxation of a CP-net N. Then, $d(u, v) \leq c(u, v)$, where c and d denote the lengths of flipping sequences from u to v under N and $R(N)$, respectively.*

Proof. By Theorem 1, every flipping sequence under N is also valid under $R(N)$. Thus, if $c(u, v)$ is the length of a shortest path π^* from u to v under N, π^* is also valid under $R(N)$, ensuring that $d(u, v) \not> c(u, v)$.

Figure 5 provides an instance where the path from u to v under $R(N)$ is strictly less than $c(u,v)$ due to additional arcs that provide a shorter path. Therefore, $d(u,v) \leq c(u,v)$. □

The fact that $d(u,v)$ is a lower bound on $c(u,v)$ suggests its use as a heuristic for dominance testing. As noted in Sect. 2, we treat DT as a search for a path from o' to o, proving that $N \models o \succeq o'$. We formalize this with a definition:

Definition 2 (Relaxed problem heuristic). *Let $h(u) = d(u,o)$ be a heuristic function that estimates the true shortest path distance $h^*(u) = c(u,o)$ from any outcome $u \in \mathcal{O}$ to the goal o under CP-net N, where $d(u,o)$ is the distance of a shortest path from u to o under relaxation $R(N)$.*

Theorem 3. *Relaxed problem heuristic h is admissible.*

Proof. A heuristic function h is said to be *admissible* if $h(u) \leq h^*(u)$ for all u, where $h^*(u) = c(u,o)$ is the true shortest path length to the goal o from u. Proof of this follows directly from Definition 2 and the bound in Theorem 2. □

We now show that h satisfies an even stronger condition [15], monotonicity, which implies consistency.

Theorem 4. *Heuristic function h is both monotone and consistent.*

Proof. A heuristic function h is said to be *monotone* if $h(u) \leq d(u,v) + h(v)$ for all u,v where v is a *successor* of u. Since $d(u,v) = 1$, we have $d(u,o) \leq 1 + d(v,o)$.

Suppose v lies on a shortest path π^* from u to o. Then $d(u,o) = 1 + d(v,o)$, and equality holds. Conversely, if v does not lie on π^*, then any path from u to o through v must be at least as long, since π^* is a shortest path. Thus, $d(u,o) \leq 1 + d(v,o)$, for every successor v of u. Therefore, h is monotone.

Since h is monotone, it follows from Pearl's result [15] that h is also *consistent*, which implies that the inequality holds for every descendant v of u, and trivially for non-descendants since then by definition $d(u,v) = \infty$. □

4.3 Decomposition Into Independent Subproblems

We now show how to decompose DT problems into independent subproblems by partitioning the CP-net, and compute the relaxed problem heuristic as the sum of the solution lengths of its constituent CP-nets.

Definition 3. *Let N be a CP-net over V. A **decomposition** of N is a relaxed partition $R(N) = (N_1, \ldots, N_K)$ of N, where each component[1] N_i is a CP-net induced by a subset $V_i \subseteq V$. The partition satisfies:*

1. *$\bigcup_{i=1}^{K} V_i = V$*
2. *no node in N_i is the parent of a node in N_j for $i \neq j$*
3. *DT-RELAX is applied to each node that is separated from a parent.*

---

[1] Since the dependency graph is directed, each N_i is a weak component of N [10].

Definition 4. *Given a decomposed CP-net $R(N)$, an **independent relaxed subproblem** $DT(N_i, o_i, o'_i)$ is a dominance testing problem involving a search for a flipping sequence π_i from o'_i to o_i under N_i, $o_i = o[V_i]$, $o'_i = o'[V_i]$.*

Consider that when $R(N)$ is decomposed into its components, each component N_i can be treated as an independent CP-net to find a shortest action sequence α_i for subproblem i. The overall shortest path from o' to o under the relaxed network $R(N)$ is constructed by concatenating these individual sequences. This decomposition lets us express the overall minimum path length $d(u, v)$ as the sum of the minimum path lengths from each constituent CP-net.

Theorem 5. *Let $R(N)$ be a decomposition of a CP-net N. Then,*

$$d(u,v) = \sum_{i=1}^{K} d_i(u_i, v_i),$$

where $d(u, v)$ is the length of a shortest path from u to v under $R(N)$ and each $d_i(u_i, v_i)$ is the length of a shortest path from $u_i = u[V_i]$ to $v_i = v[V_i]$ under N_i.

Proof. Let $u = u_1 \cdots u_k$ and $v = v_1 \cdots v_k$. Let each α_i^*, $1 \leq i \leq K$, be an optimal solution to the subproblem $DT(N_i, u_i, v_i)$, such that $|\alpha_i^*| = d_i(u_i, v_i)$.

Since each flip (X_i, x) in α_i^* depends only on variables in its respective component N_i, concatenating the solutions to the subproblems α_i^* for all i yields a valid solution α^* to the problem $DT(R(N), u, v)$. The sequence of actions α^* generates a flipping sequence from $u = u_1 \cdots u_k$ to $v = v_1 \cdots v_k$ consisting of $|\alpha^*|$ flips, where $|\alpha^*| = \sum_{i=1}^{K} |\alpha_i^*| = \sum_{i=1}^{K} d_i(u_i, v_i)$.

To see that α^* is optimal, assume for contradiction that there exists a shorter solution β. Construct subsequences β_i, each consisting of all flips (X_j, x) in β where $X_j \in V_i$, maintaining their order from β. But if $|\beta| < |\alpha^*|$, then at least one β_i must be shorter than α_i^*, contradicting our assumption that α_i^* is optimal.

Therefore, $|\alpha^*| = d(u, v) = \sum_{i=1}^{K} d_i(u_i, v_i)$, proving the theorem. □

The proof establishes that the path length in a decomposed CP-net, a relaxation of N, is the sum of the costs in its subnets. This decomposition allows us to express the heuristic function $h(u)$, the length of the solution to the relaxed problem $DT(R(N), u, o)$, as the sum of the solution lengths of the subproblems.

Definition 5 (Additive subproblem heuristic).

$$h(u) = \sum_{i=1}^{k} h_i(u_i) = \sum_{i=1}^{k} d_i(u_i, o_i) \qquad (1)$$

Here, $h(u)$ represents the solution to the relaxed problem $DT(R(N), u, o)$, and each $h_i(u)$ is the solution to the subproblem $DT(N_i, u_i, o_i)$, where $d_i(u_i, o_i)$ is the true shortest path length from u_i to o_i under N_i. Note that since the sum of solution lengths to the subproblems is an *exact calculation* of $h(u)$, the theoretical guarantees discussed in Sect. 4.2 hold here as well.

Example 3. Consider that the Relaxed CP-net $R(N)$ we saw previously in Fig. 5 is a decomposed CP-net with two components, {A,B} and {C}. Designate these as N_1 and N_2, respectively. Suppose $v = a_2 b_1 c_2$ is the goal state and we want to compute $h(u)$ for node $u = a_1 b_1 c_1$ with Eq. (1). For N_1, $u_1 = u[A,B] = a_1 b_1$ and $v_1 = v_1[A,B] = a_2 b_1$. Since $a_2 \succ a_1$, we can reach v_1 from u_1 in one flip: $a_1 b_1 \preceq a_2 b_1$. Under N_2 we can reach $v_2 = c_2$ from $u_2 = c_1$ in one flip, $c_1 \preceq c_2$, since $c_1 \sim c_2$. Thus, $h(u) = h_1(a_2 b_1) + h_2(c_2) = 1 + 1 = 2$. Additionally, $h(y) = 2 + 0 = 2$, $h(z) = 1 + 1 = 2$, and $h(v)$, as one would hope, is 0.

The relaxed subproblems also give us a way to reduce the search space by pruning paths that cannot lead to a solution. We formalize this with a theorem:

Theorem 6. *The original problem* $\mathrm{DT}(N, u, o)$ *has no solution if any of its independent relaxed subproblems* $\mathrm{DT}(N_i, u_i, o_i)$ *has no solution.*

Proof. Suppose, for contradiction, that $\mathrm{DT}(N, u, o)$ has a solution. Then, there exists a flipping sequence π from u to o under N, which remains valid under $R(N)$ by Theorem 1. From the corresponding action sequence α, extract a subsequence α_i consisting of flips (X_j, x) where $X_j \in V_i$, preserving their original order. Since each flip in α_i depends solely on variables within its component N_i, this subsequence generates a valid flipping sequence from u_i to o_i, contradicting our assumption that the relaxed subproblem has no solution. □

Finally, we note that the method can be extended to accommodate multiple overlapping partitions. By defining P such partitions and taking the maximum of their respective sums of additive heuristics, we ensure the admissibility of the overall heuristic [11,15] $h(u)$: $h(u) = \max_p \sum_{i=1}^{K} h_i^p(u)$ where $h_i^p(u)$ represents the heuristic for part i of partition p.

5 Experiments

To evaluate our method, we performed dominance testing on randomly generated CP-nets of varying configurations.[2] We generated 100 CP-nets over n binary nodes for each n from 12 to 24, and 100 CP-nets over multivalued domains ($d = 4$) for each n from 6 to 12, with bounds on indegree of 4 and 2 for the binary and multivalued problems, respectively. For each (n, d) we generated 100 pairs of outcomes uniformly at random, using a random seed for consistent assignment of problems for all CP-nets. Note that the search spaces are comparable for the two series, with $|\mathcal{O}| = 2^{24} = 4^{12}$ for the largest binary and multivalued problems.

We applied the relaxation method described in Sect. 4 to each CP-net in the test set, decomposing it into its relaxed components. For each component, we explicitly constructed the induced preference graph, using the Floyd-Warshall all-pairs shortest path (APSP) algorithm [8] to precompute optimal solution costs to all possible subproblems. Once these were cached, we could evaluate

---

[2] We used the random generation method described by Allen et al. [3].

the heuristic function in Def. 5 with table lookups and addition, amortizing the front-loaded computational costs across each set of DT problems.[3]

We evaluated three configurations of DT-Relax: (1) a single random partition with K = 3 components, (2) multiple overlapping partitions that scale with the complexity of the problem (we arbitrarily chose $P = \lfloor dn/2 \rfloor$ as the number of random partitions with $K = 3$ for each), and (3) the same, but further incorporating rank pruning [12]. As a baseline comparison, we solved the same problems with uniform cost search (UCS), which expands a node of lowest depth in the search tree without use of a heuristic. We also compared DT-Relax to the DT* method of Li et al. [13], which prioritizes nodes based on a penalty function and Hamming distance from the goal state, and with the rank pruning method of Laing et al. [12], which prioritizes nodes of highest rank. For all methods we employed suffix fixing as described by Boutilier et al. [5].

n	UCS	Penalty	Rank	DT-Relax (single)	DT-Relax (multi)	DT-Relax (multi+RP)
CP-nets over n binary variables ($d = 2$)						
12	706	53	54	80	21	11
14	2,778	187	192	228	79	35
16	11,173	770	766	600	420	127
18	42,256	3,141	3,143	2,354	446	236
20	174,882	11,581	11,923	11,692	5,574	1,491
22	682,857	47,732	48,804	52,349	26,703	6,683
24	2,680,587	195,667	202,128	282,741	99,032	30,034
CP-nets over n multivalued variables ($d = 4$)						
6	975	43	45	60	12	7
7	3,929	166	170	255	36	16
8	15,552	681	694	938	117	42
9	62,189	2,687	2,731	3,665	396	128
10	247,151	11,447	11,625	16,033	1,530	502
11	998,336	42,053	42,741	56,642	881	515
12	3,993,249	175,896	177,746	214,953	4,678	2,379

Fig. 6. Nodes expanded for CP-nets with varying configurations. Each entry is an average over 10,000 trials, rounded to the nearest whole number.

We implemented DT-Relax and all comparison methods in C++ using the g++ compiler and Boost libraries. We conducted our experiments on a Mac Studio (2022) with an Apple M1 Max processor and 64 GB of RAM. Each program was run as a single-threaded process, but we executed multiple programs simultaneously on separate cores using GNU Parallel [18]. To conserve memory, we employed a compact bitmap representation for storing outcomes. Additionally, we precompiled the rules of the CPT into an efficient hash representation,

---
[3] Korf and Felner describe these cached solutions as a *disjoint pattern database* [11].

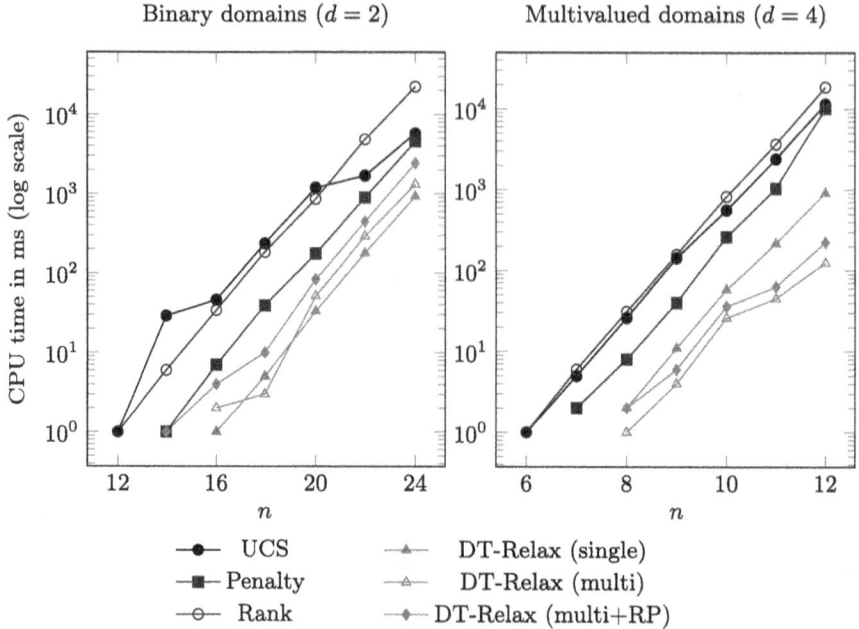

Fig. 7. CPU Time for DT problems over binary and multivalued CP-nets.

pregenerating all possible flips. A consistent random seed was used across all experiments to ensure replicability. We verified the correctness of each method by testing their answers to all problems during postprocessing.[4]

Figure 6 presents the average number of nodes expanded for DT problems of varying complexity. The data show that when an appropriate number of partitions are employed, DT-Relax significantly reduces the size of the search space, outperforming the current comparison methods. The results were particularly impressive for multivalued problems.

This improvement involves a trade-off between time saved in the search and time spent computing the heuristic function. Thus, while the number of nodes expanded is often used as a proxy for computation time, we also need to take CPU time into account.

Figure 7 plots CPU time (milliseconds) against problem size on a log scale. As expected for an NP-hard problem, all methods exhibit exponential growth. However, the DT-Relax configurations significantly outperform the others. For instance, DT problems with 12 multivalued variables that took over 10 s to solve with DT* can now be solved in just 125 ms with DT-Relax.

---

[4] Our code and data are available at https://github.com/thomas-allen-ai/dtrelax.

6 Conclusion

Perhaps the chief advantage of the relaxed subproblems approach is its flexibility. Although we solved all subproblems in advance for our experiments, the relaxation and additive subproblem heuristic that we have proposed do not depend on this. We could have instead solved subproblems on the fly using Dijkstra's algorithm, or omitted the construction of the preference graph altogether and recursively called our own solver, or another. The number of partitions and partition size can also be adjusted, and, as demonstrated, our method can be combined with other pruning methods. As an admissible heuristic, it provides a flexible, powerful approach to dominance testing in CP-nets.

Future work will involve exploring this range of capabilities further, including incorporating iterative deepening methods such as IDA* to mitigate memory issues that arise as problem size grows, and optimizing the choice of partitions to improve performance.

Acknowledgments. The authors wish to thank Michael Ramage, who contributed to the codebase, and David Toth for assisting with computational resources.

Disclosure of Interests. The authors have no competing interests to declare that are relevant to the content of this article.

References

1. Ahmed, S., Mouhoub, M.: A divide and conquer algorithm for dominance testing in acyclic CP-nets. In: Proceedings of the 31st IEEE ICTAI, pp. 392–399 (2019)
2. Alashaikh, A., Alanazi, E.: Conditional preference networks for cloud service selection and ranking with many irrelevant attributes. IEEE Access **9**, 131214–131222 (2021)
3. Allen, T.E., Goldsmith, J., Justice, H.E., Mattei, N., Raines, K.: Uniform random generation and dominance testing for CP-nets. J. Artif. Intell. Res. **59**, 771–813 (2017)
4. Bistarelli, S., Fioravanti, F., Peretti, P.: Using CP-nets as a guide for countermeasure selection. In: Proceedings of the 2007 ACM SAC, pp. 300–304. ACM (2007)
5. Boutilier, C., Brafman, R., Domshlak, C., Hoos, H., Poole, D.: CP-nets: a tool for representing and reasoning with conditional ceteris paribus preference statements. J. Artif. Intell. Res. **21**, 135–191 (2004)
6. Cadilhac, A., Asher, N., Benamara, F., Lascarides, A.: Grounding strategic conversation: Using negotiation dialogues to predict trades in a win-lose game. In: Proceedings of the EMNLP 2013, pp. 357–368 (2013)
7. Cafaro, M., Mirto, M., Aloisio, G.: Preference-based matchmaking of grid resources with CP-nets. J. Grid Comput. **11**(2), 211–237 (2013)
8. Floyd, R.: Algorithm 97: shortest path. Commun. ACM 345 (1962)
9. Goldsmith, J., Lang, J., Truszczyński, M., Wilson, N.: The computational complexity of dominance and consistency in CP-nets. J. Artif. Intell. Res. **33**(1), 403–432 (2008)

10. Knuth, D.E.: The art of computer programming, volume 4, pre-fascicle 12A: components and traversal (2022)
11. Korf, R.E., Felner, A.: Disjoint pattern database heuristics. Artif. Intell. **134**(1), 9–22 (2002)
12. Laing, K., Thwaites, P.A., Gosling, J.P.: Rank pruning for dominance queries in CP-nets. J. Artif. Intell. Res. **64**, 55–107 (2019)
13. Li, M., Vo, Q.B., Kowalczyk, R., et al.: Efficient heuristic approach to dominance testing in CP-nets. In: AAMAS, pp. 353–360 (2011)
14. Loreggia, A., Mattei, N., Rossi, F., Venable, K.B.: Value alignment via tractable preference distance, chap. 18, pp. 249–261. Chapman and Hall/CRC (2018)
15. Pearl, J.: Heuristics: Intelligent Search Strategies for Computer Problem Solving. Addison-Wesley Longman Publishing Co., Inc. (1984)
16. Rossi, F., Mattei, N.: Building ethically bounded AI. Proceedings of the AAAI Conference on Artificial Intelligence, vol. 33, no. 0101, pp. 9785–9789 (2019)
17. Rossi, F., Venable, K.B., Walsh, T.: mCP nets: representing and reasoning with preferences of multiple agents. In: Proceedings of the 19th AAAI Conference on Artificial Intelligence, pp. 729–734 (2004)
18. Tange, O.: GNU parallel: the command-line power tool. USENIX Mag. **36**(1), 42–47 (2011)
19. Tarjan, R.: Depth-first search and linear graph algorithms. SIAM J. Comput. **1**(2), 146–160 (1972)

Allocations and Matching

Core Stability and Nash Stability in k-Tiered Coalition Formation Games

Nathan Arnold[1,2]($\boxtimes$) and Judy Goldsmith[1]

[1] Department of Computer Science, University of Kentucky, 329 Rose St, Lexington, KY 40506-0633, USA
goldsmit@cs.uky.edu

[2] Department of Computer Science and Information Technology, Eastern Kentucky University, 521 Lancaster Ave., Richmond, KY 40475, USA
nathan.arnold@eku.edu

Abstract. The problem of k-tiered coalition formation games (k-TCFGs) has been considered for ranking members of a stochastic, intransitive round robin tournament, with the restriction that the ordering must have exactly k nonempty ranks for some integer k. As with other coalition formation games, an outcome of a k-TCFG may be evaluated for its stability, using the notions of Nash stability or core stability. An outcome is Nash stable if no one agent can move to a more preferable position, either by forming its own coalition or joining an existing one. An outcome is core stable if no set of agents can form a new coalition such that all agents in the set benefit. Previous research on k-TCFGs has focused on preferences derived from matchups, and has indicated that, under these matchup-oriented preferences, core stable outcomes may be significantly easier to find than Nash stable outcomes. However, the extent of this trend has not been explored. Here, we prove that for a key subset of k-TCFGs with matchup-oriented preferences, there is always at least one core stable partition. We include an illustration of the difference between Nash stabilizability and core stabilizability on an example game. We introduce a preference notation that can be used to represent any preference framework for k-TCFGs, and prove that under the subset of k-TCFGs which this notation can represent within polynomial space, the problem of determining if a game has a Nash stable list is NP-complete.

Keywords: Computational Social Choice · Game Theory · Preference Modeling

1 Introduction

The Worldwide Jousting League has hosted jousting tournaments for many years, and, in that time, has always had five ranks for its jousters.[1] Everyone who will

---

[1] See http://www.ija-usa.com/jouster-certification-ranking.html for the true part of this.

compete is assigned a rank before the year's competition begins. Most jousts will be held between members of the same rank, but a competitor from a higher rank may agree to have one from a lower rank compete on their behalf for a joust against a particular opponent. There are generally three styles of jousters: careful, swift, and powerful. Careful jousters tend to have an advantage against swift jousters, swift jousters have an advantage against powerful jousters, and powerful jousters have an advantage against careful jousters. These trends appear in every rank, although a much higher-ranking jouster may still be likely to win against a much lower-ranking opponent in spite of a stylistic disadvantage.

Sir Wright the Intrepid has been assigned the lowest rank, but has been training rigorously and wishes to be promoted by one rank. The League denies his request unless he can get the members of that rank to agree to his promotion. Meanwhile, Sir Justice the Swift has been assigned the second-highest rank, but notices that the middle rank contains many jousters with a careful style, who his opponents could nominate to compete against him in their stead, and who would have an advantage against his own style. Wishing to avoid the embarrassment of losing to a series of lower-ranking opponents, Sir Justice asks to be demoted to the second-lowest rank. That rank denied Sir Wright's initial request, but when Sir Wright and Sir Justice ask together, all of the jousters agree to the change.

As we see, a partition of jousters (or other agents) into ranks, or *tiers*, so that all are content with the partition, may be a less-than-obvious matter. In this work, we investigate notions of *stability* for *tiered coalition formation games (TCFGs)* [1], or more particularly, TCFGs with a fixed number of tiers, *(k-TCFGs)* [2]. Previous publications have suggested these problems as a way to model the ranking of power for participants in round-robin tournaments or structures where each participant has a projected probability to win against each other agent. This suggestion holds particularly for settings in which power is intransitive (as it is for the above jousters and their three styles, see also [3,4]) and in which it may be desirable for entities to share a rank despite the outcomes of individual matches among those entities.

Our focus here is on k-TCFGs, which bear particular interest not only because of their mathematical properties, but because they appear to better model real-world rankings of items with intransitive power than unrestricted TCFGs do [5]. This results in part from observations of real-world ranking systems, in which the number of ranks is often known in advance, as in our five-tier jousting example; the players are arranged to suit the ranks, rather than the other way around. Furthermore, agents in TCFGs rarely choose to share a rank if their movements are unrestricted [5], which has been noted as a shortcoming in the ability of unrestricted TCFGs to create accurate rankings [1].

A coalition formation game consists of a set of agents to be partitioned into coalitions, and a set of individual agents' preferences over partitions. There is significant work on finding optimal partitions based on these preferences, for various notions of optimal (see, e.g., [6,7] for non-utilitarian optimality notions). Other work has focused on notions of stability, as we do here. There are many computational questions that can be answered about stability: for a given notion of stability, is there always a stable partition? What is the computational complexity of finding a stable partition? What stability properties imply other properties (stability or otherwise)?[2]

Concerning k-TCFGs, it was shown in 2022 that Nash stability is not guaranteed, even when using a preference framework that guarantees Nash stable outcomes in unrestricted TCFGs [2]. However, the question was raised in 2023 of whether that same preference framework can guarantee the existence of a core stable outcome [5]. After introducing several definitions surrounding k-TCFGs and stability in Sect. 2, we present in Sect. 3 what we believe to be the first major step in the investigation of this open question of a core stability existence guarantee. Subsequently, we examine a randomly generated instance in Sect. 4, comparing the count and utility of core stable and Nash stable outcomes.

In Sect. 5, we introduce a new preference notation for TCFGs and k-TCFGs, Boolean preferences, which generalizes all preference frameworks. In Sect. 6, we prove that for the subclass of k-TCFGs with preferences representable in polynomial space using Boolean preferences, the question of if there exists a Nash stable tier list on a given instance is NP-complete.

2 Preliminaries

A *coalition formation game* (CFG) is a tuple $(N, \succeq)$, where N is a set of *agents* representing individual actors. These agents behave according to their *preferences*, which dictate the utility they derive as they and their fellow agents form and move between groups of agents, called *coalitions* [8]. At a given time, the collection of coalitions the agents are currently organized into is called a *partition*. The preferences are represented by $\succeq$, a series of binary relations; for a given agent i and two partitions A and B, the statement $A \succeq_i B$ means that A is at least as preferable as B according to i. We also use $A \succ_i B$ to denote that i strictly prefers A over B.

Typically, when examining a given CFG, we are interested in how to find stable outcomes. In many games, two stability notions are considered: Nash stability, in which a partition is stable if no agent has a move that can improve its utility, and core stability, in which a partition is stable if no coalition can form such that each agent in the coalition improves its utility. These notions are, in many games, distinct from each other, as they are in additively separable hedonic games [8,9], fractional hedonic games [10], super altruistic hedonic games [11], and role-based hedonic games [12].

A *tiered coalition formation game* (TCFG) is a CFG in which coalitions, also called tiers in this context, are strictly ordered from low to high (written in this paper as left to right), and in which the preferences of each agent are strictly determined by the set of agents in the same tier and all lower tiers, referred to as the set of *seen agents* [1]. The partitions are referred to as *tier lists*, and if an

---

[2] For instance, a Pareto optimal partition is necessarily contractually individually stable. Neither of these properties feature in this work, so they will not be defined here.

agent i sees the same set of agents in two distinct tier lists, those tier lists are equally preferable to that agent, regardless of which agents are in the same tier as i and which are in lower tiers, or the arrangement of lower-tiered agents. We use notation $Seen(i, T)$ to refer to the set of agents that an agent i sees from tier list T.

A *k-tiered coalition formation game* (k-TCFG) is a variant of TCFGs in which $k \leq |N|$ is a natural number and the only permitted tier lists are those with exactly k nonempty tiers [2,5]. Preferences are defined in the same way, but agents may not move in a manner that changes the number of tiers. In these games, we may still refer to the outcomes as tier lists, although we may also call them k-tier lists to specify that they are outcomes of a k-TCFG.

In this paper, we focus on k-TCFGs, and on the notions of Nash stability and core stability for these games.

Definition 1. A *Nash stable k-tier list* is a k-tier list T such that there is no agent i in a tier with other agents that can move to another existing tier such that it gains utility, and there is no agent j in a tier by itself that can move its tier to a different position in the order such that it gains utility. Equivalently, T is Nash stable if there is no k-tier list T' that differs from T in the tier of one agent i such that $Seen(i, T') \succ_i Seen(i, T)$.

Nash stability, in general, is concerned with whether any individual agent has an available move that improves its own utility. In the case of k-TCFGs, an individual agent may be in a tier by itself, in which case it essentially moves its tier to a different position in the order of tiers. Otherwise, the agent will seek an improving tier to join.

Definition 2 [5]**.** Let $G = (N, \succeq, k)$ be a k-TCFG and let T be a k-tier list for G. A *blocking coalition* in a k-TCFG is a nonempty subset of agents C such that:

1. There exists exactly one tier $T_A \in T$ such that $T_A \subseteq C$, and
2. $\forall i \in C : T' \succ_i T$,

where there is some $r \leq k$ such that T' is the k-tier list formed by moving all agents in C from their current position in T to a new tier at index r.

Note that a blocking coalition might be an entire tier that simply moves to a new place in the tier list, or an augmented tier (one that new agents have joined) that similarly shifts in the tier list, or it might be an augmented tier that stays where it was but acquires additional agents.

Definition 3. A *core stable k-tier list* is a k-tier list for which there exists no blocking coalition.

The publications on TCFGs and k-TCFGs so far have used a particular preference framework, called a *matchup-oriented preference framework*, which represents the preferences induced solely by projected outcomes of individual matchups between agents [1,2,5], such as in a round-robin tournament. This is a natural and easy-to-represent preference framework, and we will use it through much of this paper, as well.

Definition 4. A *Win matrix* on a set of n agents is an n by n matrix Win in which, if agent i has probability p of winning against j, $Win[i,j] = 2(p - 0.5)$ and $Win[j,i] = -Win[i,j]$.

In this definition, utilities for individual matchups are antisymmetric and bounded by $[-1, 1]$. An agent is assumed in any setting to be evenly matched against itself, and as such $Win[i,i] = 0$ for any i in any Win matrix.

Definition 5. A *matchup-oriented preference framework* for TCFGs or k-TCFGs is a framework in which, for any tier list T and any agent i, the utility i derives in T is equal to $\sum_{j \in Seen(i,T)} Win[i,j]$.

This preference framework is a relatively straightforward one, ensuring that agents seek a position from which they will see more good matchups and fewer bad matchups. Agents will, for example, rarely attempt to join a topmost tier if they do not have favorable chances to win against the majority of other agents.

Nash stability and core stability are equivalent under matchup-oriented preferences in the unrestricted TCFG setting [1], but have been demonstrated to be distinct on k-TCFGs [5].

It is unknown if all k-TCFGs admit a core stable tier list [5]. This is one of the open questions that motivates our current work. Section 3 includes a partial answer to this open question, proving that all 2-TCFGs have a core stable outcome.

By contrast, it is known that not all k-TCFGs admit a Nash stable outcome under matchup-oriented preferences. We will demonstrate this point with the following example.

Let Game 2 be a 2-TCFG on agents r, s, p representing the eponymous options in the game Rock-Paper-Scissors, where $Win[r,s] = 1$, $Win[s,p] = 1$, and $Win[p,r] = 1$. A 2-tier list on this game consists of one tier of two agents and one tier of one agent. Suppose that the tier with two agents is the lower tier. Then there is one agent in the lower tier with utility 1 and one with utility -1. The agent with utility -1 could obtain a utility of 0 by moving to the higher tier. Now suppose that the tier with two agents is the higher tier. All agents have zero utility. The agent in the higher tier which defeats the agent currently in the lower tier would gain 1 utility by moving to the lower tier. Therefore, there is no Nash stable tier list in this 2-TCFG. Note, however, that any 2-tier list is core stable, as the agent that wishes to join the singleton tier would not be welcome there.

Before proceeding to the next subtopic, we will briefly comment on the values of k considered in this paper. It perhaps goes without saying that $k = 1$ is not an interesting environment, as any 1-TCFG has only one possible outcome, and there is no avenue for deviations by agents or coalitions, nor any doubt that the outcome is stable. We also will not, in games under matchup-oriented preferences, give consideration to games in which $k = n$, where n is the number of agents. This is because at this value of k, Nash stability and core stability are equivalent (as all coalitions contain exactly one agent), and any move by an agent to improve its utility would improve the total utility of other agents, meaning that a local search in which agents are allowed to move freely always results in a stable tier list.[3] Because of this, for k-TCFGs under matchup-oriented preferences, we consider only $k \in \{2, \ldots, n-1\}$.

2.1 Exact Cover

In this paper, we will be performing a reduction from EXACT COVER, a problem which is known to be NP-complete [13].

EXACT COVER
Input: $\langle X, S \rangle$: X is a set of elements; $S = \{S_1, S_2, \ldots\}$ is a collection of subsets of X such that each element of X is in at least one S_i.
Question: Does there exist a subcollection $S^* \subseteq S$ such that every element of X is an element of exactly one subset in S^*?

3 Core Stability

Observation 1. *For a 2-tier list T, any coalition containing agents in the higher-indexed tier in T cannot block T by forming at the higher-indexed tier.*

An individual tier by itself can only block the list by choosing to swap positions with the other tier. Concerning other possible moves, if an agent is in the higher tier both before and after the move, the agent has the same set of seen agents before and after the move, and therefore cannot gain utility as a result. Thus, any blocking coalition either must be one tier choosing to swap positions with the other or must plan to form at the lower of the two positions.

Theorem 1. *Every 2-TCFG under matchup-oriented preferences admits at least one core stable 2-tier list.*

Proof. Let Game 2 be a 2-TCFG with agent set N. Let $a \in N$ be an agent such that, if u_i is the utility agent i derives from seeing all agents, then $u_a = \max_j u_j = u$.

Consider the 2-tier list $T = \{T_1, \{a\}\}$, where $T_1 = N \setminus \{a\}$. T may or may not be core stable. Assume it is not.

Because any blocking coaliton must contain exactly one tier, and because a has a nonnegative sum of utility against all other agents, T_1 cannot block this list, and any blocking coalition must contain a. If the coalition forms at the higher tier, the utility of a does not change, and so the coalition is not a blocking coalition. Therefore, any blocking coalition must form at the lower tier.

As T is not core stable by assumption, let C be the largest set of agents such that $C \cup \{a\}$ blocks T, and let $D = T_1 \setminus C$. In order for this deviation to

---

[3] This property at $k = n$ has not been explicitly stated for k-TCFGs in previous work, but follows trivially from results found by Siler in the paper that introduced TCFGs [1].

be valid, a must derive a negative utility $-v$ from seeing only the agents in D, and must derive a positive utility w from seeing only the agents in C, where $w - v = u$. Furthermore, each agent c in C must have preferences satisfying $(\sum_{d \in D} Win[c, d]) - Win[c, a] < 0$.

This deviation results in the 2-tier list $T' = \{C \cup \{a\}, D\}$. From T', the lower tier cannot benefit from swapping positions with the higher tier. Any blocking coalition cannot contain D, as it is indifferent to growth and, as a whole, has a positive sum of matchups against each agent beneath it. A coalition which contains $C \cup a$ and some number of agents in D, and which would form at the lower tier, could only block T' if each agent would gain utility from this move. However, this would mean that this coalition, which contains more agents than $C \cup \{a\}$, would also have blocked T, resulting in a contradiction. Therefore, T' is core stable, and, if T is not core stable, this method must produce a core stable T'. □

4 Illustrating the Prevalence of Stability

In the interest of exploring the differences between core stability and Nash stability, we examine an example matchup-oriented preference matrix, and the differences in prevalence between core stability and Nash stability across the k-TCFGs in this matrix. This example should not be confused with a systematic survey, but nevertheless will highlight differences between core stability and Nash stability in this setting.

Let Environment 1 be the randomly generated Win matrix of ten agents $\{i_1, \ldots, i_{10}\}$ described in Table 1, where row i_p, column i_q denotes $Win[i_p, i_q]$.

Table 1. Environment 1

	i_1	i_2	i_3	i_4	i_5	i_6	i_7	i_8	i_9	i_{10}
i_1	0	−1	−1	1	−1	1	1	−1	−1	1
i_2	1	0	−1	1	1	1	1	1	1	1
i_3	1	1	0	1	1	1	−1	−1	1	−1
i_4	−1	−1	−1	0	−1	−1	−1	−1	1	1
i_5	1	−1	−1	1	0	−1	−1	−1	−1	1
i_6	−1	−1	−1	1	1	0	−1	1	−1	1
i_7	−1	−1	1	1	1	1	0	1	−1	1
i_8	1	−1	1	1	1	−1	−1	0	−1	1
i_9	1	−1	−1	−1	1	1	1	1	0	−1
i_{10}	−1	−1	1	−1	−1	−1	−1	−1	1	0

We use this ten-agent matrix to define eight k-TCFGs, where in each one k is assigned a different integer value from 2 to 9 inclusive. Within each of these k-TCFGs, we iterated over every possible k-tier list and evaluated if it was core stable and if it was Nash stable. The results of this examination are in Table 2.

Table 2. Stability on Environment 1

k	# of lists	# core stable	# Nash stable
2	1022	732	1
3	55980	12112	1
4	818520	38525	1
5	5103000	51189	2
6	16435440	36883	3
7	29635200	15029	6
8	30240000	3159	11
9	16329600	287	8

In this instance, there is a dramatic difference between Nash stabilizability and core stabilizability demonstrated by these data. At every value of k, the number of core stable k-tier lists vastly exceeds the count of Nash stable lists. We note that at $k = 2$, a majority of possible lists are core stable; this is, perhaps, unsurprising in light of Observation 1. The proportion of core stable lists decreases as k increases, and the raw number of core stable tier lists is highest at $k = 5$. Meanwhile, it appears that there is a generally positive correlation between the value of k and Nash stabilizability, at least in this environment.

We also recorded the highest and lowest total utilities of Nash and core stable tier lists, and the highest total utility of any tier list, for each value of k. These results indicate the potential penalty to stability for finding an optimal, stable outcome, as well as the potential penalty to stability for simply finding a stable outcome.[4] These measurements are recorded in Table 3.

Table 3. Minimum and maximum utility of core and Nash stable lists, compared to overall maximum utility, by k

k	core min	core max	Nash min	Nash max	Overall max
2	−11	15	13	13	15
3	−7	22	11	11	22
4	−6	24	14	14	24
5	−3	26	16	22	26
6	2	27	17	23	27
7	3	28	17	27	28
8	7	29	18	28	29
9	12	30	26	30	30

---

[4] Such penalties are typically recorded as a price of stability [14], price of anarchy [15], core price of stability [16], or core price of anarchy, but these measurements are designed for games in which stable outcomes cannot have negative total utilities, which is not the case for k-TCFGs.

Evidently, if one simply finds a core stable tier list in this game, the result may have much lower than optimal utility, with some utility values of core stable lists in the negatives. However, the maximum utility for core stable tier lists and tier lists as a whole is equal at each value of k. By contrast, there is always a penalty to utility for Nash stability except at $k = 9$, but the utility of a Nash stable list is no less than half the maximum utility for each value of k in k-TCFGs on this environment.

These results indicate that simply finding a core stable outcome may not be sufficient for finding an outcome with sufficiently high utility. However, even if the conjecture holds that all k-TCFGs have at least one core stable tier list, we are interested not only in finding *a* core stable tier list, but in finding one of high or optimal utility, preferably in polynomial time.

We noticed that the maximum-utility core stable lists at each k met the overall maximum utility, and so conducted a follow-up investigation on the same instance, counting the number of tier lists with exactly the maximum utility, and determining if all of these lists are core stable. The summary of this investigation is in Table 4.

Table 4. Results of examining maximum-utility tier lists

k	Max utility	Count of max lists	Are all maximum utility lists stable?
2	15	1	Yes
3	22	1	Yes
4	24	2	Yes
5	26	2	Yes
6	27	30	Yes
7	28	70	Yes
8	29	56	Yes
9	30	18	Yes

We found that, indeed, for any given k, all lists with the maximum utility are core stable. This result is intriguing, especially as it was proven in 2023 that locally maximal utility does not guarantee core stability [5]. Further investigation would be needed to find if there is a game in which a maximum-utility tier list is not core stable, or perhaps to prove that a maximum-utility tier list is always core stable.

5 Boolean Preferences for TCFGs and k-TCFGs

We introduce a preference notation for use in TCFGs and k-TCFGs: Boolean preferences. In this preference notation, each agent's preferences consist of a series of Boolean formulas over other agents being or not being in the set of seen agents, and an associated scalar value for each formula, representing the change in utility if that formula is satisfied.

We will define $i \triangleright j$ as a Boolean variable that evaluates to true if i sees j and false otherwise. We will also use the form $i \triangleright A$ for agent i and subset of agents A, which is equivalent to a conjunction of each $i \triangleright j$ for each $j \in A$.

Definition 6. A *seen formula* for agent i is a Boolean formula in which Boolean variables are of the form $i \triangleright j$ for any number of other agents j.

Definition 7. A *Boolean preference framework* for TCFGs or k-TCFGs is a preference framework in which an agent i has preferences consisting of a series of pairs (SF, c), where SF is a seen formula for i, and where c is a real number denoting the change in utility for i if SF is satisfied by the set of agents seen by i.

Example 1. Consider a game with agents i, j, k, in which i has preferences $(i \triangleright j, 2), (\neg(i \triangleright k), -1)$. Then, i has utility 2 if it sees j and k, utility 1 if it sees j and not k, utility 0 if it sees k and not j, and utility -1 if it sees neither of the other agents.

Example 2. Consider a game with agents a, b, c, in which a has preferences $(a \triangleright b, -2), (a \triangleright b \wedge a \triangleright c, 5)$. If a sees b, it loses 2 utility. However, if a sees c in addition to b, it gains 5 utility for a total of 3. If a does not see b, none of the seen formulas in its preferences are satisfied, and it has 0 utility.

Remark 1. Matchup-oriented preferences (in which each agent assigns a utility value to seeing each other agent, and utilities are antisymmetric) are a subclass of Boolean preferences in which the formulas consist of a single variable.

Matchup-oriented preferences have been used in the previous literature on these problems [1,2,5], and remain perhaps the most natural form of preferences for these games. Representing them as Boolean preferences means simply converting the utility c that i gains if it sees j to the pair $(i \triangleright j, c)$ for each i, j.

Observation 2. *All preference frameworks for TCFGs and k-TCFGs can be expressed in terms of Boolean preferences.*

Agent preferences in these games are determined solely by the set of seen agents. Any preferences for an agent i can be expressed as a mapping from each possible set of seen agents to the utility that agent derives from seeing exactly that set of agents. Subsequently, for a given set of seen agents, a seen formula can be derived; namely, a conjunction of i seeing each agent in that set and not seeing each agent not in that set.

6 Nash Stabilizability of k-TCFGs Under Boolean Preferences

Here, we use Boolean preferences to show that, within the subset of k-TCFGs with compact preferences (that is, the preferences are represented in an amount of space polynomial on the number of agents) with no disjunctions, there are games where the existence of a Nash stable tier list cannot be determined in polynomial time unless $P = NP$.

kTCFG-NASH-BOOLEAN-POLYSIZE
Input: $(N, \succeq)$: N is a set of agents; $\succeq$ is a collection of Boolean preferences for N with size polynomial in $|N|$ and without disjunctions.
Question: Is there a Nash stable k-tier list of N?

There is a distinct version of this problem for each value of k. We begin with the somewhat straightforward claim that for a given k, the Nash stabilizability decision problem is at least as hard for any $k+1$ as it is for k.

Proposition 2. *There is a reduction from kTCFG-NASH-BOOLEAN-POLYSIZE to $(k+1)$TCFG-NASH-BOOLEAN-POLYSIZE.*

Proof. Let G be a k-TCFG on agent set N. Let M be the agent set constructed by $N \cup \{a\}$ and let H be a $(k+1)$-TCFG on M, in which a has the preferences $(a \blacktriangleright M, 1)$ and each remaining agent b has the same preferences as it did in G, with the addition of the pair $(b \blacktriangleright a, -\infty)$. Then for any tier list T that is Nash stable in G, the tier list T' constructed by appending $\{a\}$ as the new highest tier to T is stable. If a tier list T is not Nash stable, the tier list T' likewise constructed is not Nash stable. Lastly, only tier lists in which the highest tier consists solely of a are stable, as other agents lose utility from seeing a, and as a only obtains positive utility by being in the highest tier. Therefore, H is Nash stabilizable if and only if G is Nash stabilizable. □

What follows is the reduction from EXACT COVER to 3TCFG-NASH-BOOLEAN-POLYSIZE. In order to claim that the reduction is correct, we will prove that the reduction can be computed in time polynomial in the size of the EXACT COVER instance, and that for given EXACT COVER input I and its corresponding input $f(I)$ to 3TCFG-NASH-BOOLEAN-POLYSIZE, 3TCFG-NASH-BOOLEAN-POLYSIZE($f(I)$) is true iff EXACT COVER(I) is true.

Theorem 3. *There is a reduction from* EXACT COVER *to 3TCFG-NASH-BOOLEAN-POLYSIZE.*

Proof. Let $\langle X, S \rangle$ be an instance of EXACT COVER. That is, let X be a set of elements $\{x_0, \ldots, x_n\}$, and let S be a collection of subsets of X, $\{S_0, \ldots, S_m\}$, where every element of X is an element of at least one subset in S. EXACT COVER on this problem is true if there exists a subcollection $S^* \subseteq S$ such that every element of X is an element of exactly one subset in S^*, and false if such a subcollection does not exist.

Construct a 3-TCFG with a set of agents and their preferences as follows. The agent set N consists of agent subsets A, T, Y, Z. First, A contains agents a, b, c. The preferences of a are $(a \blacktriangleright b, -2), (a \blacktriangleright b \wedge \neg(a \blacktriangleright c), 1)$. The preferences of b are $(b \blacktriangleright c, -2), (b \blacktriangleright a, 1)$. The preferences of c are $(c \blacktriangleright N, 1)$.

Before continuing with the construction, we argue that in such a game where agents are allowed to move one at a time according to their preferences, the equilibrium state for agents in A is that a will be in tier 1, b in tier 2, and c in tier 3. First, as all tiers must be nonempty at all times, c will always have 1 utility at tier 3 and will never have positive utility at any other tier, and therefore will always move to or remain in tier 3. Subsequently, b will always move to be below c. If b is in tier 2, a will move to tier 1. As long as no other agent in A moves from this position, no agent in A will ever gain utility from moving, and this subset of agents is thus at equilibrium. If, instead, b is in tier 1, then a has -2 utility at tier 1 and at tier 3, but -1 utility at tier 2, and so will move to tier 2. If a is at tier 2 and c is at tier 3, b will move to tier 2. Subsequently, a will move to tier 1, and these agents once again will be at the equilibrium state.

Because of this, a stable tier list must have these placements of the agents in A, so we will assume without loss of generality that these movements occur first.

Next, for every element x_i in X, there is corresponding agent y_i in Y and agent z_i in Z, and for every subset S_j in S, there is one corresponding agent t_j in T. Every agent y_i in Y has $(y_i \blacktriangleright b \wedge \neg(y_i \blacktriangleright z_i), 2), (y_i \blacktriangleright c, 1)$. Every agent z_i in Z has $(z_i \blacktriangleright y_i, 2), (z_i \blacktriangleright b, 1), (z_i \blacktriangleright c, -1)$. Every agent t_j in T has $(t_j \blacktriangleright N, 1)$ and a set of agents NBR_j associated with it, which will be constructed as we process S.

For each S_j containing agent x_i such that $S_1, \ldots, S_n$ are the subsets other than S_j containing x_i, agent y_i gains $(y_i \blacktriangleright t_j \wedge \neg(y_i \blacktriangleright t_1) \wedge \ldots \wedge \neg(y_i \blacktriangleright t_n), 5)$, and $NBR_j = NBR_j \cup \{t_1, \ldots, t_n\}$. After this process, each t_j in T gains $(\neg(t_j \blacktriangleright b) \wedge \forall r \in NBR_j : \neg(t_j \blacktriangleright r), 5)$.

There are $2|X| + |S| + 3$ agents. The agents in A and Z have constant preferences, while agents in T may have preferences taking up to $O(|X| * |S|)$ space, and agents in Y have preferences taking up to $O(|S|)$ space. The spacial complexity of this instance is therefore $O(|X| * |S|^2)$, and it can be computed in time polynomial on the size of the original instance of EXACT COVER.

Suppose for a given instance of EXACT COVER that there exists a subcollection S^* which is an exact cover. Under this reduction, the following 3-tier list is Nash stable: tier 1 contains a, all agents in T corresponding to a subset in S^*, and all agents in Y; tier 2 contains b and all agents in Z; tier 3 contains c and all agents in T corresponding to a subset not in S^*. For a given $S_j \in S^*$, t_j sees no other agents in T corresponding to a subset in S that shares an element in common with S_j. As a result, it has 5 utility at tier 1 and can gain no utility from moving. Likewise, for an element $x_i \in s_j$, the agent y_i sees t_j and no other agent in T corresponding to a subset containing x_i, and so x_i has 5 utility and can gain no utility from moving. Agents in Z can see b and agents in Y, and so have 3 utility and would have lower utility at any other tier. Agents in T not corresponding to a subset in S^* gain 1 utility from seeing all agents, and cannot move in such a way that their preference formula that would provide 5 utility can be satisfied. Therefore, no agent can move in a profitable manner, and the list is Nash stable.

Now suppose that for a given instance of EXACT COVER, there exists no exact cover by subsets in S. The constructed 3-TCFG has the property that agent t_j in T will either move to tier 1 if the formula containing $\neg t_j \blacktriangleright b$ is satisfied by such a move, or move to tier 3 if that formula is not satisfied. The agents will move until agents in T and in tier 3 all have at least one mutually exclusive neighbor in tier 1; that is, $\forall t_j$ in tier 3, there exists at least one t_h in tier 1 such that S_j and S_h share an element in common. The elements in T and in tier 1 in a given equilibrium state for agents in T correspond to a partial or complete cover of X, as the corresponding subsets are mutually exclusive and may or may not constitute an exact cover of all agents.

Subsequently, for each t_j in tier 1, each y_i corresponding to an $x_i \in S_j$ moves to tier 1, as it gains 5 utility from seeing exactly one t_j corresponding to an S_j containing x_i, which outweighs the 1 utility it loses by failing to see c. As we have assumed that the exact cover does not exist in the original problem, for any collection of mutually exclusive subsets, there must be at least one element x_i not appearing in any subset. Therefore, for any subset of T currently in tier 1, the corresponding y_i does not have incentive to move to tier 1. Further, y_i gains 1 utility by seeing c, so it will move to tier 3 unless incentivized to by the set of seen elements from another tier. With this arrangement, z_i would have 0 utility at tier 1, 1 utility at tier 2, and 2 utility at tier 3, so it will move to tier 3. With z_i in tier 3, y_i could see b without seeing z_i by moving to tier 2, which gives it a utility of 2 over its current 1, so it will move to tier 2. This causes z_i to move to tier 2, where it now has a utility of 3. But because y_i once again sees z_i, the formula $b \wedge \neg z_i$ is no longer satisfied, and y_i moves to tier 3, creating a loop of agent movements. There is therefore no Nash stable tier list. □

From this, we can conclude that kTCFG-NASH-BOOLEAN-POLYSIZE is NP-complete for $k \geq 3$. The complexity class remains unknown for $k = 2$, but this result indicates the difficulty of finding a Nash stable k-tier list in a general case.

This reduction does not make use of matchup-oriented preferences. However, we conjecture that the introduction of matchup-oriented preferences does not make this decision problem easier.

7 Conclusion

We have proven that on matchup-oriented preferences, all 2-TCFGs admit a core stable outcome. This represents a significant step in answering the question of whether all k-TCFGs with these preferences have a nonempty core. Future efforts may address specific values of k, such as 3 or $n-1$, or may even yield a proof that applies to all values of k. We presented an example that illustrates the difference between core stability and Nash stability in this environment, and that highlights a possible avenue of investigation for the open question on core stability. We introduced a preference notation that can generalize both matchup-oriented preferences and any preference frameworks introduced in the future for these games. We presented a reduction from EXACT COVER that proves the NP completeness of finding Nash stabilizability for a subset of k-TCFGs in polynomial space.

The reduction presented here does not prove the complexity of finding if a Nash stable tier list exists in a k-TCFG with matchup-oriented preferences. Indeed, while the proof applies to the consideration of complexity for a broad subset of k-TCFGs, there is a possibility that a subclass within that subset can be proven (without first proving that P = NP) to have a polynomial-time solution for this problem. We conjecture that the problem remains NP-complete for matchup-oriented preferences, but we recognize a potential topic for future research in finding subclasses of k-TCFG with a P-time solution to determining Nash stabilizability.

Declarations.
- **Materials availability:** A copy of the code used in the investigation in Section 4 is available at: https://github.com/naroarnold/ktcfg-nash-core.

References

1. Siler, C.: Tiered coalition formation games. In: The International FLAIRS Conference Proceedings, vol. 30, pp. 210–214 (2017)
2. Arnold, N., Goldsmith, J., Snider, S.: Extensions to tiered coalition formation games. In: The International FLAIRS Conference Proceedings, vol. 35 (2022). https://doi.org/10.32473/flairs.v35i.130708
3. Akin, E.: Generalized intransitive dice: mimicking an arbitrary tournament. J. Dyn. Games **8**(1), 1–20 (2020)
4. Saarinen, S., Tovey, C.A., Goldsmith, J.: A model for intransitive preferences. In: Workshops at the Twenty-Eighth AAAI Conference on Artificial Intelligence (2014)
5. Arnold, N., Snider, S., Goldsmith, J.: Socially conscious stability for tiered coalition formation games. Ann. Math. Artif. Intell. (2023). https://link.springer.com/article/10.1007/s10472-023-09897-4
6. Bullinger, M.: Pareto-optimality in cardinal hedonic games. In: AAMAS, vol. 20, pp. 213–221 (2020)
7. Waxman, N., Kraus, S., Hazon, N.: On maximizing egalitarian value in kcoalitional hedonic games. arXiv preprint arXiv:2001.10772 (2020)
8. Bogomolnaia, A., Jackson, M.O.: The stability of hedonic coalition structures. Games Econom. Behav. **38**(2), 201–230 (2002)
9. Banerjee, S., Konishi, H., Sonmez, T.: Core in a simple coalition formation game. Soc. Choice Welfare **18**, 135–153 (2001)
10. Brandl, F., Brandt, F., Strobel, M.: Fractional hedonic games: individual and group stability. In: Proceedings of the 2015 International Conference on Autonomous Agents and Multiagent Systems, pp. 1219–1227 (2015)
11. Schlueter, J., Goldsmith, J.: Super altruistic hedonic games. In: The International FLAIRS Conference Proceedings, vol. 33, pp. 160–165 (2020)

12. Spradling, M., Goldsmith, J.: Stability in role based hedonic games. In: The International FLAIRS Conference Proceedings, vol. 28, pp. 85–90 (2015)
13. Karp, R.M.: Reducibility among combinatorial problems. In: Complexity of Computer Computations (1972). https://doi.org/10.1007/978-1-4684-2001-2_9
14. Anshelevich, E., Dasgupta, A., Kleinberg, J., Tardos, É., Wexler, T., Roughgarden, T.: The price of stability for network design with fair cost allocation. SIAM J. Comput. **38**(4), 1602–1623 (2008)
15. Koutsoupias, E., Papadimitriou, C.: Worst-case equilibria. In: Meinel, C., Tison, S. (eds.) STACS 1999. LNCS, vol. 1563, pp. 404–413. Springer, Heidelberg (1999). https://doi.org/10.1007/3-540-49116-3_38
16. Monaco, G., Moscardelli, L., Velaj, Y.: Stable outcomes in modified fractional hedonic games. Auton. Agent. Multi-Agent Syst. **34**(1), 4 (2020)

Envy-Free and Efficient Allocations for Graphical Valuations

Neeldhara Misra and Aditi Sethia(✉)

Indian Institute of Technology, Gandhinagar, Gandhinagar, India
{neeldhara.m,aditi.sethia}@iitgn.ac.in

Abstract. We consider the complexity of finding envy-free allocations for the class of graphical valuations. Graphical valuations were introduced by Christodoulou et al. [14] as a structured class of valuations that admit allocations that are envy-free up to any item(EFX). These are valuations where every item is valued by two agents, lending a (simple) graph structure to the utilities, where the agents are vertices and are adjacent if and only if they value a (unique) common item. Finding envy-free allocations for general valuations is known to be computationally intractable even for very special cases: in particular, even for binary valuations, and even for identical valuations with two agents. We show that, for binary graphical valuations, the existence of envy-free allocations can be determined in polynomial time. In contrast, we also show that allowing for even slightly more general utilities $\{0, 1, d\}$ leads to intractability even for graphical valuations. This motivates other approaches to tractability, and to that end, we exhibit the fixed-parameter tractability of the problem parameterized by the vertex cover number of the graph when the number of distinct utilities is bounded. We also show that, all graphical instances that admit EF allocations also admit one that is non-wasteful. Since EFX allocations are possibly wasteful, we also address the question of determining the price of fairness of EFX allocations. We show that the price of EFX with respect to utilitarian welfare is one for binary utilities, but can be arbitrarily large $\{0, 1, d\}$ valuations. We also show the hardness of deciding the existence of an EFX allocation which is also welfare-maximizing and of finding a welfare-maximizing allocation within the set of EFX allocations.

Keywords: Fair Division · Utilitarian Welfare · Price of Fairness · Graphical Valuations

1 Introduction

Given a set of n agents and m items, dividing all the items among the agents in a manner considered *fair* for every agent is an important assignment problem. The *fairness* aspect can have different interpretations and has been studied in the literature with various lenses: *envy-freeness* [16,18], *equitability* [20], *share-based notions* [10,26] *and their approximations* [10,17,19,23].

Finding allocations that are envy-free is a gold standard in such assignment problems. This entails that no agent should feel envious of any other agent under the allocation. That is, every agent should value its own allocated bundle at least as much as it values anyone else's bundle. The problem with such envy-free allocations is two-fold: existential and computational. That is, they might not exist for many instances (say, when there are more agents than items), and deciding whether they exist is computationally intractable even for very special and structured instances. In particular, it is NP-complete even for binary valuations (where agents value items at either 0 or 1) [3] and weakly NP-Complete for two agents and identical valuations [22].

Motivated by these issues, we focus on a recently introduced class of structured valuations, called graphical valuations introduced by Christodoulou et al. [14] as a class of valuations that admit allocations that are envy-free up to any item (EFX)[1]. These are valuations where every item is valued by exactly two agents, lending a (simple) graph structure to the utilities, where the agents are associated with vertices and items with edges. Two agent-vertices are adjacent if and only if they value a (unique) common edge-item, represented by the edge between them. Such valuations may arise in scenarios where agents only value the items that are geographically closer. For instance, in real estate allocation, potential buyers might only be interested in properties within a certain distance from their workplace or amenities; employees might value office spaces closer to their teams and likewise [14].

Our Contributions. We highlight our main contributions below and put them in context with the already-known results.

- We show that an EF allocation if it exists, can be found efficiently for graphical valuations where agents have binary ($\{0,1\}$) valuations over the items (Theorem 2). This is in contrast to the intractability of EF allocation for binary utilities in general.
- We show that if we allow for even slightly more general valuations than binary, for instance, $\{0, 1, d\}$-valuations for some constant d, the problem again becomes intractable (Theorem 3).
- The above hardness motivates a parameterized approach towards tractability and towards that, we present a *fixed-parameter tractable*[2] algorithm for finding EF allocations for graphical instances with bounded number of distinct utilities, where the parameterization is in terms of the minimum vertex cover of the associated graph G (Theorem 5).
- We show that if there is an EF allocation for any graphical instance, then there is also an EF allocation that does not 'waste' any item, that is, it does not assign an item to an agent who derives 0 value from it. This shows that if there is an EF allocation, then there is an EF orientation of the graph G (Theorem 1). This result stands in contrast to the fact that an EFX allocation

---
[1] We refer the reader to next section for the definition of EFX.
[2] An algorithm that runs in time $f(k)poly(n,m)$ where f is some computable function of the parameter k.

always exists but an EFX orientation may not exist [14]. In terms of the price of EF, this implies that for $\{0,1\}$-graphical valuations, there is no loss in the welfare while achieving EF allocations, whenever they exist.
- Christodoulou et al. [14] showed that EFX allocations not only always exist but can be found efficiently for graphical valuations. But this comes with a sacrifice in terms of welfare. In particular, there are cases where any EFX allocation must assign items to agents for which they are irrelevant (0-valued). In this work, we quantify the loss of welfare while achieving EFX allocations and show that for $\{0,1\}$-graphical instances, the price of EFX for Utilitarian (sum of agents' utilities) welfare is 1 (Theorem 6). That is, restricted to binary graphical valuations, there is no loss in any of the welfare notions and an EFX allocation that maximizes the respective welfare can be found efficiently. On the other hand, we show that for slightly general valuations than binary, that is, for $\{0,1,d\}$-valuations, there are instances with a huge loss in the utilitarian welfare and consequently, price of EFX shoots up to ∞ (Theorem 7).
- On the computational side, we show that for general graphical valuations, finding EFX allocations that also maximize utilitarian welfare is NP-Hard (Theorem 8). It follows that finding a welfare-maximizing allocation within the set of EFX allocations is also hard.

Additional Related Work. Several special cases and approximations have been extensively studied in the fair division literature to understand the extent of tractability of EFX allocations: binary valuations [9]; bounded number of agents [1,13,25]; and bounded number of unallocated items [5,11]. Graphs have also been associated with fair division in various contexts and models. Allocations, where items allocated to each agent form a connected subgraph in a provided item graph, have been studied [8,15]. In a different model, Payan et al. [24] looked at graph-EFX which requires that an agent, represented by a vertex, satisfy EFX only against its adjacent vertices. Our work is closely aligned with that of Christodoulou et al. [25] who introduced graphical valuations and showed the hardness of deciding the existence of an EFX orientation. Following this, Zeng and Mehta [29] characterized that graphs with chromatic number at most 2 admit EFX orientations for any given valuations, while graphs with chromatic number strictly greater than 3 may not admit such orientations for all valuations. They also characterized EFX orientability for binary valuations.

The quantification of welfare loss that is inevitable due to the fairness constraint has also been of interest in the literature. To capture this, the notion of *price of fairness* was proposed in the works of Bertsimas et al. [6] and Caragiannis et al. [12]. Since then, various works have given bounds for the price of proportionality, envy-freeness, EF1, EFX, equitability, EQ1, maximum Nash welfare, and more [2,4,7,27,28].

2 Preliminaries

Model. A fair allocation instance $\mathcal{I} := (N, M, \mathcal{U})$ consists of a set N of n agents, a set M of m items and a set of valuation functions $\mathcal{U} = \{u_1, \ldots u_n\}$ such that $u_i := 2^M \to \mathbb{R}_{\geq 0}$. Each u_i captures the utility that agent i derives from a set of items in M. A valuation function u_i is said to be additive if the value of a bundle is the sum of the values of the items in the bundle. In this work, we assume that all the valuation functions are additive. An allocation $\Phi := \{\Phi_1, \ldots \Phi_n\}$ is a partition of M items into n bundles Φ_i, one for each agent.

Graphical Allocation Instance. A graphical fair allocation instance $\mathcal{I} = \{G = (V, E), \mathcal{U}\}$ takes as input an undirected, simple graph G and a valuation function $\mathcal{U}$. The set of vertices V in G corresponds to n agents and the set of edges E in G corresponds to m items to be allocated. We will often use the terms "items" and "edges" interchangeably because of this correspondence. Every agent only values a subset of the incident edge-items. Also, note that every edge is valued by exactly two agents, and every pair of agents value at most one edge together, the one which is incident on both of them. A $\{0, 1\}$-graphical instance is one such that $u_i \in \{0, 1\} \; \forall \; i \in N$. Given a graph G, an *orientation* O_G is an allocation with the additional property that every edge is assigned to one of the two endpoints. A directed graph that directs the edges of G towards the vertex that receives the edges is called an *orientation graph* of G. Note that every orientation corresponds to a complete allocation. An allocation is an orientation if it assigns the edges to the incident vertices. We say that an orientation satisfies a property if the corresponding allocation satisfies that property.

Fairness Notions. An allocation Φ is said to be envy-free (EF) if every agent values its bundle at least as much as it values any other allocated bundle. That is, $u_i(\Phi_i) \geq u_i(\Phi_j) \; \forall \; i, j \in N$. Note that an EF allocation may not exist so we resort to approximations. An allocation is said to be EFX if the envy between a pair of agents can be eliminated by removal of *any* item from the envied bundle. That is, $u_i(\Phi_i) \geq u_i(\Phi_j \setminus \{g\}) \; \forall \; g \in \Phi_j \; \& \; i, j \in N$. Further, we say that an allocation is non-wasteful if it allocates the items to agents who value them. An item is said to be wasted if it is allocated to an agent who derives 0 utility from it. Any allocation that consists of wasted items is said to be wasteful.

Welfare Notions and Price of Fairness. The welfare notion that we consider in this work is the Utilitarian social welfare, which is the sum of the individual agent utilities. We say that an allocation is Utilitarian maximal (UM) if it maximizes the utilitarian welfare. (We defer the discussion for other welfare notions like Nash and Egalitarian to the appendix). The price of a fairness notion F with respect to a welfare notion W is the supremum over all fair division instances with n agents and m items of the ratio of maximum welfare under any allocation to the maximum welfare under the fair allocation. In particular, the price of EFX with respect to utilitarian welfare is:

$$PoF_{UM} = \sup_{I \in \mathcal{I}} \frac{\max_{\Phi^* \in UM(I)} \sum_i v_i(\Phi_i^*)}{\max_{\Phi \in EFX(I)} \sum_i v_i(\Phi_i)}$$

3 Envy-Free Graphical Allocations

Although it is known that EFX allocations always exist on graphical valuations [14], an EF allocation may not exist on graphical instances as well, as illustrated by a simple example of a graph consisting of only one edge. Whichever incident vertex receives the edge, the other one is bound to be envious. We show that it is possible to determine if an EF allocation exists in polynomial time for $\{0,1\}$-graphical valuations, and in the event that the instance admits an EF allocation, such an allocation can be found in polynomial time. Before that, we present a series of structural results. The following result is in contrast to the EFX fairness, where the existence of an EFX allocation does not guarantee an EFX orientation but any EF allocation does guarantee an EF orientation.

Theorem 1. *Given a graphical allocation instance, there is an EF allocation if and only if there is an EF orientation.*

Proof. An orientation is EF if the corresponding allocation is EF, so the reverse direction holds. We argue the forward direction.

Suppose there is an EF allocation Φ for the given instance, which does not correspond to any EF orientation. We assume that everyone values at least one item, otherwise the agent can be removed from the instance. Since Φ is not an orientation, there must be some edges allocated to vertices that are not incident on them. All such edges are allocated wastefully as an agent does not value an edge that is not incident on itself. Consider the re-allocation Φ' such that all such wastefully allocated edges are re-allocated to one of their incident vertices, chosen arbitrarily. Say, edge $e = (ij)$ which was previously wastefully allocated to vertex k is now re-allocated to i, WLOG.

Under Φ', an agent who loses an item can not envy anyone, as its utility does not decrease. Any agent can potentially be envious of only those agents that are incident on it. Indeed, if i is not incident to agent k, then $v_i(\Phi_k) = 0$ as k only receives the edges incident on itself, none of which are valued by i. Moreover, suppose j is envious of i under Φ' as i receives the edge $e = (ij)$ that is also valued by j. Notice that e is the only item that is valued by j in the bundle Φ_i since it is the unique item valued by both i and j. Therefore, if j is envious of i, we have $u_j(\Phi_i') = u_j(e) > u_j(\Phi_j') \geq u_j(\Phi_j)$. The last inequality holds as no agent's utility decreases under the re-allocation Φ'. This implies that j valued e more than the bundle it got under the EF allocation Φ. But then, $u_j(\Phi_j) < u_j(e) \leq u_j(\Phi_k)$, where k is the recipient of e under Φ. This implies that j was envious of k in the allocation Φ, which is a contradiction to the fact that Φ was EF. Therefore, all the agents are EF under Φ', and Φ' assigns edges to only incident vertices. Therefore, Φ' corresponds to an EF orientation. □

Lemma 1. *Given any graphical allocation instance, suppose v_i^{max} is the maximum value any agent i has for any item. Then, a non-wasteful allocation is EF if and only if i gets a utility of at least v_i^{max} $\forall\, i \in [n]$.*

Proof. Let Φ be any EF allocation. Let v_i^{max} be the maximum value an agent i has for an edge e. Suppose $v_i(\Phi_i) < v_i^{max}$, then clearly, $e \notin \Phi_i$. Let $e \in \Phi_j$ for some agent j. Then $v_i(\Phi_i) < v_i(e) = v_i(\Phi_j)$. Therefore, i is envious of j which is a contradiction to the fact that Φ is EF. Therefore, every agent i must get a utility of at least v_i^{max} under an EF allocation Φ. Conversely, suppose every agent gets a utility of at least v_i^{max} under a non-wasteful allocation Φ. Since Φ is a non-wasteful allocation, it corresponds to an orientation in G. So every agent receives a subset of edges that are incident on it. Consider an agent i. We have $u_i(\Phi_i) \geq v_i^{max}$. Consider any other agent j incident on i. If the edge $e = (ij) \in \Phi_j$, then $u_i(\Phi_j) = u_i(e) \leq v_i^{max}$, else $u_i(\Phi_j) = 0$, as i does not value any edge incident on j except e. Also, for any agent j not incident on i, $u_i(\Phi_j) = 0$ as i does not value any edge which is not incident on itself. Therefore, we have that $u_i(\Phi_i) \geq u_i(\Phi_j)$ for all $1 \leq i \neq j \leq n$ and hence the orientation is EF. □

This gives us the following corollary.

Corollary 1. *For graphical instances, if an agent i gets a utility of at least v_i^{max} under a partial orientation O_P, then i remains EF under any extension of O_P.*

In particular for binary valuations, there is an EF allocation where every item is allocated to an agent who values it at 1, so, we have the following result.

Corollary 2. *For $\{0,1\}$-graphical instances, the price of EF with respect to utilitarian social welfare is 1.*

Note that the above result is not true for binary valuations in general. Consider the instance in Table 1. It is not a graphical instance as a_1 and a_3 value 3 items positively. An EF allocation must allocate at least one item from $\{g_1, g_3, g_4\}$ wastefully. Indeed, if all of them are allocated non-wastefully, then the agent who ends up receiving two of them is envied by the other one.

Table 1. An EF allocation that allocates an item wastefully.

	g_1	g_2	g_3	g_4
a_1	(1)	0	1	1
a_2	0	(1)	0	(0)
a_3	1	0	(1)	1

Theorem 2. *For $\{0,1\}$-graphical instances, an EF allocation can be found efficiently, if it exists.*

Proof. Consider an instance $\mathcal{I}$ of $\{0,1\}$-graphical valuations. Since an EF allocation exists if and only if there is an EF orientation (Theorem 1), we will construct an EF orientation if it exists. For all asymmetric edges $e = (ij)$, we orient them towards the incident agent who values e at 1, say i. This does not create any envy in the graph as the only agent who values e is i. We call such vertices i as special vertices since they remain envy-free under any completion of the allocation and are not envied by anyone else. Once we orient all the asymmetric edges, we remove them from the graph. The edges which are valued at 0 by both end-points are oriented arbitrarily and removed from the graph. This gives us a collection of connected subgraphs $H = \{H_1, H_2, \ldots H_k\}$ such that all edges in H are symmetric and valued at 1 by both the end-points. For each $H_i \in H$, we consider the following cases:

1. H_i is a tree. Then, $V(H_i) = E(H_i) + 1$. By pigeonholing, at least one agent, say i, does not receive any edge item from $E(H_i)$. Such a vertex i is always envious under any allocation unless i is already a special vertex. In the former case, there is no complete EF allocation. Otherwise, if there is a special agent i, then we root H_i either on i and construct an orientation such that every vertex gets an edge item from its parent. This way, everyone except i receives a utility of at least 1 from the edges in H_i and hence is EF in any complete orientation. Also, i is EF since it is a special vertex.

2. H_i contains a cycle, say $C = \{v_1, v_2, \ldots v_c, v_1\}$. We orient the edges (v_i, v_{i+1}) towards v_i and (v_c, v_1) towards v_c. Then, every vertex in the cycle is EF as $v_i(\Phi_i) \geq 1$ and remains EF in any completion of this orientation (Corollary 1). Therefore, the edges inside the cycle can be oriented arbitrarily. We now remove the cycle C from H_i, replace it with a vertex c, and construct a spanning tree of H_i rooted at c. We then construct an orientation that allocates every vertex in the spanning tree, except c, an edge from its parent. This implies that every agent in the spanning tree except the root c ends up with a utility of at least 1. All agents corresponding to the root c already had a utility of at least 1. Since all the agents in H_i now have utility at least 1, therefore everyone is EF in any completion of the partial orientation. Therefore, the remaining edges in H_i can be oriented arbitrarily, and hence we get an EF allocation for H_i.

The algorithm loops over every H_i in H and if there is an EF allocation for every H_i, it corresponds to a complete EF allocation (since vertices across components do not envy each other). Else, if there is at least one H_i for which there is no EF allocation, then the algorithm outputs that no complete EF allocation exists. This is true because an envious agent in H_i can not be made EF by any of the edges in the other components, as it does not value them. This settles our claim. □

We now show in the following result that if we slightly generalize from binary to $\{0, 1, d\}$ graphical valuations, it becomes hard to decide if the graphical instance admits an EF allocation.

Theorem 3. *Deciding whether an EF allocation exists is NP-Hard even for symmetric $\{0, 1, d\}$-graphical valuations.*

Proof. We present a reduction from MULTI-COLORED INDEPENDENT SET (MCIS), where given a regular graph $G = (V_1 \uplus \cdots \uplus V_k, E)$, the problem is to decide if there exists a subset $S \subseteq V(G)$ such that $G[S]$ is an independent set and $|V_i \cap S| = 1$ for all $i \in [k]$. We construct the graphical instance as follows. All vertices in $V(G)$ correspond to agents and all edges in $E(G)$ to items. Every agent $v \in V(G)$ values its incident edges at 1. That is, all edges in G are symmetric with a weight of 1. For every vertex partition V_i, we add a vertex-agent w_i adjacent to all the vertices in V_i. Every edge $\{(w_i, v) : v \in V_i\}$ is a symmetric edge such that w_i and v value it at d, where d is the degree of any vertex in the (regular) graph G. This completes the construction. A schematic of this construction is shown in Fig. 1. We now argue the equivalence.

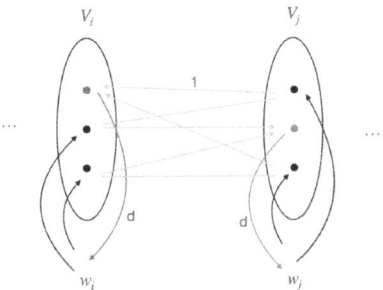

Fig. 1. A schematic of reduced instance in the proof of Theorem 3.

Forward Direction. Suppose MCIS is a Yes-instance and there is an independent set $S = \{s_1, s_2, \ldots s_k\} \subseteq V(G)$ such that $G[S]$ is an independent set and $|V_i \cap S| = 1$. Then, we do the following orientation of $E(G)$ to get an EF allocation.

- $\{(s_i, w_i) : i \in [k]\}$ are oriented towards w_i.
- $\{(v, w_i) : v \in V_i \setminus \{s_i\}\}$ are oriented towards v.
- $\{(s_i, v) : v \in N(s_i) \setminus \{w_i\}\}$ are oriented towards s_i.
- All the remaining edges are oriented arbitrarily.

Let Φ be the allocation corresponding to the above orientation. Then, $u_{w_i}(\Phi_{w_i}) = d$. Note that since all the other edge-items valued at d by w_i are allocated to distinct agents in V_i, hence $u_{w_i}(\Phi_j) \leq d$ for any $j \in V(G)$, so all agents $\{w_i : i \in [k]\}$ are envy-free. Similarly all $v \in V_i \setminus \{s_i\}$ have a utility of d each for Φ_v and a utility of at most d for any other bundle. So, all such agents are envy-free. The remaining agents $\{s_i : i \in [k]\}$ get a utility of d from the d distinct edge-items ($|N(s_i) \setminus w_i| = d$) valued at 1 each in their respective bundles. Note that all s_i also value any other bundle at atmost d and hence $\{s_i : i \in [k]\}$ are envy-free. This implies that Φ is an EF allocation.

Reverse Direction. Suppose there is an EF allocation Φ in the reduced instance. Under Φ, each of the $w_i's$ must get at least one incident edge-item to be envy-free. Otherwise, $u_{w_i}(\Phi_i) = 0$ but w_i values every bundle that ends up with any edge-item $\{(w_i, v) : v \in V(G)\}$ at d, and hence is envious. Also, since there are only $|V_i| - 1$ edge-items valued at d by all the $|V_i|$ agents in V_i, so by pigeon-holing, there is at least one agent in every partition V_i, say $s_i \in V_i$, which does not end up with a d-valued item. Since $u_{s_i}(\Phi_{w_i}) = d$, therefore, s_i must get a utility of at least d from the remaining items in order to be envy-free. This is feasible only if all the agents $\{s_i : i \in [k]\}$ get the respective d edge-items incident on them in the original graph G. This implies that $\{s_1, s_2, \ldots s_k\}$ must form an independent set in the original graph G. This settles the reverse direction. □

Given the hardness of finding EF allocation for $\{0, 1, d\}$-graphical valuations, we consider the parameterized tractability in this context. On a positive note, we show that the problem admits an FPT algorithm parameterized by the vertex cover number of the associated graph, which is the size of the smallest vertex cover (a set of vertices that includes at least one endpoint of every edge) of the graph. We will use the following classical result by Lenstra.

Theorem 4 ([21]). *An integer linear programming (ILP) instance of size L with p variables can be solved using $\mathcal{O}\left(p^{2.5p+o(p)} \cdot (L + \log M_x) \log (M_x M_c)\right)$ arithmetic operations and space polynomial in $L + \log M_x$, where M_x is an upper bound on the absolute value a variable can take in a solution, and M_c is the largest absolute value of a coefficient in the vector c.*

Theorem 5. *Given a graphical allocation instance with a bounded number of distinct utilities, the problem of finding an EF allocation admits an FPT algorithm parameterized by the Vertex Cover Number of the associated graph G.*

Proof. We formulate an ILP where the number of variables is bounded by a function of the size of the minimum vertex cover of G. We will show that the ILP is feasible if and only if there is an EF orientation in the allocation instance. Then, we invoke Theorem 4 to get a feasible solution of the ILP, if it exists, and hence, get the desired FPT algorithm parameterized by minimum vertex cover number.

Let B be the (bounded) set of distinct utilities. Let S be a minimum Vertex Cover of G and $|S| = k$. We have that $I = V(G) \setminus S$ is an independent set. We say that two vertices in I are of the 'same class' C_i if they are incident to the same subset of vertices in S. This partitions I into at most 2^k classes, corresponding to the subsets of S. That is, $I = \{C_1, C_2, \ldots C_{2^k}\}$. Further, for each class C_i, we say that two vertices have the 'same signature' σ_i if they value the subset in S in the same manner. That is, $\{v_1, v_2, \ldots v_s\} \in C_i$ have the same signature if their common neighborhood $\{s_1, s_2, \ldots s_t\} \in S$ is valued by all of them at $\{u_1, u_2, \ldots u_t\}$ such that $u_i \in B$. Since the degree of every vertex in I is at most k, this gives us at most $|B|^k$ many signatures for every class. All vertices of the same signature in a class are said to be of the same type. In aggregate, we have at most $2^k \cdot |B|^k$ many types of vertices in I.

For each vertex v in a type T, there are 2^d possible orientations of the edges incident on v, where d is the degree of v in G. Note that $d \leq k$, so there are at most 2^k such orientations. We say that an orientation is 'good' for the vertex v if it orients at least one of the highest-valued edges of v towards it. We denote the set of good orientations as O.

Towards formulating the ILP, for every type T and a good orientation o, we create the variables $x(T, o)$, which denote the number of vertices in the type T that are oriented according to the orientation o. Note that these are $f(k) = (2^k \cdot |B|^k) \cdot 2^k = 4^k \cdot |B|^k$ many variables.

We first describe the constraints to ensure the envy-freeness of the vertices in the independent set I. Let n_T be the number of vertices in the type T. Any vertex is EF if and only if it gets its highest valued edge oriented towards it (Corollary 1). Therefore, if the vertex v of type T ends up in a good orientation, it is EF. To ensure this, we add the constraints as described in Equation (1). Note that LHS of Eq. (1) equals n_T only when every vertex in the type T is oriented according to some good orientation o. Indeed, if any vertex fails to end up in a good orientation, then it is not counted in the sum and hence RHS is strictly less than n_T, which then violates the constraint.

Now consider a vertex i in S. Let $u(i, T, o)$ denote the utility that an agent $i \in S$ gets when a vertex in type T is oriented according to the orientation o. Note that for a fixed orientation, $u(i, T, o)$ is a constant. If $x(T, o)$ many vertices in type T are oriented according to o, then the utility that agent i derives from the edge items across S and I from type T is precisely $x(T, o) \cdot u(i, T, o)$. To capture the utility that i gets from edges in $E(S)$, we can do a brute-force search on which edges are allocated to i (since there are at most $\binom{k}{2}$ edges in $E(S)$). To that end, we create binary variables x_{ie} which take value 1 if the edge $e \in E(S)$ is allocated to i, otherwise 0. These are at most $g(k) = k \cdot k^2$ many variables. And, the utility that i derives from $E(S)$ is precisely $\sum_{e \in E(S)} u_i(e) x_{ie}$. Equation (2) ensures that every edge in S is allocated to at most 1 agent in S, while Eq. (3) ensures that every edge in $E(S)$ is allocated. Finally, for i to be EF, it must get at least v_i^{max} utility under any allocation. This is captured by the constraints in Eq. (4).

$$\sum_{o \in O} x(T, o) = n_T \quad \forall\, T \tag{1}$$

$$\sum_{i \in S} x_{ie} = 1 \quad \forall\, e \in E(S) \tag{2}$$

$$\sum_{i \in S} \sum_{e \in E(S)} x_{ie} = |E(S)| \tag{3}$$

$$\sum_{e \in E(S)} u_i(e) x_{ie} + \sum_{T,o} x(T, o) \cdot u(i, T, o) \geq v_i^{max} \quad \forall\, i \in S \tag{4}$$

$$x(T, o) \geq 0 \quad \forall\, T\, \&\, o \in O \tag{5}$$

$$x_{ie} \in \{0, 1\} \quad \forall\, i \in S\, \&\, e \in E(S) \tag{6}$$

In aggregate, the number of variables created is $f(k) + g(k)$. We now argue the correctness of the ILP. Let O be the orientation that corresponds to the values that $x(T, o)$ takes in some feasible solution of the ILP. For every vertex in the independent set I, Eq. (1) ensures that it ends up in a good orientation, and therefore gets one of its highest valued edges oriented towards itself under O. This ensures the envy-freeness of vertices in I. The envy-freeness of vertices in S is ensured in Eq. (4) via a brute-force search of an orientation that gives every vertex in S its highest valued edge. Therefore, a feasible solution to the ILP corresponds to an envy-free allocation of the original instance.

Conversely, suppose there is an EF allocation Φ in the original instance. Then, by Theorem 1, there is an EF orientation O. Let O_I and O_S be the restrictions of O for vertices in I and S respectively. Since O is EF, we have that both O_I and O_S are good orientations. This implies that there exist orientations under which every vertex ends up being in a good orientation. Hence, the constraints 1 and 4, which loop over all the good orientations, are satisfied when the variables $x(T, o)$ and x_{ie} correspond to the orientations O_I and O_S respectively. This implies that the ILP is feasible and this settles our claim. □

4 EFX and Welfare-Maximization

In this section, we discuss the price of EFX on graphical instances. Every graph may not admit an EFX orientation but does admit an EFX allocation, so it must be the case that some welfare is lost in the process of achieving EFX. We quantify this loss with respect to Utilitarian welfare.

Theorem 6. *For $\{0, 1\}$-graphical instances, a non-wasteful EFX allocation always exists and can be found in polynomial time. Therefore, the Price of EFX with respect to Utilitarian welfare is 1.*

Proof. Consider an instance $\mathcal{I}$ of $\{0, 1\}$-graphical allocations. For all the asymmetric edges $e = (ij)$, we orient them towards the incident agent who values e at 1, say i. We call i a special vertex, since i is now envy-free in any completion of this partial allocation and is also not envied by anyone else. Once we orient all the asymmetric edges, we remove them from the graph. The edges that are valued at 0 by both end-points can be allocated to the non-envied agents arbitrarily at the end of the algorithm, so for now, we remove them from the graph and consider a collection of connected subgraphs $H = \{H_1, H_2, \ldots H_k\}$ such that all edges in H are symmetric and valued at 1 by both the end-points. For each $H_i \in H$, we consider the following cases:

1. H_i is a Tree. Suppose there is a special agent i, then proceed as in the Case 1 of Theorem 2 and hence arrive at an EF (hence, EFX) orientation on H_i. Suppose there is no special vertex in the tree H_i. Then there is no complete EF allocation. To find an EFX allocation, we root H_i on any vertex, say the least degree vertex i, and construct an orientation such that every vertex gets an edge item from its parent. Note that i leaves empty-handed and is envious

of its neighbors in H_i. Since every envied agent (precisely, the children of i in the tree H_i) gets exactly one edge item (precisely the edge from i), the allocation is EFX.
2. H_i contains a cycle, say $C = \{v_1, v_2, \ldots v_c, v_1\}$. This case is the same as Case 2 in the proof of Theorem 2 and therefore a non-wasteful EF (hence EFX allocation exists) such that every agent receives a utility of at least 1.

Therefore, we get a non-wasteful EFX allocation Φ. For $\{0, 1\}$-valuations, a non-wasteful allocation is also utilitarian optimal, and hence Φ is also utilitarian optimal. Therefore, the price of EFX is 1 in this case. □

Theorem 7. *The price of EFX with respect to Utilitarian welfare is ∞ even for $\{0, 1, d\}$-graphical valuations.*

Proof. We construct an instance where the price of fairness is a function of the highest degree of a vertex in the graph. Consider a star graph G rooted at the vertex r which is incident to d many leaf vertices. The root vertex r values each of the d incident edges at d. All the leaf vertices value their incident edge at 1. A utilitarian welfare maximizing allocation gives all the edges to r, generating a welfare of d^2. Clearly, this allocation is not EFX since the envied agent r has multiple items and every leaf agent violates EFX. Under any EFX allocation, r can not receive more than 1 item, otherwise, the corresponding leaf vertex whose incident edge is allocated to r, violates EFX. Therefore, the maximum welfare under an EFX allocation is $d + (d-1)$, where one d-valued edge is allocated to r and the rest all $(d-1)$ edges are allocated to their corresponding leaf vertices, valued at 1 by each of them. Therefore, $PoF_{UM} = \frac{d^2}{d+(d-1)} > \frac{d^2}{2d} \approx d$. This implies that welfare loss can be as high as possible, and hence PoF is ∞. □

Theorem 8. *Given an instance of graphical valuations, deciding the existence of a utilitarian welfare-maximizing and EFX allocation (UM+EFX) is NP-Hard.*

Proof. We present a reduction from MULTI-COLORED INDEPENDENT SET (MCIS), where given a regular graph $G = (V_1 \uplus \cdots \uplus V_k, E)$ with degree d, the problem is to decide if there exists a subset $S \subseteq V(G)$ such that $G[S]$ is an independent set and $|V_i \cap S| = 1$ for all $i \in [k]$. We construct the graphical instance as follows. All vertices in $V(G)$ correspond to agents and all edges in $E(G)$ to items. Every agent $v \in V(G)$ values its incident edges at 1. That is, all edges in G are symmetric with a weight of 1. For every vertex partition V_i, we add a path of three edges and four vertex-agents $\{w_i^1, w_i^2, w_i^3, w_i^4\}$ such that w_2^i is adjacent to all the vertices in V_i. All edges from w_i^2 to V_i are valued symmetrically at d by both end-points. The edge (w_i^1, w_i^2) is valued at 0 by w_i^1 and at 1 by w_i^2. The edge (w_i^2, w_i^3) is valued at d by both w_i^2 and w_i^3. Finally, the edge (w_i^3, w_i^4) is valued at d by w_i^3 and at 0 by w_i^4. This completes the construction. A schematic of this construction is shown in the appendix. We now argue the equivalence.

Forward Direction. Suppose MCIS is a Yes-instance and there is an independent set $S = \{s_1, s_2, \ldots s_k\} \subseteq V(G)$ such that $|V_i \cap S| = 1$. Then, we do the following orientation of $E(G)$ to get an allocation that is welfare-maximizing and EFX.

- $\{(s_i, w_i^2)\}$ are oriented towards w_i $\forall\, i \in [k]$.
- $\{(v, w_i) : v \in V_i \setminus \{s_i\}\}$ are oriented towards v $\forall\, i \in [k]$.
- $\{(s_i, v) : v \in N(s_i) \setminus \{w_i\}\}$ are oriented towards s_i $\forall\, i \in [k]$.
- $\{w_i^1, w_i^2\}$ are oriented towards w_i^2 $\forall\, i \in [k]$.
- $\{w_i^2, w_i^3\}$ & $\{w_i^3, w_i^4\}$ are oriented towards w_i^3 $\forall\, i \in [k]$.

Let Φ be the allocation corresponding to the above orientation. Then, by construction, every edge is allocated to an agent who values it the most. Therefore, Φ is a (utilitarian) welfare-maximizing allocation. We now argue that Φ also satisfies EFX. The agents w_i^1 and w_i^4 do not value any item, so even though they are empty-handed under Φ, they do not envy any agent. All the agents in V_i except s_i get a utility of d each and they value every other bundle at most d, hence are envy-free. Likewise, w_i^2 is envy-free as it gets a utility of $d + 1$ and values every other bundle no more than d. Also, w_i^3 gets all the edges it values, so there is no envy on its part. Lastly, each s_i gets d of its incident edges valued at 1 each, deriving a value of d, and hence they are envy-free. This implies that Φ is EF and hence, EFX.

Reverse Direction. Suppose there is a welfare-maximizing allocation Φ which also satisfies EFX. Then because Φ maximizes welfare, it must satisfy the following partial allocation: w_i^3 must receive both its incident edges $\{w_i^2, w_i^3\}$ & $\{w_i^3, w_i^4\}$ as it values them highly at d, and $\{w_i^1, w_i^2\}$ must be allocated to w_i^2 as a utilitarian welfare maximizing allocation is also non-wasteful. This forces w_i^2 to be envious of w_i^3 even after one item is removed from the envied bundle. Therefore, w_i^2 must receive at least one item that it values at d incident to the partition V_i. This in turn forces at least one vertex from V_i to violate EFX with respect to w_i^2, hence it must receive at least d utility from the remaining items. This is feasible only when it is allocated all its d incident edges. Since this is true for at least one vertex in all V_i such that $i \in [k]$, it must be the case that all these k vertices form an independent set in G. This implies that MCIS is a yes-instance. This concludes the argument. □

We now present a polynomial-time reduction from deciding the existence of a welfare-maximizing and EFX allocation (UM+EFX) to finding a welfare-maximizing allocation within the set of EFX allocations (UM/EFX). Let w^* be the maximum utilitarian welfare (w^* can be computed in linear time by giving each item to an agent who values it the most). Now suppose the latter problem can be solved in polynomial time. Then, let w be the maximum welfare within EFX allocations. If $w = w^*$, we have a "yes" instance of UM+EFX; else if $w \neq w^*$, we have a "no" instance. Therefore, we get the following result.

Corollary 3. *Given an instance of graphical valuations, finding a utilitarian welfare maximizing allocation within the set of EFX allocations (UM/EFX) is NP-Hard.*

5 Concluding Remarks

We studied the complexity of finding envy-free allocations for graphical valuation and quantified the loss of welfare in the process of achieving approximate (i.e., EFX) envy-freeness, which was the original motivation for the study of the class of graphical valuations. We believe there are several directions of interest for future work that build on our preliminary line of inquiry here. For instance, for parameterized results, one could consider structural parameters that are smaller than vertex cover. Extending the PoF discussion beyond binary setting for other welfare notions would also be of interest. Finally, one might also generalize the class of graphical valuations in many ways. One generalization is to allow graphs with multiedges which then corresponds to instances where an item is liked by at most two agents but a pair of agents together can derive positive value from more than one item. The restricted setting of hypergraphs where every edge corresponds to multiple but the same number of vertices is also an interesting direction to pursue.

References

1. Akrami, H., Alon, N., Chaudhury, B.R., Garg, J., Mehlhorn, K., Mehta, R.: EFX: a simpler approach and an (almost) optimal guarantee via rainbow cycle number. In: Proceedings of the 24th ACM Conference on Economics and Computation, EC 2023, p. 61. Association for Computing Machinery, New York (2023)
2. Aumann, Y., Dombb, Y.: The efficiency of fair division with connected pieces. ACM Trans. Econ. Comput. **3**(4) (2015)
3. Aziz, H., Gaspers, S., Mackenzie, S., Walsh, T.: Fair assignment of indivisible objects under ordinal preferences. Artif. Intell. **227**, 71–92 (2015)
4. Bei, X., Lu, X., Manurangsi, P., Suksompong, W.: The Price of Fairness for Indivisible Goods. Theory Comput. Syst. **65**(7), 1069–1093 (2021)
5. Berger, B., Cohen, A., Feldman, M., Fiat, A.: Almost full EFX exists for four agents. In: Proceedings of the AAAI Conference on Artificial Intelligence, vol. 36, no. 5, pp. 4826–4833 (2022)
6. Bertsimas, D., Farias, V.F., Trichakis, N.: The Price of Fairness. Oper. Res. **59**(1), 17–31 (2011)
7. Bhaskar, U., Misra, N., Sethia, A., Vaish, R.: The price of equity with binary valuations and few agent types. In: Deligkas, A., Filos-Ratsikas, A. (eds.) SAGT 2023. LNCS, vol. 14238, pp. 271–289. Springer, Cham (2023). https://doi.org/10.1007/978-3-031-43254-5_16
8. Bilò, V., et al.: Almost envy-free allocations with connected bundles. Games Econom. Behav. **131**, 197–221 (2022)
9. Bu, X., Song, J., Yu, Z.: EFX allocations exist for binary valuations. In: Li, M., Sun, X., Wu, X. (eds.) IJTCS-FAW 2023. LNCS, vol. 13933, pp. 252–262. Springer, Cham (2023). https://doi.org/10.1007/978-3-031-39344-0_19
10. Budish, E.: The combinatorial assignment problem: approximate competitive equilibrium from equal incomes. J. Polit. Econ. **119**(6), 1061–1103 (2011)
11. Caragiannis, I., Gravin, N., Huang, X.: Envy-freeness up to any item with high nash welfare: the virtue of donating items. In: Proceedings of the 2019 ACM Conference on Economics and Computation, pp. 527–545 (2019)

12. Caragiannis, I., Kaklamanis, C., Kanellopoulos, P., Kyropoulou, M.: The efficiency of fair division. Theory Comput. Syst. **50**(4), 589–610 (2012)
13. Chaudhury, B.R., Garg, J., Mehlhorn, K.: EFX exists for three agents. In: Proceedings of the 21st ACM Conference on Economics and Computation, EC 2020, pp. 1–19. Association for Computing Machinery, New York (2020)
14. Christodoulou, G., Fiat, A., Koutsoupias, E., Sgouritsa, A.: Fair allocation in graphs. In: Proceedings of the 24th ACM Conference on Economics and Computation, EC 2023, pp. 473–488. Association for Computing Machinery, New York (2023)
15. Deligkas, A., Eiben, E., Ganian, R., Hamm, T., Ordyniak, S.: The parameterized complexity of connected fair division. In: IJCAI, pp. 139–145 (2021)
16. Foley, D.: Resource allocation and the public sector. Yale Econ. Essays 45–98 (1967)
17. Freeman, R., Sikdar, S., Vaish, R., Xia, L.: Equitable allocations of indivisible goods. In: Proceedings of the 28th International Joint Conference on Artificial Intelligence, pp. 280–286 (2019)
18. Gamow, G., Stern, M.: Puzzle-Math (1958)
19. Garg, J., Taki, S.: An improved approximation algorithm for maximin shares. Artif. Intell. **300**, 103547 (2021)
20. Gourvès, L., Monnot, J., Tlilane, L.: Near fairness in matroids. In: Proceedings of the 21st European Conference on Artificial Intelligence, pp. 393–398 (2014)
21. Lenstra, H.W., Jr.: Integer programming with a fixed number of variables. Math. Oper. Res. **8**(4), 538–548 (1983)
22. Lipton, R.J., Markakis, E., Mossel, E., Saberi, A.: On approximately fair allocations of indivisible goods. In: Proceedings of the 5th ACM Conference on Electronic Commerce, EC 2004, pp. 125–131. Association for Computing Machinery, New York (2004)
23. Lipton, R.J., Markakis, E., Mossel, E., Saberi, A.: On approximately fair allocations of indivisible goods. In: Proceedings of the 5th ACM Conference on Electronic Commerce, pp. 125–131 (2004)
24. Payan, J., Sengupta, R., Viswanathan, V.: Relaxations of envy-freeness over graphs. In: Proceedings of the 2023 International Conference on Autonomous Agents and Multiagent Systems,AAMAS 2023, pp. 2652–2654. International Foundation for Autonomous Agents and Multiagent Systems, Richland (2023)
25. Plaut, B., Roughgarden, T.: Almost envy-freeness with general valuations. SIAM J. Discrete Math. **34**(2), 1039–1068 (2020)
26. Steinhaus, H.: The problem of fair division. Econometrica **16**(1), 101–104 (1948)
27. Sun, A., Chen, B., Doan, X.V.: Equitability and welfare maximization for allocating indivisible items. Auton. Agents Multi-Agent Syst. **37**(8) (2023)
28. Sun, A., Chen, B., Vinh Doan, X.: Connections between fairness criteria and efficiency for allocating indivisible chores. In: Proceedings of the 20th International Conference on Autonomous Agents and MultiAgent Systems, pp. 1281–1289 (2021)
29. Zeng, J.A., Mehta, R.: On the structure of envy-free orientations on graphs (2024)

Manipulation With(out) Money in Matching Market

Sushmita Gupta[1](✉) and Pallavi Jain[2]

[1] The Institute of Mathematical Sciences, HBNI, Chennai, India
sushmitagupta@imsc.res.in
[2] Indian Institute of Technology Jodhpur, Jodhpur, India
pallavi@iitj.ac.in

Abstract. The issue of manipulation in the stable marriage game is well-known and have been studied for many decades. The question of weighted manipulation where each manipulative action is charged individually and the question is to decide if a favorable outcome can be attained within a pre-specified budget has only recently been considered by Boehmer et al. [SAGT'20]. They considered several manipulative actions with uniform cost. In this paper, we generalise that model and consider arbitrary cost functions. Moreover, we study an additional question where given a manipulative action and an agent, the goal is to match the agent under a stable matching with the cost not exceeding the budget. We formally address some questions raised by Boehmer et al. and in the process show that in this extended model, all problems under consideration are intractable and even exhibit parameterized hardness. The most intriguing aspect of the analysis is that we are able to identify a common underlying structural property that makes each of the problems hard despite the fact that the manipulative action undertaken and/or the desired outcomes are very different from each other. Moreover, Boehmer et al.'s work revealed several dichotomies–be it classical or parameterized–and therefore it is apriori not obvious why in the weighted setting all problems must be $W[1]$-hard with respect to the combined parameter of budget and the number of unmatched vertices in a stable matching before manipulation. We discuss our analysis by way of presenting a metagadget that is at the heart of each hardness result, and show how to enrich it in different ways to yield hardness result for each of the individual problems.

Keywords: stable matching · manipulation · bribery · control

1 Introduction

Matching under preferences is an extensively studied area of theoretical and empirical research that has a wide range of applications in economics and social sciences. The most popular and standard problem in this field is the *stable marriage problem* (SMP), introduced by Gale and Shapley [15], where we have two parties; a set of men and a set of women. Each man has a preference list that orders women according to his preference, and similarly each woman also has a preference ordering on the set of men. The goal is to find a *stable marriage*: a set of disjoint man-woman pairs, viewed as matching partners that do not contain a *blocking pair*, defined to be a man and a woman who prefer each other to their current matching partners. The idea being that existence of blocking pairs pose a threat to the "stability of the marriage". Among the many applications of the stable marriage problem, college admission and matching medical residents to hospitals are perhaps the best known ones.

Manipulation by individual agents in stable matching has a long history; see [30] for theoretical context and discussions. Specifically, *truncation* and *permutation* strategies have been studied quite extensively for many decades, [12,29,31]. The more general phenomenon of strategic behavior extends beyond that of the individual player to the central agency/organizers who may want to manipulate the outcome to the benefit or detriment of certain individuals by adding or deleting agents. These mechanisms have been studied quite extensively in the context of voting and election, where the former is known as *constructive control* and the latter as *destructive control* [2,13].

In the context of the college admissions problem, there have been reports from college admission systems in China, Bulgaria, Moldova, and Serbia, that bribery has been employed to gain desirable admissions [19,20]. Thus, in addition to studying the feasibility of a manipulative action, it is natural to study the "cost" associated with it and if one can attain the stated goal within a prespecified budget. This article is a computational study of that question: *Given a particular manipulative action, and a stated goal, what is the cost of attaining that goal via that manipulative action? And more specifically, can it be attained within the allowable budget?*

This particular question was considered by Boehmer et al. [3] in the presence of unit and uniform cost functions, i.e., each manipulative action costs one irrespective of the agent and the manipulative action. We extend their model by studying cost functions that may vary with individuals and the manipulative action. More formally, we consider the possible manipulative actions and their associated costs[1].

---

[1] We will refer to an agent with the female pronoun in order to be consistent with literature which has studied manipulation from the women's side when using the man-proposing Gale Shapley algorithm.

ACTIONS

REORDER: Agent(s) can be bribed to reorder their preferences arbitrarily.

SWAP: Agent(s) can be bribed for swapping a pair of agents who occupy consecutive positions in their preference list.

ADD: A pre-specified subset of agents are "passive" in the sense that even though they are present in the preference list of other agents they are not considered while computing the matching. The passive agents can be bribed to participate in the matching market (algorithm).

DELETE: Agent(s) can be bribed to leave the market, i.e., be removed from every individual's preference list to which she currently belongs.

DELETE ACCEPTABILITY: A pair of agents can be bribed to delete each other from their respective preference lists.

The goals is to change the SMP instance using a manipulative action within the given budget to attain a certain target as listed below.

MANIPULATION GOALS

CONSTRUCTIVE-PARTNER EXISTS: manipulate so that a designated agent is **saturated** (i.e., matched) in a stable matching of the manipulated instance.

DESTRUCTIVE-PARTNER EXISTS: manipulate so that a designated agent is **unsaturated** in a stable matching of the manipulated instance.

CONSTRUCTIVE-EXISTS: manipulate so that a designated man-woman pair **belongs to** any stable matching of the manipulated instance.

DESTRUCTIVE-EXISTS: manipulate so that a designated man-woman pair **does not belong to** any stable matching of the manipulated instance.

EXACT UNIQUE: manipulate so that a given matching is the **unique** stable matching of the manipulated instance.

For the $\mathcal{Y}$ manipulation action and $\mathcal{X}$ manipulation goal, we refer to the corresponding problem as $\mathcal{Y}$-$\mathcal{X}$. The goal of the $\mathcal{Y}$-$\mathcal{X}$ problem is to decide if it is possible to manipulate the original instance using the manipulative action $\mathcal{Y}$ to attain the goal $\mathcal{X}$ within the budget. Depending on the manipulation action and the goal, the input of the instance varies. The classical SMP takes as an instance, a bipartite graph $G = (M \cup W, E)$, where M and W denote the set of vertices representing the agents on the two sides and E denotes the set of edges representing acceptable matches between vertices on different sides, and a preference list $\mathcal{L}$ of every vertex in G over its neighbors. The input instance of the $\mathcal{Y}$-$\mathcal{X}$ problem is as follows.

> **INPUT**
>
> For action $\mathcal{Y} \in \{\text{REORDER}, \text{SWAP}, \text{ADD}, \text{DELETE}, \text{DELETE ACCEPTABILITY}\}$ the input consists of $I_\mathcal{Y}$, a designated agent s for goal $\mathcal{X} \in \{\text{CONSTRUCTIVE-PARTNER EXISTS}, \text{DESTRUCTIVE-PARTNER EXISTS}\}$ (or a designated edge for goal $\mathcal{X} \in \{\text{CONSTRUCTIVE-EXISTS}, \text{DESTRUCTIVE-EXISTS}\}$ or a matching for EXACT UNIQUE), and a budget $b \in \mathbb{N}$. Specifically,
>
> - for $\mathcal{Y} \in \{\text{REORDER}, \text{ADD}, \text{DELETE}\}$, $I_\mathcal{Y} = (G = (V, E), \mathcal{L}, c : V \to \mathbb{N})$;
> - $I_{\text{SWAP}} = (G = (V, E), \mathcal{L}, \{c_v : V \times V \to \mathbb{N}\}_{v \in V})$;
> - $I_{\text{DELETE ACCEPTABILITY}} = (G = (V, E), \mathcal{L}, c : V \times V \to \mathbb{N})$;
>
> where $\mathcal{L}$ is the set of preference lists, c is the cost function.

Our Contributions. In this paper, we carry forward the future research directions proposed by Boehmer et al. [3]. In particular, we consider weighted manipulation and destructive manipulation. Additionally, we propose a new manipulation goal, CONSTRUCTIVE-PARTNER EXISTS, in which the objective is to manipulate the given STABLE MARRIAGE instance using manipulative action of some given type such that a designated agent is matched in the new instance.

- We show that CONSTRUCTIVE-PARTNER EXISTS is NP-hard and in fact is also W[1]-hard, for each of the manipulation actions, reorder, swap, add, delete, and delete acceptability, with respect to the combined parameter $b + n_{\text{unmatch}}$, where b denotes the budget and n_{unmatch} denotes the number of unmatched agents in the stable matching of the original instance.
- Boehmer et al. [3] showed that CONSTRUCTIVE-EXISTS is NP-hard and W[1]-hard with respect to the budget, for the manipulative actions reorder, swap, add, and delete acceptability, while it is polynomial time solvable for the delete, when the cost of deleting every agent is unit. They specified that their polynomial algorithm does not work for the weighted manipulation [3]. We show that in fact, CONSTRUCTIVE-EXISTS is NP-hard and W[1]-hard with respect to the parameter $b + n_{\text{unmatch}}$ for the manipulative actions reorder, swap, add, and delete acceptability.
- We further show that, in polynomial time, CONSTRUCTIVE-PARTNER EXISTS can be reduced to DESTRUCTIVE-PARTNER EXISTS for the manipulation actions reorder, swap, add, delete, and delete acceptability. Thus, DESTRUCTIVE-PARTNER EXISTS and DESTRUCTIVE-EXISTS are also NP-hard and W[1]-hard with respect to $b + n_{\text{unmatch}}$, for all the considered actions.
- Using rotation, the fundamental object behind the mathematics of stable matching lattice, and rotation elimination properties we are able to design a polynomial time algorithm for SWAP-EXACT UNIQUE for preference list lengths of size at most two. When size is at least four, we show that the problem remains NP-hard, via a reduction from VERTEX COVER. We further design an XP algorithm with respect to b for SWAP-EXACT UNIQUE, however, an open question regarding FPT with respect to b for this problem still remains open.

The hardness results for CONSTRUCTIVE-PARTNER EXISTS and CONSTRUCTIVE-EXISTS fall under a common theme that underlines all the seemingly different manipulative actions and varied outcomes. To highlight the structural similarity among the problems we will present the hardness proofs, reductions from MULTICOLORED CLIQUE piece by piece. Specifically, the presentation of the gadget is broken into layers at the foundation of which is the *metagadget*.

Next, we describe and explain the intuition behind its design. Following that in Fig. 2 we will describe schematic idea of how we add specific components to extend that construction towards a gadget for each of the problems under consideration.

Our Methodology. To show the intractability result of $\mathcal{Y}$-CONSTRUCTIVE-PARTNER EXISTS, where $\mathcal{Y} \in \{$REORDER, SWAP, ADD, DELETE, DELETE ACCEPTABILITY$\}$, we give a poly-time reduction from MULTICOLORED CLIQUE, in which we are given a graph $G = (V, E)$ and a partition of V into k parts, $V_1, \ldots, V_k$, and the objective is to decide if there exists a subset $S \subseteq V$ such that $|S \cap V_i| = 1$, for each $i \in [k]$, and the induced subgraph $G[S]$ is a complete graph. The MULTICOLORED CLIQUE problem is known to be NP-hard and W[1]-hard with respect to the solution size [11,16]. We first present a metagadget, specifically for CONSTRUCTIVE-PARTNER EXISTS, and then we will add the gadget on the top of it to show the intractability for each manipulation action. For the hardness of CONSTRUCTIVE-EXISTS, instead of designated vertex d_1, we have designated edge $d_1 d_2$.

Metagadget: Let (G, k) be an instance of MULTICOLORED CLIQUE. For any $\{i, j\} \subseteq [k]$, $i < j$, let E_{ij} denote the set of edges between sets V_i and V_j. We construct a graph G' as follows. We call E_{ij} as an *edge class*. Figure 1 describes this metagadget.

- For each vertex $u \in V(G)$, we have four vertices in G', denoted by $\{u_i \colon i \in [4]\}$ that are connected via a path: (u_1, u_2, u_3, u_4).
- For each edge $e \in E(G)$, we add vertices e and $\tilde{e}$ to $V(G')$ and the edge $e\tilde{e}$.
- For each $\{i, j\} \subseteq [k]$, $i < j$, we add a vertex q_{ij} to G'. Additionally, for each edge $e(= uv) \in E_{ij}$, we add three edges $q_{ij}e$, eu_3, and ev_3 to G'.
- We add four *dummy vertices* to G', denoted by $\{d_i \colon i \in [4]\}$ that are connected via a path: (d_1, d_2, d_3, d_4). For each $\{i, j\} \subseteq [k]$, $i < j$, we add the edge $d_4 q_{ij}$ to G' as well.
- The designated vertex is d_1.

Next, we describe the preference list of every vertex in G'. For every vertex $u \in V(G)$, we define $\mathcal{E}_u = \{e \in V(G') \colon e(= uv) \in E(G)\}$, the set of vertices in G' that corresponds to an edge incident to u in G. For every $\{i, j\} \subseteq [k]$, $i < j$, we define $\mathcal{E}_{q_{ij}} = \{e \in V(G') \colon e \in E_{ij}\}$, the set of vertices corresponding to the edges in the edge class E_{ij}. Further, we define $\mathcal{Q}_{d_4} = \{q_{ij} \in V(G') \colon \{i, j\} \subseteq [k], i < j\}$, the set of neighbors of d_4 except d_3. Table 1 presents the preference lists.

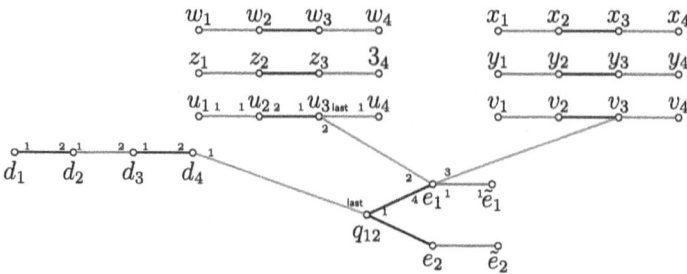

Fig. 1. An illustration of graph G' in metagadget. Here, $k = 2$ for an instance of MULTICOLORED CLIQUE, and d_1 is the designated vertex to whom we want to match after manipulation.

Table 1. Preference lists in the metagadget, where $[\cdot]$ represents an arbitrary order.

For each $i \in [k]$ and each $u \in V_i$, we have the following preference lists:

u_1: $\langle u_2 \rangle$
u_2: $\langle u_1, u_3 \rangle$
u_3: $\langle u_2, [\mathscr{E}_u], u_4 \rangle$
u_4: $\langle u_3 \rangle$

For each edge $e \in E_{ij}$ with endpoints $u \in V_i$ and $v \in V_j$, where $1 \leq i < j \leq k$, we have the following preference lists:

e: $\langle \tilde{e}, u_3, v_3, q_{ij} \rangle$
$\tilde{e}$: $\langle e \rangle$

For each $\{i, j\} \subseteq [k]$, $i < j$, we have the following preference lists:

q_{ij}: $\langle [\mathscr{E}_{q_{ij}}], d_4 \rangle$

For the dummy vertices, we have the following preference lists:

d_1: $\langle d_2 \rangle$
d_2: $\langle d_3, d_1 \rangle$
d_3: $\langle d_4, d_2 \rangle$
d_4: $\langle \mathscr{Q}_{d_4}, d_3 \rangle$

Intuition: We discuss here the idea behind our gadget. Since d_1 is a designated vertex whom we wish to match, we would like to manipulate the original instance such that $d_1 d_2$ belongs to some stable matching. To avoid the blocking edge $d_2 d_3$, d_3 needs to be matched with d_4. Similarly, to avoid the blocking edges between d_4 and q_{ij}s in $\mathscr{Q}_{d_4}$, every vertex in $\mathscr{Q}_{d_4}$, i.e., q_{ij}s, needs to be matched to some vertex in $\mathscr{E}_{q_{ij}}$, i.e., some vertex $e \in V(G')$ corresponding to an edge $e \in E_{ij}$. We trigger this effect by appropriately setting the cost functions and budget such that manipulating the preference list of any dummy vertex and any vertex in $\mathscr{E}_{q_{ij}}$, where $1 \leq i < j \leq k$, is not possible within the budget. Note that in the original instance $e\tilde{e}$ belongs to every stable matching. However, we manipulate the original instance, according to the manipulative action, so that $q_{ij} e$ is part of the stable matching. In the MULTICOLORED CLIQUE solution, we will take those edges whose corresponding vertex in G' is matched to some q_{ij}. Note that the clique solution must also contain endpoint of these edges. So, we want to trigger this effect of manipulation to the gadget corresponding to vertices, which we call the *vertex gadget*. Note that edges of type u_3e in G', i.e., the edge between u_3 and the vertex in G' corresponding to an edge incident to u in G, acts as a "bridge" between the vertices in G' (corresponding to vertices and edges in G). We encode the property that if we are breaking the stable matching edge $e\tilde{e}$ by manipulating the instance, we also break the stable matching edge u_3u_4. Thus, we ensure that we do not change the preference list of e so that eu_3 is a blocking edge if u_3u_4 is a matching edge and thereby trigger the manipulation action to the vertex gadget.

For each manipulative action, we will extend this gadget by adding more vertices, and set the cost functions appropriately. Figure 2 describes the overview of extending the metagadget for each manipulative action.

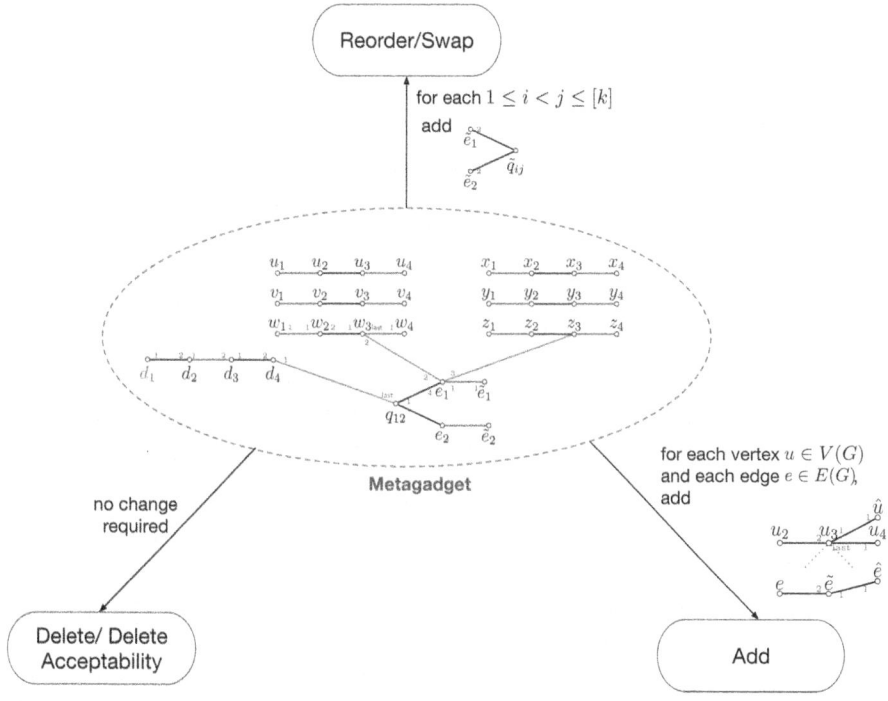

Fig. 2. An illustration of extending metagadget to various manipulation action for CONSTRUCTIVE-PARTNER EXISTS.

Related Work. In addition to the works on stable marriage and its strategic questions we have mentioned in the Introduction, it is notable that the approach and problems studied in this paper are closely aligned with the study of manipulation, bribery and control in voting literature [13]. Among the most common forms of manipulation are the addition and/or deletion of candidates, and the addition and/or deletion of voters, hence they deserve special attention. These forms of manipulation give rise to two well-studied families of computational problems, termed *constructive control* and *destructive control*. In the former, the objective of the manipulator is to ensure that a distinguished candidate wins the election; in the latter a particular candidate does not win the election.

For over three decades, since the seminal work of Bartholdi et al. [1,2], computational problems of constructive control have been extensively studied and are nowadays relatively well understood. While initially, the computational problems (of adding/deleting candidates/voters) were defined for a single distinguished candidate [1,2], the objective was soon generalized to settings that capture the case of a set of distinguished candidates [23–26]. Having a shorter history, but introduced more than a decade ago by Conitzer et al. [6] and Hemaspaandra et al. [18], are the computational problems of destructive control.

On the topic of strategic questions studied in the realm of the stable marriage problem, numerous work has appeared over the decades, a detailed accounting can be found in [22, Chapter 2.9]. Our work can be seen as a way to modify the input profile such that the set of stable marriages change as a result of the manipulative action, recent work on *robust stable matching* [5,21] can be viewed as finding a stable marriage where the manipulator cannot make it unstable by manipulation. But more directly, CONSTRUCTIVE-PARTNER EXISTS can be seen as manipulation towards making the given vertex a part of a *stable pair* (a pair that appears in some stable marriage) and CONSTRUCTIVE-EXISTS as manipulation towards making the given edge a stable pair.

Cseh and Hegger [7] studied the complexity of forcing an edge to a stable solution, however, they did not consider manipulation actions. Bredereck et al. [4] studied the complexity of change in stable matching if the profile is changed.

Overview of the Paper. In light of the space constraint and the presentation of the metagadget, we have chosen to focus the short version of the paper to one hardness result and the exposition of the results for SWAP-EXACT UNIQUE: has a polytime algorithm when preference lists are of length two and an XP algorithm with respect to b in general. We will present the construction and proof details of all the other hardness results in an extended journal version which will be made available in arXiv.

Preliminaries. In the preference list of a vertex u, if v appears before w, then we say that u prefers v more than w. We call an edge in the graph as a *fixed edge* if it belongs to every stable matching of the graph. We will be using terminologies on graph theory and parameterized algorithms from [9] and [8,10,14], respectively.

2 Reorder-Constructive-Partner Exists

In this section, we show that REORDER-CONSTRUCTIVE-PARTNER EXISTS is NP-hard and W[1]-hard with respect to $b + n_{\mathsf{unmatch}}$, where b is the budget and n_{unmatch} is the number of unmatched agents in the stable matching in the original instance. In particular, we prove the following result.

Theorem 1. REORDER-CONSTRUCTIVE-PARTNER EXISTS *is* NP-*hard and* W*[1]-hard when parameterized by* $b + n_{\text{unmatch}}$.

To prove Theorem 1, we give a reduction from MULTICOLORED CLIQUE.

Construction. Let (G, k) be an instance of MULTICOLORED CLIQUE. We first construct the graph G' using the metagadget (Fig. 1). Next, we do as follows.

- For each $\{i, j\} \subseteq k$, $i < j$, we add a vertex $\tilde{q}_{ij}$ to $V(G')$.
- For each $e \in E_{ij}$, we add the edge $\tilde{e}\tilde{q}_{ij}$ to $E(G')$.

Table 2 presents the preference list of every vertex in G'. Note that the stable matching of G' contains the following edges: $u_1 u_2$, $u_3 u_4$, for every $u \in V(G)$, $e\tilde{e}$, for every $e \in E(G)$, $d_2 d_3$, and $d_4 q_{ij}$, for some $1 \leq i < j \leq k$, where q_{ij} is the most preferred vertex of d_4. Thus, the designated vertex d_1 is unsaturated in any stable matching of G' due to Rural Hospital theorem [27,28].

Table 2. Preference lists for REORDER-CONSTRUCTIVE-PARTNER EXISTS.

For each $\{i, j\} \subseteq [k]$, $i < j$, we have the following preference list of $\tilde{q}_{ij}$:

$\tilde{q}_{ij}$: $\langle [N(\tilde{q}_{ij})] \rangle$

For each $\{i, j\} \subseteq [k]$, $i < j$, $e \in E_{ij}$, we have the following preference list of the corresponding vertex $\tilde{e}$:

$\tilde{e}$: $\langle e, q_{ij} \rangle$

For all the remaining vertices $w \in V(G')$, the preference list is same as in Table 1.

Next, we define the cost function $c \colon V(G) \to \mathbb{N}$. For each $e \in E_{ij}$, $c(\tilde{e}) = 1$ and for each $u \in V(G)$, $c(u_2) = 1$. For the remaining vertices $w \in V(G')$, $c(w) = k + \binom{k}{2} + 1$. We set the budget $b = k + \binom{k}{2}$. Clearly, we can only reorder the preference list of $\tilde{e}$, where $e \in E(G)$, and u_2, where $u \in V(G)$, as the cost of reordering any other vertex exceeds the budget.

Claim 1. G' *is a bipartite graph.*

Intuition: The idea is that since we need to match d_1 to d_2 (as d_2 is the unique neighbor of d_1 in G'), the most preferred vertex of d_2, i.e., d_3, also needs to be matched with d_4. But, d_4 prefers all the vertices in $\mathcal{Q}_{d_4}$ more than d_3 and the preference list of d_4 cannot be reordered within the budget. So, all q_{ij}s needs to be matched to some vertex in $\mathcal{E}_{q_{ij}}$ and this triggers the effect that for some $e \in E_{ij}$, the corresponding vertex $\tilde{e}$ reorder the preference list. Since $q_{ij}e$ is a potential matching edge for some vertex $e \in V(G')$ corresponding to the edge $e(= uv) \in E_{ij}$, to avoid blocking edge incident to u_3, v_3, and e, u_3 and v_3 needs to matched with u_2 and v_2, respectively (later on we will formally argue that u_3 or v_3 cannot be matched to any vertex in $\mathcal{E}_{u_3}$ or $\mathcal{E}_{v_3}$, respectively). Thus, when we match e with q_{ij}, it triggers the effect that u_2 and v_2 also change their preference list. Thus, corresponds to a clique in G.

Next, we formally prove the equivalence between the instance (G, k) of MULTICOLORED CLIQUE and the instance $(G', \mathcal{L}, d_1, c, b)$ of REORDER-CONSTRUCTIVE-PARTNER EXISTS in the following lemma.

Lemma 1. (G, k) *is a yes-instance of* MULTICOLORED CLIQUE *if and only if* $(G', \mathcal{L}, d_1, c, b)$ *is a yes-instance of* REORDER-CONSTRUCTIVE-PARTNER EXISTS.

Proof (Theorem 1). Given an instance (G, k) of MULTICOLORED CLIQUE, we first construct an instance of REORDER-CONSTRUCTIVE-PARTNER EXISTS using Construction on page 2. Note that the stable matching of G' under $\mathcal{L}$ is

$$\eta = \{u_1 u_2, u_3 u_4 \colon u \in V(G)\} \cup \{e\tilde{e} \colon e \in E(G)\} \cup \{d_2 d_3\} \cup$$
$$\{d_4 q_{ij} \colon \{i, j\} \subseteq [k], i < j, q_{ij} \text{ is most preferred vertex of } d_4\}$$

Therefore, there are only $2\binom{k}{2}$ unsaturated vertex in η, i.e., $n_{\text{unmatch}} = 2\binom{k}{2}$. Furthermore, since $b = k + \binom{k}{2}$, due to Lemma 1, the theorem follows. □

3 Swap-Exact Unique

Let $\mathcal{I} = (G, \mathcal{L}, (c_v)_{v \in V}, \mu, b)$ denote an instance of SWAP-EXACT UNIQUE, where the vertex set of G is $A \cup B$, and $\mathcal{L}$ is the set of preference lists, called the *preference profile*. For an agent $v \in V(G)$, let p_v denote a permutation of the set of neighbors of v in $\mathcal{I}$. We will use $\mathcal{L} \oplus p_v$ to denote the preference profile obtained by changing the preference list of v to p_v. For a set S of preference lists of agents $V' \subseteq V$, we use $\mathcal{L} \oplus S$ to define the preference profile obtained by replacing the preference lists of the agents in V' with those in S.

Rotations: For a stable matching μ in $\mathcal{L}$, and a vertex $m \in A$ who is matched in μ, we define $s_\mu(m)$ to be the first B-vertex in m's preference list who prefers m to its μ-partner. If no such vertex exists, then $s_\mu(m) = \emptyset$. A A-*rotation exposed* in a stable matching μ is a sequence $(m_0, w_0), \ldots, (m_{r-1}, w_{r-1})$ such that for each $i \in \{0, \ldots, r-1\}$, it holds that $(m_i, w_i) \in \mu$ and $w_{i+1} = s_\mu(m_i)$, where index i is taken modulo r. Analogously, we can define a B-*rotation exposed* in a stable matching μ by reversing the roles of A and B-vertices.

For the purpose of our analysis in this section, where we are only dealing with one matching and the goal is to make it the unique stable matching by changing the preference profile, we will not refer to the underlying stable matching explicitly and instead, refer to the underlying preference profile in which the rotation is exposed. Thus, instead of $s_\mu(m) = w$ as above we will use $s_\mathcal{L}(m)$ to refer to the first vertex who appears after $\mu(m)$ in the preference list of m in $\mathcal{L}$ who prefers m to its μ-partner according to its preference list in $\mathcal{L}$. Our analysis in this section is inspired by the following result.

Proposition 1 *[17]. Matching μ is the unique stable matching in a given instance iff there is neither A nor B-rotation exposed by μ in that instance.*

Rotation Detection: To detect if there is A-rotation exposed by μ in $\mathcal{L}$, we need to create a Hasse diagram, denoted by $H(\mathcal{L})$, based on the function $s_{\mathcal{L}}(m)$ for each $m \in A$. Specifically, for each $m \in A$, consider the μ-partner of $s_{\mathcal{L}}(m)$, denoted by $n_{\mathcal{L}}(m)$. Add a directed arc from m to $n_{\mathcal{L}}(m)$. If $s_{\mathcal{L}}(m) = \emptyset$, then add arc from m to a special vertex t. Note that every vertex in this digraph has out-degree at most one. It is easy to see that a A-rotation is exposed by μ in $\mathcal{L}$ iff this digraph has a directed cycle.

Reduction Rules. In the exposition of our algorithm, we will use the tool of a *reduction rule*, defined as a set of rules applied to the given instance of a problem to produce another instance of the same problem. A reduction rule is said to be *safe* when the input instance is a yes-instance iff the reduced instance is a yes-instance. We say that a reduction rule is applicable on an instance if the output instance is different from the input instance. We apply the rules exhaustively, that is whenever applicable, and we stop when none of them are applicable.

We begin with polynomial-time algorithm for SWAP-EXACT UNIQUE.

Theorem 2. SWAP-EXACT UNIQUE *is polynomial-time solvable when all preference lists are of size at most two.*

For the ease of exposition we will present the analysis for unit cost, however as we point out in the closing analysis, the algorithm can be extended to arbitrary cost function as well with appropriate modifications to the reduction rules followed by a simple branching algorithm at the end.

Let $\mathcal{I} = (G, \mathcal{L}, (c_v)_{v \in V}, \mu, b)$ denote an instance of SWAP-EXACT UNIQUE, where each preference list in $\mathcal{L}$ is of size at most two. Note that a solution S for $\mathcal{I}$ can be viewed as a set of preference lists (of size two) corresponding to a subset of vertices $V_S \subseteq V$ such that the ordering is the reverse of the one in $\mathcal{L}$. Thus, it follows that μ is the unique stable matching in the preferences profile given by $\mathcal{L} \oplus S$. Moreover, for a solution S to be minimal, every vertex in V_S should be part of a blocking pair or a rotation in $\mathcal{L}$.

Note that the definition of a rotation is constructive and the structure may exist even if the underlying matching is not stable. In what follows, we will not make explicit reference to the underlying matching being stable. The reduction rules are applied by detecting a rotation via the Hasse-diagram as described above.

We will prove the above theorem via a series of reduction rules. We begin noting that since the preference lists are of size two, the following must hold.

Observation 1. *Any two blocking pairs in $\mathcal{I}$ are vertex disjoint, i.e., for any two blocking pairs (m, w) and (m', w'), we have $m \neq m'$ and $w \neq w'$.*

The following result follows from the fact that when two rotations are exposed in the same matching, then they must be vertex disjoint cycles in the Hesse diagram $H(\mu)$.

Proposition 2 *[17]. The set of rotations exposed by a matching μ must be vertex-disjoint.*

Reduction Rule 1 *(Passive vertices).* Let $\mathcal{I} = (G, \mathcal{L}, (c_v)_{v \in V}, \mu, b)$ denote an instance of SWAP-EXACT UNIQUE. Suppose that for an edge $(m, w) \in E \setminus \mu$, m prefers its μ-partner over w and w prefers its μ-partner over m. Then, we define new preference lists $p'_m : \langle \mu(m) \rangle$ and $p'_w : \langle \mu(w) \rangle$, and the reduced instance is $\mathcal{I} = (G \setminus \{(m,w)\}, \mathcal{L} \oplus \{p'_m, p'_w\}, (c_v)_{v \in V}, \mu, b)$.

Reduction Rule 2. Let $\mathcal{I} = (G, \mathcal{L}, (c_v)_{v \in V}, \mu, b)$ denote an instance of SWAP-EXACT UNIQUE. Let ρ denote an exposed rotation in $\mathcal{I}$. Then, do as follows.

- if ρ is a A-rotation, then for some $w \in B$ who is in $V(\rho)$, we define p' to be the preference list of w after swapping;
- else, if ρ is a B-rotation, then for some $m \in A$ who is in ρ, we define p' to be the preference list of m after swapping.

The new instance is $\mathcal{I}' = (G, \mathcal{L} \oplus p', (c_v)_{v \in V}, \mu, b-1)$.

We will next handle the blocking pairs in the instance. We note that in order to *fix* a blocking pair (m, w), i.e., ensure that (m, w) is not a blocking pair in the resulting instance, it is enough to swap the preference list of either m or w. However, by swapping a A-vertex's (B-vertex's) preference list we may inadvertently create a $A(B)$-rotation in the resulting instance. Hence, to kill that we may have to swap both the preference lists of m and w. The following definition allows us to distinguish between these different types of rotations.

Definition 1. *Let (m, w) be a blocking pair in μ. Then, in $\mathcal{I}$, the preference lists are $p_m : \langle w, \mu(m) \rangle$ and $p_w : \langle m, \mu(w) \rangle$. Let $p'_m : \langle \mu(m), w \rangle$ and $p'_w : \langle \mu(w), m \rangle$, be the lists obtained after swapping.*

If the matching pairs $(m, \mu(m))$ and $(\mu(w), w)$ are part of an A-rotation in $\mathcal{L} \oplus p'_m$ and are also part of a B-rotation in $\mathcal{L} \oplus p'_w$, then we call (m, w) a type II blocking pair. Else, it is type I.

The next rule is about type I blocking pairs. For these it is enough to swap the preference list of one of the members.

Reduction Rule 3. Let $\mathcal{I} = (G, \mathcal{L}, (c_v)_{v \in V}, \mu, b)$ denote an instance of SWAP-EXACT UNIQUE. Let (m, w) be a type I blocking pair with respect to μ in $\mathcal{I}$. Let $p'_m : \langle \mu(m), w \rangle$ and $p'_w : \langle \mu(w), m \rangle$ denote the swapped preference lists.

Then, for the pairs $(m, \mu(m))$ and $(\mu(w), w))$ we do as follows:

- if they are not part of an A-rotation in $\mathcal{L} \oplus p'_m$, then the reduced instance is $\mathcal{I}' = (G, \mathcal{L} \oplus p'_m, (c_v)_{v \in V}, \mu, b-1)$;
- else, if they are not part of a B-rotation in $\mathcal{L} \oplus p'_w$, then the reduced instance is $\mathcal{I}' = (G, \mathcal{L} \oplus p'_w, (c_v)_{v \in V}, \mu, b-1)$

Intuitively speaking, when a type II blocking pair is fixed, it creates a rotation in the new instance, and that rotation will have to be killed in a subsequent step to meet the goal of the problem.

Reduction Rule 4. *Let $\mathcal{I} = (G, \mathcal{L}, (c_v)_{v \in V}, \mu, b)$ denote an instance of* SWAP-EXACT UNIQUE. *Let (m, w) be a type II blocking pair with respect to μ in $\mathcal{I}$, where $m \in A$ and $w \in B$. We denote $p'_m : \langle \mu(m), w \rangle$ and $p'_w : \langle \mu(w), m \rangle$ the swapped preference lists of m and w. Then, the reduced instance is $\mathcal{I}' = (G, \mathcal{L} \oplus \{p'_m, p'_w\}, (c_v)_{v \in V}, \mu, b-2)$.*

We apply each of these rules exhaustively and whenever applicable. We stop when for the instance $\mathcal{I}$ one of the following conditions has been reached:

(i) none of the reduction rules are applicable and $b \geq 0$;
(ii) budget $b < 0$; or
(iii) Reduction Rules 2–4 are applicable but the budget $b \leq 0$.

If condition (1) holds, then there are no blocking pair and no rotations are exposed in the instance. Thus, μ is the unique stable matching and we exit by saying "yes". For the other conditions we exit by saying "no".

When Costs are Non-uniform: We note that Reduction Rule 1 can be applied without any consideration for cost. However, there are cost-version of Reduction Rule2, where we swap the preference list of the cheapest B-vertex if ρ is an A-rotation. The reason behind this is that by choosing an A-vertex m to kill ρ, even if cheaper will end up costing more since the preference list of the B-agent $s_{\mathcal{L}}(m)$ will have to be swapped as well to fix the blocking pair formed by swapping the preference list of m. Hence, instead one can just swap the preference list of the B-vertex in ρ who has the minimum cost. Analogously, we can argue that if ρ is a B-rotation, then we swap the preference list of the cheapest A-vertex in ρ. From here, we note that we can have both type I or type II blocking pairs and due to variable cost, it may be that for a type I blocking pair (m, w), swapping p_m does not create rotation but it costs more than swapping p_w which creates a rotation. In this case, we have to compare the cost of m vs the sum of the cost of w and any other A-agent in the resulting B-rotation. The lesser of the two options will lead to the blocking pair (m, w) being fixed and any rotation that results from it being killed. Similarly, for a type II blocking pair (m, w) such that swapping p_m in profile $\mathcal{L}$ creates a A-rotation ρ, we will have to compare the cost of $c_m + c_w$ vs $c_m + \min_{w' \in \rho_B}\{c_{w'}\}$ where ρ_B denotes the B-members of the A-rotation ρ created by swapping p_m. Symmetrically, if swapping p_w in profile $\mathcal{L}$ creates a B-rotation ρ, we will have to compare the cost of $c_m + c_w$ vs $c_w + \min_{m' \in \rho_A}\{c_{m'}\}$ where ρ_A denotes the A-members of the B-rotation ρ created by swapping p_w.

Next, we design an XP algorithm with respect to the budget, b.

Theorem 3. SWAP-EXACT UNIQUE *is* XP *with respect to b.*

Proof. Let $\mathcal{I} = (G = (V, E), \mathcal{L}, (c_v)_{v \in V}, \mu, b)$ be an instance of SWAP-EXACT UNIQUE. We will prove the following technical result

Lemma 2. *Suppose that $\mathcal{I}$ is a yes-instance. Then, there exists a solution in which the manipulated preference list, for any agent $v \in V$, only differs from*

the original preference list in at most $2b$ positions. Specifically, all swaps are contained within the sublist obtained by restricting the preference list of v in $\mathcal{L}$ to b positions to the left of $\mu(v)$ and b positions to the right of $\mu(v)$.

Using the above result we can design our algorithm as follows: Consider a subset $S \subseteq V$ of size at most b. For each agent $v \in S$, construct the $2b$ length sublist from the preference list of v as described above. Guess the permutations that would have led to the solution for the instance $\mathcal{I}$. For each "guess" check if μ is indeed the unique stable matching in the resulting instance. If any one guess yields a solution, then we are done. Else, we can return "no". The time complexity is due to $\sum_{i=0}^{b} \binom{n}{i}((2b)!)^i \leq \mathcal{O}((n(2b)!)^b)$, and the correctness follows from the above lemma.

4 Outlook

In this work we showed that the considered problems are W[1]-hard with respect the number of agents unmatched in a stable marriage of the original instance. Given that exhaustive search yields an FPT algorithm with respect to the total number of agents, an interesting question is: what is the parameterized complexity of the problems with respect to the number of agents matched in a stable marriage of the original instance? One direction of research would be to study approximation algorithms, a so far unexplored avenue for manipulation in matching. Apart from these, it would be interesting to study manipulation effects on other notions of "optimal" matching, e.g., max/min cardinality stable matching, egalitarian matching, sex-equal stable matching, min regret stable matching, popular matching, etc.

References

1. Bartholdi-III, J.J., Tovey, C.A., Trick, M.A.: The computational difficulty of manipulating an election. Soc. Choice Welf. **6**(3), 227–241 (1989)
2. Bartholdi-III, J.J., Tovey, C.A., Trick, M.A.: How hard is it to control an election? Math. Comput. Model. **16**(8–9), 27–40 (1992)
3. Boehmer, N., Bredereck, R., Heeger, K., Niedermeier, R.: Bribery and control in stable marriage. J. Artif. Intell. Res. **71**, 993–1048 (2021)
4. Bredereck, R., Chen, J., Knop, D., Luo, J., Niedermeier, R.: Adapting stable matchings to evolving preferences. In: AAAI 2020, pp. 1830–1837 (2020)
5. Chen, J., Skowron, P., Sorge, M.: Matchings under preferences: strength of stability and trade-offs. In: Proceedings of the 2019 ACM Conference on Economics and Computation (EC), pp. 41–59 (2019)
6. Conitzer, V., Sandholm, T., Lang, J.: When are elections with few candidates hard to manipulate? J. ACM **54**(3) (2007)
7. Cseh, Á., Heeger, K.: The stable marriage problem with ties and restricted edges. Discret. Optim. **36**, 100571 (2020)
8. Parameterized Algorithms. Springer, Cham (2015). https://doi.org/10.1007/978-3-319-21275-3_15

9. Diestel, R.: Graph Theory, 4th Edition, volume 173 of Graduate texts in mathematics. Springer, Cham (2012). https://doi.org/10.1007/978-3-319-49475-3.
10. Fundamentals of Parameterized Complexity. TCS, Springer, London (2013). https://doi.org/10.1007/978-1-4471-5559-1_35
11. Downey, R.G., Fellows, M.R.: Fixed-parameter tractability and completeness ii: on completeness for w [1]. Theor. Comput. Sci. **141**(1–2), 109–131 (1995)
12. Ehlers, L.: Truncation strategies in matching markets. Math. Oper. Res. **33**(2), 327–335 (2008)
13. Faliszewski, P., Rothe, J.: Handbook of Computational Social Choice, Chapter Control and Bribery in Voting. Cambridge Univ, Press, Cambridge (2016)
14. Parameterized Complexity Theory. TTCSAES, Springer, Heidelberg (2006). https://doi.org/10.1007/3-540-29953-X_9
15. Gale, D., Shapley, L.S.: College admissions and the stability of marriage. Am. Math. Monthly **69**, 9–15 (1962)
16. Garey, M.R., Johnson, D.S.: Computers and Intractability, vol. 174. Freeman, San Francisco (1979)
17. Gusfield, D., Irving, R.W.: The Stable Marriage Problem-Structure and Algorithm. The MIT Press, Cambridge (1989)
18. Hemaspaandra, E., Hemaspaandra, L.A., Rothe, J.: Anyone but him: the complexity of precluding an alternative. Artif. Intell. **171**(5–6), 255–285 (2007)
19. Heyneman, S., Anderson, K., Nuraliyeva, N.: The cost of corruption in higher education. Comp. Educ. Rev. **52**(1), 1–25 (2008)
20. Liu, Q., Peng, Y.: Corruption in college admissions examinations in china. Int. J. Educ. Dev. **41**, 104–111 (2015)
21. Mai, T., Vazirani, V.V.: Finding stable matchings that are robust to errors in the input. In: Proceedings of the 26th Annual European Symposium on Algorithms (ESA), pp. 60:1–60:11 (2018)
22. Manlove, D.: Algorithmics of Matching Under Preferences, vol. 2. World Scientific, Singapore (2013)
23. Meir, R., Procaccia, A.D., Rosenschein, J.S., Zohar, A.: Complexity of strategic behavior in multi-winner elections. J. Artif. Intell. Res. **33**, 149–178 (2008)
24. Meir, R., Procaccia, A.D., Rosenschein, J.S.: A broader picture of the complexity of strategic behavior in multi-winner elections. In: 7th AAMAS, pp. 991–998 (2008)
25. Obraztsova, S., Zick, Y., Elkind, E.: On manipulation in multi-winner elections based on scoring rules. In: 12th AAMAS, pp. 359–366 (2013)
26. Procaccia, A.D., Rosenschein, J.S., Zohar, A.: Multi-winner elections: complexity of manipulation, control and winner-determination. In: 16th IJCAI, vol. 7, pp. 1476–1481 (2007)
27. Roth, A.E.: The evolution of the labor market for medical interns and residents: a case study in game theory. J. Polit. Econ. **92**(6) (1984)
28. Roth, A.E.: On the allocation of residents to rural hospitals: a general property of two-sided matching markets. Econom. J. Econ. Soc. **54**(2), 425–427 (1986)
29. Roth, A.E., Rothblum, U.G.: Truncation strategies in matching markets-in search of advice for participants. Econometrica **67**(1), 21–43 (1999)
30. Roth, A.E., Sotomayor, M.: Two-Sided Matching: A Study in Game Theoretic Modeling and Analysis. Cambridge Univ, Press, Cambridge (1990)
31. Teo, C.-P., Sethuraman, J., Tan, W.-P.: Gale-Shapley stable marriage problem revisited: strategic issues and applications. Manag. Sci. **47**(9), 1252–1267 (2001)

Abstracts

Metric Distortion Under Probabilistic Voting

Sahasrajit Sarmsarkar[iD] and Mohak Goyal[(✉)][iD]

Stanford University, Stanford, CA 94305, USA
{sahasras,mohakg}@stanford.edu

Abstract. Metric distortion in social choice theory provides a framework for evaluating how well voting rules minimize social cost when both voters and candidates are situated within a shared metric space, with voters submitting rankings and the rule outputting a single winner. Traditional studies focus on deterministic voting, where voters' submitted rankings are always consistent with their distances from the candidates. This model sometimes leads to counterintuitive results, such as the Random Dictator (RD) Rule outperforming the Copeland Rule despite the latter's adherence to the Condorcet criterion.

In this paper, we extend the metric distortion framework to incorporate probabilistic voting models, where voters' rankings are generated according to a probability distribution that depends on the underlying metric distances. Our study includes a broad class of probabilistic models, such as the Plackett-Luce (PL) model and a newly introduced Pairwise Quantal Voting (PQV) model inspired by quantal response theory. We define three axiomatic properties for these models: scale-freeness with distances, pairwise order probabilities being independent of other candidates, and strict monotonicity of pairwise order probabilities in distances.

Our results apply to a broad class of probabilistic models. We provide distortion bounds for numbers of voters $n \geq 3$ and number of candidates $m \geq 2$. For large elections ($n \to \infty$ and $m \ll n$), we establish matching upper and lower bounds on the distortion of Plurality, an upper bound for Copeland, and a lower bound for RD. The distortion of Plurality grows linearly in m; Copeland's upper bound is constant, while RD's lower bound increases sublinearly in m, depending on the probabilistic model. For example, in the PL model with candidate strength inversely proportional to the square of their metric distance, Copeland's distortion is at most 2, while RD's is $\Omega(\sqrt{m})$.

Our technique is as follows: for the problem of the "adversary" to maximize the distortion, we establish a critical threshold for the expected fraction of votes on pairwise comparisons along all edges on a directed path from a winner to the "true optimal" candidate for Copeland and Plurality. This path involves one or two hops for Copeland and one for Plurality. We then formulate a linear-fractional program incorporating this threshold, linearize it via the sub-level sets technique, and find a feasible solution to the dual problem. Concentration inequalities on this

---

S. Sarmsarkar and M. Goyal—Contributed equally.

© The Author(s), under exclusive license to Springer Nature Switzerland AG 2025
R. Freeman and N. Mattei (Eds.): ADT 2024, LNAI 15248, pp. 291–292, 2025.
https://doi.org/10.1007/978-3-031-73903-3

solution provide an upper bound on the distortion, and we find a matching lower bound for Plurality by construction. The lower bound for RD is also by construction.

Keywords: Metric Distortion · Probabilistic Voting · Social Choice

A Linear Theory of Multi-Winner Voting

Lirong Xia[✉]

Rutgers University and DIMACS, New Brunswick, USA
xialirong@gmail.com

Abstract. We initiate the development of a linear theory of multi-winner voting by representing multi-winner rules and proportionality axioms as linear systems. We show that many existing multi-winner rules, including Thiele methods and their sequential variants are linear, and many proportionality axioms, including JR, EJR, PJR, EJR+, PJR+, and Core stability, are linear. This allows us to leverage previous work to obtain general results on axiomatic satisfaction and computation for multi-winner voting. As an example, we prove that under a large class of *Impartial Culture* distributions for approval preferences, all k-committees are core-stable and satisfy EJR+ with high probability that converges to 1 at an exponential rate in the number of voters. This shows that, surprisingly, while whether the core is always non-empty is still an important open question, it is very likely that all ABC rules are core-stable, which implies that weaker proportionality axioms, e.g., PJR+, EJR, PJR, JR, are satisfied with high probability as well.

Group Fairness in Multi-period Mobile Facility Location Problems

Haris Aziz[1], Hau Chan[2], Xingchen Sha[3](✉), Toby Walsh[1], and Lirong Xia[4]

[1] UNSW Sydney, Sydney, Australia
{haris.aziz,t.walsh}@unsw.edu.au
[2] University of Nebraska-Lincoln, Lincoln, USA
hchan3@unl.edu
[3] Columbia University, New York, USA
xingchen.sha@columbia.edu
[4] Rutgers University, New Brunswick, USA
lirong.xia@rutgers.edu

Abstract. We study the group-fair multi-period mobile facility location problems in offline and online settings, where agents from different groups are located on a real line and arrive in different periods with complete and no arrival information of the agents, respectively. The problems capture several real-world scenarios such as locating mobile blood donation centers, vaccination vehicles, and medical clinics. Our goal is to locate a mobile facility at each period to serve the arriving agents in order to minimize the maximum total group-fair cost and the maximum average group-fair cost objectives that measure the costs or distances of groups of agents to their corresponding facilities across all periods. We first consider the problems from the algorithmic perspective and show that computing optimal facility locations for both group-fair cost objectives is polynomially solvable for both settings. We then consider the problems from the mechanism design perspective, where the agents' locations are private, for the two settings. For both settings and objectives, we design deterministic strategyproof mechanisms to elicit the agents' locations truthfully while optimizing the group-fair cost objectives. We show that the designed mechanisms obtain several approximation guarantees and have almost tight approximation ratios for certain periods and settings. We refer readers to the full paper for details.

Keywords: Facility location · Group fairness · Mechanism design · Optimization · Multiple periods · Online · Offline

Extending Myerson's Optimal Auctions to Correlated Bidders via Neural Network Interpolation

Mingyu Guo[1,2(✉)], Jiayuan Liu[2], and Vincent Conitzer[2]

[1] University of Adelaide, Adelaide, Australia
mingyu.guo@adelaide.edu.au
[2] Carnegie Mellon University, Pittsburgh, USA
{mingyugu,jiayuan4}@andrew.cmu.edu, conitzer@cs.cmu.edu

Abstract. We aim to design revenue-maximizing single-item auctions that are deterministic, strategy-proof and ex post individually rational. Myerson's seminal work on optimal auction design [Myerson 81] solved this problem for independent bidders. Myerson introduced the novel concept of virtual valuation and showed that revenue maximization is equivalent to virtual valuation maximization. Coincidentally, by greedily allocating the item to the bidder with the highest (ironed) virtual valuation, the resulting allocation is guaranteed to be monotone – a necessary and sufficient condition for strategy-proofness.

For correlated bidders, Myerson's greedy allocation no longer guarantees monotonicity/strategy-proofness. We propose a simple yet empirically effective approach for designing near-optimal auctions for correlated bidders. We train a neural network to mimic Myerson's greedy allocation, which is to allocate to whoever has the highest *marginal profit*, a variant of virtual valuation proposed in [Papadimitriou and Pierrakos 11]. We use a multilayer perceptron (MLP) with ReLU activation to model the allocation function. This architecture allows us to *exactly verify* the monotonicity of a trained allocation using mixed-integer-programming (MIP). Given a correlated distribution, we can use our approach to check whether Myerson's optimal auction, or more accurately, the *neural network interpolation version* of it, remains monotone (strategy-proof) despite the correlation. The more important and more general use case of our approach is that even when the greedy allocation is not 100% monotone, it is empirically possible to derive a neural network interpolation version of it, which is close to the original and is verifiably monotone. This is via several monotonicity seeking techniques, including 1) repeated trials; 2) counterexample-guided training; and 3) post-processing monotonicity fix. The latter two are enabled by the MLP+ReLU architecture. We performed an extensive suite of experiments on 59 correlated distributions. Our mechanisms outperform all baselines, including a verifiably strategy-proof variant of RegretNet [Dütting et al. 19]. The revenue gap between the achieved revenue using our approach and the upper bound is maximally 2.1% under an evolutionary computation generated adversarial distribution.

No-Regret Learning for Stackelberg Equilibrium Computation in Newsvendor Pricing Games

Larkin Liu[1(✉)] and Yuming Rong[2]

[1] Technische Universität München, Munich, Germany
ryrong@google.com
[2] Google, Mountain View, USA

Abstract. We introduce the application of online learning in a Stackelberg game pertaining to a system with two learning agents in a dyadic exchange network, consisting of a supplier and retailer, specifically where the parameters of the demand function are unknown [2]. In this game, the supplier is the first-moving leader, and must determine the optimal wholesale price of the product. Subsequently, the retailer who is the follower, must determine both the optimal procurement amount and selling price of the product. In the perfect information setting, this is known as the classical *price-setting Newsvendor* problem [3], and we prove the existence of a unique Stackelberg equilibrium [4] when extending this to a two-player pricing game. In the framework of online learning, the parameters of the reward function for both the follower and leader must be learned, under the assumption that the follower will best respond with optimism under uncertainty. A novel algorithm based on contextual linear bandits with a measurable uncertainty set is used to provide a confidence bound on the parameters of the stochastic demand [1]. Consequently, optimal finite time regret bounds on the Stackelberg regret, along with convergence guarantees to an approximate Stackelberg equilibrium, are provided.

References

1. Abbasi-Yadkori, Y., Pál, D., Szepesvári, C.: Improved algorithms for linear stochastic bandits. In: Advances in Neural Information Processing Systems, vol. 24 (2011)
2. Cesa-Bianchi, N., et al.: Learning the Stackelberg equilibrium in a newsvendor game. In: Proceedings of the 2023 International Conference on Autonomous Agents and Multiagent Systems, pp. 242–250 (2023)
3. Mills, E.S.: Uncertainty and Price Theory. Q. J. Econ. **73**(1), 116–130 (1959)
4. Zhu, B., et al.: The sample complexity of online contract design. In: arXiv preprint arXiv:2211.05732 (2022)

Equilibria of Data Marketplaces with Privacy-Aware Sellers under Endogenous Privacy Costs

Diptangshu Sen[✉], Jingyan Wang, and Juba Ziani

Georgia Institute of Technology, Atlanta, GA30332, USA
dsen30@gatech.edu

In today's digital era, tremendous amounts of data are generated and processed daily. Some organizations hold vast amounts of user or customer data, while others continually seek to augment their datasets by acquiring and purchasing data from external sources. A critical issue that arises when user data is collected and transacted is then data privacy. Users are increasingly aware of how their data can be shared or leaked, and privacy considerations are now an important factor for many when it comes to engaging with new applications or platforms. These concerns are exacerbated by numerous real-world instances of online and social media platforms disclosing private user data to third parties, sometimes without explicit consent.

In this work, we aim to model how user privacy costs may not just be exogenous and fixed, but a function of *how widely their data is shared and transacted*. We study this phenomenon in two-sided online platforms: the platform can collect data from participating users, and can share or resell this data to potentially numerous downstream data buyers. Our main modeling novelty is that we explicitly consider how users' privacy costs can depend on *how many third parties acquire their data*, in particular treating user privacy costs as *endogenous* and a function of the platform's data pricing and re-selling decisions.

Our main technical contribution is to characterize user participation equilibria as a function of platform price P under the endogenous privacy cost model. We consider two variants of our model that correspond to different utility functions for the users: i) when each user gets a constant benefit for joining the platform and ii) when each user's benefit is linearly increasing in the number of other users that participate. In both variants, we see the emergence of multiple equilibrium regimes depending on the magnitude of the benefit offered, and highlight the existence of counter-intuitive equilibria as well as equilibrium multiplicity. In all cases, we demonstrate that equilibria in our setting are fundamentally different from the case when privacy costs are exogenous, the most significant point of difference being that under exogenous privacy costs, the user-side participation rate is completely independent of the platform's price P and the buyer-side decisions and thus can never be improved without investing in improving the quality of service offered. Finally, we provide experiments expanding our results to more general distributions of user costs and more general utility functions.

Learning Linear Utility Functions From Pairwise Comparison Queries

Luise Ge[(✉)], Brendan Juba, and Yevgeniy Vorobeychik

Washington University in St. Louis, St. Louis, MO 63130, USA
g.luise@wustl.edu

Abstract. We study learnability of linear utility functions from pairwise comparison queries. In particular, we consider two learning objectives. The first objective is to predict out-of-sample responses to pairwise comparisons, whereas the second is to approximately recover the true parameters of the utility function. We show that in the passive learning setting, linear utilities are efficiently learnable with respect to the first objective, both when query responses are uncorrupted by noise, and under Tsybakov noise when the distributions are sufficiently "nice". In contrast, we show that utility parameters are not learnable for a large set of data distributions without strong modeling assumptions, even when query responses are noise-free. Next, we proceed to analyze the learning problem in an active learning setting. In this case, we show that even the second objective is efficiently learnable, and present algorithms for both the noise-free and noisy query response settings. Our results thus exhibit a qualitative learnability gap between passive and active learning from pairwise preference queries, demonstrating the value of the ability to select pairwise queries for utility learning.

Keywords: Learning from pairwise comparisons · RLHF and Value Alignment

The Fairness-Quality Trade-Off in Clustering

Rashida Hakim[1](✉), Ana-Andreea Stoica[2], Mihalis Yannakakis[1], and Christos H. Papadimitriou[1]

[1] Columbia University, New York, USA
rashida.hakim@columbia.edu
[2] MPI for Intelligent Systems, Tübingen, Germany

Fairness in clustering has been a topic of extensive research; however, the trade-off between fairness and quality has seldom been systematically explored. This study addresses this gap by introducing novel algorithms designed to trace the complete trade-off curve, or Pareto front, between quality and fairness in clustering problems. The Pareto front represents the set of clusterings that are not outperformed in both objectives by any other clustering, providing a comprehensive view of the trade-offs involved. Unlike traditional approaches that optimize one objective subject to a constraint on the other, our algorithms allow for a flexible exploration without predefined bounds. This is particularly valuable when decision makers do not know a priori how the two objectives affect one another in a particular dataset.

A General Framework. Unlike previous studies that focus on specific objectives for quality and fairness, our approach encompasses all objectives within two general classes that include several special cases previously considered in the group fairness literature that are concerned with how points of different demographics are represented across clusters.

Novel Algorithms. Clustering is itself an NP-hard problem. Even if we fix the centers to make the problem easier (called the assignment problem), introducing fairness constraints again makes the problem NP-hard, even to compute just a single point on the Pareto front. We present dynamic programming algorithms for computing the *whole* front that run in exponential time in the number of clusters k (but not in the number of points n) for a general set of fairness objectives. We also present a polynomial-time algorithm for scenarios where cluster centers are fixed and the fairness objective involves minimizing the sum of imbalances between two groups within each cluster, which may be a sensible fairness notion when two groups are equally represented in the dataset.

Empirics. We also implement our algorithms on several different datasets to explore real-world trade-off curves. We show empirical benefit to the tight approximation in the fairness objective that our exponential time algorithms are able to achieve. We do this by showing that our Pareto front approximation lies strictly below the Pareto front approximation computed via repeated application of an algorithm from previous work.

Author Index

A
Allen, Thomas E. 225
Arnold, Nathan 243
Auriau, Vincent 207
Aziz, Haris 294

B
Bassani, Matteo 160
Belahcène, Khaled 207

C
Caballero, William N. 128
Camacho, J. M. 98
Chan, Hau 294
Conitzer, Vincent 295

D
Dima, Simon 113
Dogan, Vedat 144

F
Fischer, Simon 113

G
Galliera, Raffaele 160
Garvardt, Jaroslav 82
Ge, Luise 299
Goldsmith, Judy 243
Goyal, Mohak 291
Guo, Mingyu 295
Gupta, Sushmita 273

H
Hakim, Rashida 300
Heitzig, Jobst 113
Herin, Margot 191

J
Jain, Pallavi 273
Jiang, Liu 225
Juba, Brendan 299

K
Karh Bet, Garo 3
Karia, Neel 18
Komusiewicz, Christian 82

L
Lang, Jérôme 18
Liu, Jiayuan 295
Liu, Larkin 297
Lorke, Berthold Blatt 82

M
Malherbe, Emmanuel 207
Markakis, Evangelos 48
Mikšaník, David 33
Misra, Neeldhara 258
Mousseau, Vincent 207

N
Naveiro, Roi 98, 128

O
O'Sullivan, Barry 144
Oliver, Joss 113

P
Papadimitriou, Christos H. 300
Papasotiropoulos, Georgios 48
Pennock, David 174
Perny, Patrice 191
Prestwich, Steven 144

R

Ríos Insua, David 98, 128
Rong, Yuming 297
Rothe, Jörg 3
Rudich, Avi 67
Rudich, Isaac 67
Rue, Rachel 67

S

Sarmsarkar, Sahasrajit 291
Schestag, Jannik 82
Schvartzman, Ariel 33, 174
Sen, Diptangshu 298
Sethia, Aditi 258
Sha, Xingchen 294
Sokolovska, Nataliya 191
Soukup, Jan 33
Stoica, Ana-Andreea 300
Suri, Niranjan 160

V

Venable, Kristen Brent 160
Vorobeychik, Yevgeniy 299

W

Walsh, Toby 294
Wang, Jingyan 298

X

Xia, Lirong 293, 294
Xue, Eric 174

Y

Yannakakis, Mihalis 300

Z

Ziani, Juba 298
Zorn, Roman 3

SPRINGER NATURE

GPSR Compliance

The European Union's (EU) General Product Safety Regulation (GPSR) is a set of rules that requires consumer products to be safe and our obligations to ensure this.

If you have any concerns about our products, you can contact us on ProductSafety@springernature.com

In case Publisher is established outside the EU, the EU authorized representative is:

Springer Nature Customer Service Center GmbH
Europaplatz 3
69115 Heidelberg, Germany

The manufacturer's authorised representative in the EU is Springer Nature Customer Service Centre GmbH, Europaplatz 3, 69115 Heidelberg, Germany. If you have any concerns regarding our products, please contact ProductSafety@springernature.com

Printed and bound by CPI Group (UK) Ltd, Croydon, CR0 4YY

25/03/2026

02078190-0009